About Island Press

ISLAND PRESS, a nonprofit organization, publishes, markets, and distributes the most advanced thinking on the conservation of our natural resources—books about soil, land, water, forests, wildlife, and hazardous and toxic wastes. These books are practical tools used by public officials, business and industry leaders, natural resource managers, and concerned citizens working to solve both local and global resource problems.

Founded in 1978, Island Press reorganized in 1984 to meet the increasing demand for substantive books on all resource-related issues. Island Press publishes and distributes under its own imprint and offers these services to other nonprofit organizations.

Support for Island Press is provided by Apple Computers Inc., Mary Reynolds Babcock Foundation, Geraldine R. Dodge Foundation, The Charles Engelhard Foundation, The Ford Foundation, Glen Eagles Foundation, The George Gund Foundation, William and Flora Hewlett Foundation, The Joyce Foundation, The John D. and Catherine T. MacArthur Foundation, The Andrew W. Mellon Foundation, The Joyce Mertz-Gilmore Foundation, The New-Land Foundation, The J. N. Pew, Jr., Charitable Trust, Alida Rockefeller, The Rockefeller Brothers Fund, The Florence and John Schumann Foundation, The Tides Foundation, and individual donors.

CLIMATE CHANGE

The IPCC Response Strategies

CLIMATE CHANGE

The IPCC Response Strategies

World Meteorological Organization/United Nations Environment Program

INTERGOVERNMENTAL PANEL
ON CLIMATE CHANGE

ISLAND PRESS

Washington, D.C. ☐ *Covelo, California*

Library of Congress Cataloging-in-Publication Data

Climate change: the IPCC response strategies
  p.   cm.
    ISBN 1-55963-103-1. — ISBN 1-55963-102-3 (pbk.)
    1. Climatic changes—International cooperation.
  OC981.8.C5C516   1991
  363.73'92—dc20   90-26690
                                              CIP

        ISBN 1-55963-102-3 (paperback)
              1-55963-103-1 (cloth)

    Printed on recycled, acid-free paper

  Manufactured in the United States of America

      10 9 8 7 6 5 4 3 2 1

CLIMATE CHANGE

The IPCC Response Strategies

World Meteorological Organization / United Nations Environment Program

INTERGOVERNMENTAL PANEL
ON CLIMATE CHANGE

Note to the Reader

The report of the Response Strategies Working Group (RSWG) was compiled through an unprecedented international cooperative effort to deal with the many climate change response strategies. The chairs of the four RSWG subgroups and the coordinators of the five RSWG topic areas took the responsibility for completing their individual reports. Along with their respective governments, they contributed generously of their time and resources to that end.

It was not possible or intended to review each subgroup report and topic paper in plenary session. The RSWG report thus constitutes a series of independently prepared underlying documents which attempt to analyze as thoroughly as possible the issues addressed in each subgroup or topic area. The synthesis of the concepts from these underlying reports is the RSWG Policymakers Summary, on which consensus was reached at the RSWG's Third Plenary Session in Geneva on June 9, 1990.

—F. M. B.

The Intergovernmental Panel on Climate Change (IPCC) was jointly established by us in 1988. Prof. B. Bolin is the Chairman of the Panel. The Panel's charge was to:

a) assess the scientific information that is related to the various components of the climate change issue, such as emissions of major greenhouse gases and modification of the Earth's radiation balance resulting therefrom, and that is needed to enable the environmental and socio-economic consequences of climate change to be evaluated; and
b) formulate realistic response strategies for the management of the climate change issue.

The Panel began its task by establishing its Working Goups I, II, and III respectively to:

a) assess available scientific information on climate change;
b) assess the environmental and socio-economic impacts of climate change; and
c) formulate response strategies.

The Panel also established a Special Committee on the Participation of Developing Countries to promote, as quickly as possible, the full participation of the developing countries in its activities.

The Panel has completed its First Assessment Report (FAR). The FAR consists of

• the Overview;
• the policymakers summaries of the IPCC Working Groups and the Special Committee;
• the reports of the IPCC Working Groups.

The Overview and the policymakers summaries are to be found in a single volume. The reports of the Working Groups are being published individually.

The present volume is based upon the findings of Working Group III. In order to appreciate the linkages among the various aspects of the climate change issue, it is recommended that it be read in the context of the full IPCC First Assessment Report.

The Chairman of Working Group III, Dr. F. M. Bernthal, and his Secretariat have succeeded beyond measure in mobilizing the cooperation and enthusiasm of literally hundreds of experts from all over the world. They have produced a volume of remarkable depth and breadth and a policymakers summary that is a model of writing aiming at explaining complex issues to the non-specialist.

We take this opportunity to congratulate and thank Dr. Bernthal for a job well done.

G. O. P. OBASI
Secretary-General, WMO

M. K. TOLBA
Executive Director, UNEP

OFFICERS OF WORKING GROUP III

CHAIRMAN:
F. Bernthal (U.S.A.)

CO-CHAIRS:
E. Dowdeswell (Canada)
J. Luo (China)
D. Attard (Malta)
P. Vellinga (Netherlands)
R. Karimanzira (Zimbabwe)

LEAD CONTRIBUTORS

RSWG SUBGROUP CO-CHAIRS

ENERGY AND INDUSTRY:
K. Yokobori (Japan)
Shao-Xiong Xie (China)

AGRICULTURE, FORESTRY,
AND OTHER HUMAN ACTIVITIES:
D. Kupfer (Germany, Fed. Rep.)
R. Karimanzira (Zimbabwe)

COASTAL ZONE MANAGEMENT:
J. Gilbert (New Zealand)
P. Vellinga (Netherlands)

RESOURCE USE AND MANAGEMENT:
R. Pentland (Canada)
J. Theys (France)
I. Abrol (India)

TASK A (EMISSIONS SCENARIOS) COORDINATORS

D. Tirpak (U.S.A.)
P. Vellinga (Netherlands)

TASK B (IMPLEMENTATION MECHANISMS) COORDINATORS

PUBLIC EDUCATION AND INFORMATION:
G. Evans (U.S.A.)
Ji-Bin Luo (China)

TECHNOLOGY DEVELOPMENT
AND TRANSFER:
K. Madhava Sarma (India)
K. Haraguchi (Japan)

ECONOMIC (MARKET) MEASURES:
J. Tilley (Australia)
J. Gilbert (New Zealand)

FINANCIAL MEASURES:
J. Oppeneau (France)
P. Vellinga (Netherlands)
A. Ibrahim (Egypt)

LEGAL AND INSTITUTIONAL
MECHANISMS:
R. Rochon (Canada)
D. Attard (Malta)
R. Beetham (U.K.)

Acknowledgments

The work of the Response Strategies Working Group was organized and coordinated by a network of officers from all over the world. I take this opportunity to thank the RSWG Vice-Chairs for their dedication in ensuring that Working Group III was able to fulfill the mission assigned to it. Special credit should go to the cochairs of the four RSWG subgroups and the coordinators of Task A (emissions scenarios) and Task B (implementation measures), who guided the efforts of the hundreds of experts in scores of countries around the world who contributed to the preparation of the specific sections of this report.

In addition, I would like to express my gratitude to the staff of the Office of Global Change in the U.S. Department of State, (Daniel Reifsnyder, Frances Li, Stephanie Kinney, Robert Ford, and Granville Sewell), for their support as Secretariat to the RSWG. Special thanks are due the staff of the U.S. National Science Foundation who, under the supervision of Dr. Li and Dr. Beverly Fleisher, oversaw the publication of this entire compendium. Finally, the Response Strategies Working Group owes an enormous debt of gratitude to IPCC Chairman, Prof. Bert Bolin, for his guidance and counsel, and to the IPCC Secretariat, Dr. N. Sundararaman and his staff.

—FREDERICK M. BERNTHAL

Contents
Climate Change
The IPCC Response Strategies

I

POLICYMAKERS SUMMARY OF THE RESPONSE STRATEGIES WORKING GROUP OF THE INTERGOVERNMENTAL PANEL ON CLIMATE CHANGE (WORKING GROUP III)

Contents
Policymakers Summary
Formulation of Response Strategies

Chairman's Introduction

The First Plenary meeting of Working Group III of the IPCC, the Response Strategies Working Group (RSWG), was held in Washington, January 30–February 2, 1989. This meeting was largely organizational, and it was not until after a subsequent RSWG Officers Meeting in Geneva, May 8–12, 1989, that the real work by the four RSWG subgroups, the Emissions Scenarios Task Force (Task A), and "Implementation Measures" Topic Coordinators (Task B) began.

The Second RSWG Plenary Session was held in Geneva, from October 2–6, 1989, to discuss the implementation measures: (1) public education and information; (2) technology development and transfer; (3) financial measures; (4) economic measures; and (5) legal measures, including elements of a framework climate convention. A consensus was reached on five topical papers dealing with these measures, with the understanding that they would be "living documents" subject to further modification as new information and developments might require.

The Third Plenary Meeting of RSWG, held in Geneva, June 5–9, 1990, achieved three objectives:

1) It reached consensus on the attached "policymakers summary," the first interim report of the RSWG.

2) It completed final editing and accepted the reports of the four RSWG subgroups, of the coordinators of Task A, and of the coordinators of the five Task B topical papers. These documents comprise the underlying material for the consensus report of this meeting, the policymakers summary; they are not themselves the product of a RSWG plenary consensus, although many governments participated in their formulation.

Finally,

3) The Working Group agreed to submit comments on its suggested future work programme to the RSWG Chairman by July 1, 1990, for transmission to the Chair of the IPCC. There was general agreement that the work of the RSWG should continue.

The primary task of the RSWG was, in the broad sense, technical, not political. The charge of IPCC to RSWG was to lay out as fully and fairly as possible a set of response policy options and the factual basis for those options.

Consistent with that charge, it was *not* the purpose of the RSWG to select or recommend political actions, much less to carry out a negotiation on the many difficult policy questions that attach to the climate change issue, although clearly the information might tend to suggest one or another option. Selection of options for implementation is appropriately left to the policymakers of governments and/or negotiation of a convention.

The work of RSWG continues. The Energy and Industry Subgroup has, since the June RSWG Plenary Meeting, held additional meetings in London (June 1990) and Paris (September 1990), the results of which are not reflected in this report.

It should be noted that quantitative estimates provided in the report regarding CFCs, including those in Scenario A ("Business as Usual"), generally do not reflect decisions made in June 1990 by the

Parties to the Montreal Protocol. Those decisions accelerate the timetable to phase out production and consumption of CFCs, halons, carbon tetra-chloride, and methyl chloroform.

It should further be noted that quantitative estimates of forestry activities (e.g., deforestation, bio-mass burning, including fuel wood, and other changes in land-use practices), as well as agricultural and other activities provided in the Report continue to be reviewed by experts.

Two specific items of unfinished business submitted to RSWG by the Ministers at the November 1989 meeting in Noordwijk are the consideration of the feasibility of achieving: (1) targets to limit or reduce CO_2 emissions, including, e.g., a 20 percent reduction of CO_2 emission levels by the year 2005; (2) a world net forest growth of 12 million hectares a year in the beginning of the next century.

The subgroup chairs and topic coordinators took the responsibility for completing their individual reports and, along with their respective governments, contributed generously of their time and resources to that end.

The RSWG Policymakers Summary is the culmination of the first year of effort by this body. The RSWG has gone to considerable lengths to ensure that the summary accurately reflects the work of the various subgroups and tasks. Given the very strict time schedule under which the RSWG was asked to work, this first report can be only a beginning.

—FREDERICK M. BERNTHAL
Chairman
Response Strategies Working Group
Intergovernmental Panel on
Climate Change

EXECUTIVE SUMMARY

Working Group III (Response Strategies Working Group) was tasked to formulate appropriate response strategies to global climate change. This was to be done in the context of the work of Working Group I (Science) and Working Group II (Impacts), which concluded that:

We are certain emissions resulting from human activities are substantially increasing the atmospheric concentrations of the greenhouse gases: carbon dioxide, methane, chlorofluoro-carbons (CFCs), and nitrous oxide. These increases will enhance the greenhouse effect, resulting on average in an additional warming of the Earth's surface.

The longer emissions continue at present-day rates, the greater reductions would have to be for concentrations to stabilize at a given level.

The long-lived gases would require immediate reductions in emissions from human activities of over 60 percent to stabilize their concentrations at today's levels.

Based on current model results, we predict under the IPCC "Business-as-Usual" emissions of greenhouse gases, a rate of increase of global mean temperature during the next century of about 0.3°C per decade (with an uncertainty range of 0.2°C to 0.5°C per decade), greater than that seen over the past 10,000 years; under the same scenario, we also predict an average rate of global mean sea level rise of about 6 cm per decade over the next century (with an uncertainty range of 3–10 cm per decade).

There are many uncertainties in our predictions particularly with regard to the timing, magnitude, and regional patterns of climate change.

Ecosystems affect climate, and will be affected by a changing climate and by increasing carbon dioxide concentrations. Rapid changes in climate will change the composition of ecosystems; some species will benefit while others will be unable to migrate or adapt fast enough and may become extinct. Enhanced levels of carbon dioxide may increase productivity and efficiency of water use of vegetation.

In many cases, the impacts will be felt most severely in regions already under stress, mainly the developing countries.

The most vulnerable human settlements are those especially exposed to natural hazards, e.g., coastal or river flooding, severe drought, landslides, severe storms and tropical cyclones.

Any responses will have to take into account the great diversity of different countries' situations and responsibilities and the negative impacts on different countries, which consequently would require a wide variety of responses. Developing countries, for example, are at widely varying levels of development and face a broad range of different problems. They account for 75 percent of the world population and their primary resource bases differ widely. Nevertheless, they are most vulnerable to the adverse consequences of climate change because of limited access to the necessary information, infrastructure, and human and financial resources.

MAIN FINDINGS

1) Climate change is a global issue; effective responses would require a global effort that may have a considerable impact on humankind and individual societies.

2) Industrialized countries and developing countries have a common responsibility in dealing with problems arising from climate change.

3) Industrialized countries have specific responsiblities on two levels:

   (a) a major part of emissions affecting the atmosphere at present originates in industrialized countries where the scope for change is greatest. Industrialized countries should adopt domestic measures to limit climate change by adapting their own economies in line with future agreements to limit emissions;

   (b) to cooperate with developing countries in international action, without standing in the way of the latter's development, by contributing additional financial resources, by appropriate transfer of technology, by engaging in close cooperation concerning scientific observation, by analysis and research, and finally by means of technical cooperation geared to forestalling and managing environmental problems.

4) Emissions from developing countries are growing and may need to grow in order to meet their development requirements and thus, over time, are likely to represent an increasingly significant percentage of global emissions. Developing countries have the responsibility, within the limits feasible, to take measures to suitably adapt their economies.

5) Sustainable development requires the proper concern for environmental protection as the necessary basis for continuing economic growth. Continuing economic development will increasingly have to take into account the issue of climate change. It is imperative that the right balance between economic and environmental objectives be struck.

6) Limitation and adaptation strategies must be considered as an integrated package and should complement each other to minimize net costs. Strategies that limit greenhouse gas emissions also make it easier to adapt to climate change.

7) The potentially serious consequences of climate change on the global environment give sufficient reasons to begin by adopting response strategies that can be justified immediately even in the face of significant uncertainties.

8) A well-informed population is essential to promote awareness of the issues and provide guidance on positive practices. The social, economic, and cultural diversity of nations will require tailored approaches.

A FLEXIBLE AND PROGRESSIVE APPROACH

Greenhouse gas emissions from most sources are likely to increase significantly in the future if no response measures are taken. Although some controls have been put in place under the Montreal Protocol for CFCs and halons, emissions of CO_2, CH_4, N_2O, and other gases such as several CFC substitutes will grow. Under these scenarios, it is estimated that CO_2 emissions will increase from approximately 7 billion* tonnes carbon (BTC) in 1985 to between 11–15 BTC by 2025. Similarly, man-made methane emissions are estimated to increase from about 300 teragrams (Tg) to over 500 Tg by the year 2025. Based on these projections, Working Group I estimated that global warming of 0.3°C/decade could occur.

The climate scenario studies of Working Group I further suggest that control policies on emissions can indeed slow global warming, perhaps from 0.3°C/decade to 0.1°C/decade. The social, economic, and environmental costs and benefits of these control policies have not been fully assessed. It must be emphasized that implementation of measures to reduce global emissions is very difficult, as

* 1 billion = 1000 million

energy use, forestry, and land use patterns are primary factors in the global economy. To take maximum advantage of our increasing understanding of scientific and socio-economic aspects of the issue, a flexible and progressive approach is required. Subject to their particular circumstances, individual nations may wish to consider taking steps now to attempt to limit, stabilize, or reduce the emission of greenhouse gases resulting from human activities and prevent the destruction and improve the effectiveness of sinks. One option that governments may wish to consider is the setting of targets for CO_2 and other greenhouse gases.

Because a large, projected increase in world population will be a major factor in causing the projected increase in global greenhouse gases, it is essential that global climate change strategies include strategies and measures to deal with the rate of growth of the world population.

SHORTER-TERM

The Working Group has identified measures at the national, regional, and international levels as applicable, which, while helping to tackle climate change, can yield other benefits.

LIMITATION

- *Improved energy efficiency* reduces emissions of carbon dioxide, the most significant greenhouse gas, while improving overall economic performance and reducing other pollutant emissions and increasing energy security.
- *Use of cleaner energy sources and technologies* reduces carbon dioxide emissions, while reducing other pollutant emissions that give rise to acid rain and other damaging effects.
- *Improved forest management* and, where feasible, *expansion of forest areas* as possible reservoirs of carbon.
- *Phasing out of CFCs under the Montreal Protocol*, thus removing some of the most powerful and long-lived greenhouse gases, while also protecting the stratospheric ozone layer.
- Agriculture, forestry, and other human activities are also responsible for substantial quantities of greenhouse gas emissions. In the short

term, reductions can be achieved through *improved livestock waste management, altered use and formulation of fertilizers,* and other *changes to agricultural land use,* without affecting food security, as well as through improved management in landfill and wastewater treatment.

ADAPTATION

- Developing *emergency and disaster preparedness* policies and programmes.
- Assessing areas at risk from sea level rise and developing *comprehensive management plans* to reduce future vulnerability of populations and coastal developments and ecosystems as part of coastal zone management plans.
- Improving the *efficiency of natural resource use,* research on control measures for desertification and enhancing adaptability of crops to saline regimes.

LONGER-TERM

Governments should prepare for more intensive action, which is detailed in the report. To do so, they should undertake now:

- *Accelerated and coordinated research programmes* to reduce scientific and socio-economic uncertainties with a view toward improving the basis for response strategies and measures.
- Development of *new technologies* in the fields of energy, industry, and agriculture.
- Review planning in the fields of energy, industry, transportation, urban areas, coastal zones, and resource use and management.
- Encourage beneficial behavioral and structural (e.g., transportation and housing infrastructure) changes.
- Expand the global ocean observing and monitoring systems.

It should be noted that no detailed assessments have been made as of yet of the economic costs and benefits, technological feasibility, or market potential of the underlying policy assumptions.

INTERNATIONAL COOPERATION

The measures noted above require a high degree of international cooperation, with due respect for national sovereignty of states. The international negotiation on a framework convention should start as quickly as possible after the completion of the IPCC First Assessment Report. This, together with any additional protocols that might be agreed upon, would provide a firm basis for effective cooperation to act on greenhouse gas emissions and adapt to any adverse effects of climate change. The convention should, at a minimum, contain general principles and obligations. It should be framed in such a way as to gain the adherence of the largest possible number and most suitably balanced range of countries, while permitting timely action to be taken.

Key issues for negotiation will include the criteria, timing, legal form and incidence of any obligations to control the net emissions of greenhouse gases, how to address equitably the consequences for all, any institutional mechanisms that may be required, the need for research and monitoring, and in particular, the request of the developing countries for additional financial resources and for the transfer of technology on a preferential basis.

FURTHER CONSIDERATION

The issues, options, and strategies presented in this document are intended to assist policymakers and future negotiators in their respective tasks. Further consideration of the summary and the underlying reports of Working Group III should be given by every government, as they cut across different sectors in all countries. It should be noted that the scientific and technical information contained in the policymakers summary and the underlying reports of Working Group III do not necessarily represent the official views of all governments, particularly those that could not participate fully in all Working Groups.

FORMULATION OF RESPONSE STRATEGIES
BY WORKING GROUP III

1. SOURCES OF ANTHROPOGENIC GREENHOUSE GASES

A wide range of human activities result in the release of greenhouse gases, particularly CO_2, CH_4, CFCs, and N_2O, into the atmosphere. Anthropogenic emissions can be categorized as arising from energy production and use, non-energy industrial activities (primarily the production and use of CFCs), agricultural systems, and changes in land-use patterns (including deforestation and biomass burning). The relative contributions of these activities to radiative forcing during the 1980s are discussed in the text and shown in Figure 1 (see Working Group I report for further explanation of the radiative forcing of the various greenhouse gases; see also the Chairman's introduction regarding the quantitative estimates of the contributions to radiative forcing from these activities).

IPCC Working Group I calculated that the observed increases in the atmospheric concentrations of CO_2, CH_4, CFCs, and N_2O during the 1980s, which resulted from human activities, contributed to the enhanced radiative forcing by 56 percent, 15 percent, 24 percent, and 5 percent, respectively.

ENERGY

The single largest anthropogenic source of radiative forcing is energy production and use. The consumption of energy from fossil fuels (coal, petro-

leum, and natural gas, excluding fuelwood) for industrial, commercial, residential, transportation, and other purposes results in large emissions of CO_2 accompanied by much smaller emissions of CH_4 from coal mining and the venting of natural

FIGURE 1: Estimated Contribution of Different Human Activities to the Change in Radiative Forcing During the Decade from 1980 to 1990*

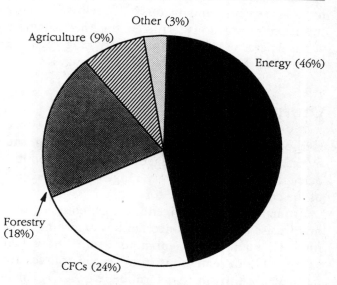

* Percentages derived from estimated greenhouse gas concentrations in the atmosphere and the Global Warming Potentials of these greenhouse gases given by Working Group I on pages 11 and 12 of that group's Policymakers Summary.

gas; the energy sector accounts for an estimated 46 percent (with an uncertainty range of 38–54 percent) of the enhanced radiative forcing resulting from human activities.

Natural fluxes of CO_2 into the atmosphere are large (200 Bt/yr*), but inputs from man-made sources are large enough to significantly disturb the atmospheric balance.

INDUSTRY

The production and use of CFCs and other halocarbons in various industrial processes comprise about 24 percent of the enhanced radiative forcing.

FORESTRY

Deforestation, biomass burning including fuelwood, and other changes in land-use practices release CO_2, CH_4, and N_2O into the atmosphere and together comprise about 18 percent (with an uncertainty range of 9–26 percent) of the enhanced radiative forcing.

AGRICULTURE

Methane releases from rice cultivation and from livestock systems, and nitrous oxide released during the use of nitrogenous fertilizers together comprise about 9 percent (with an uncertainty range of 4–13 percent) of the enhanced radiative forcing.

OTHER SOURCES

Carbon dioxide from cement manufacturing and methane from landfills together comprise about 3 percent (with an uncertainty range of 1–4 percent) of the enhanced radiative forcing.

Estimates of current greenhouse gas emissions are not precise because of uncertainties regarding both total emissions and emissions from individual sources. Global emissions from certain sources are particularly difficult to determine, e.g., CO_2 emission from deforestation, CH_4 emission from rice cultivation, livestock systems, biomass burning,

* Billion (or 1000 million) tons per year

coal mining and venting of natural gas, and N_2O emissions from all sources. The range of such estimates can be quite large, typically, a factor of 1.5 for methane from livestock, a factor of 4 for CO_2 from deforestation, and up to a factor of 7 for rice cultivation.

2. FUTURE EMISSIONS OF GREENHOUSE GASES

Greenhouse gas emissions from most sources are likely to increase significantly in the future if no policy measures are taken. As economic and population growth continues, in particular in the developing countries, there is expected to be an increase in energy use, industrial and agricultural activity, deforestation, and other activities which result in a net increase of greenhouse gas emissions. Although some controls have been put in place under the Montreal Protocol for certain CFCs and halons, emissions of CO_2, methane, nitrous oxide, and other greenhouse gases are likely to increase under current patterns of economic activity and growth.

However, because of the inherent limitations in our ability to estimate future rates of population and economic growth, etc., there is some uncertainty in the projections of greenhouse gas emissions, individual behavior, technological innovation, and other factors that are crucial for determining emission rates over the course of the next century. This lends uncertainty to projections of greenhouse gas emissions over several decades or longer. Reflecting these inherent difficulties, the RSWG's work on emissions scenarios are the best estimates at this time covering emissions over the next century, but further work needs to be done.

The RSWG used two methods to develop scenarios of future emissions as discussed in Sections 2.1 and 2.2. One method used global models to develop four scenarios which were subsequently used by Working Group I to develop estimates of future warming. The second method used studies of the energy and agriculture sectors submitted by over 21 countries and international organizations to estimate emissions. These latter studies were aggregated into a reference scenario. Both approaches show that emissions of CO_2 and CH_4 will increase in the future. Both approaches indicate that CO_2

emissions will grow from approximately 7 BTC to between 11 and 15 BTC by the year 2025.

2.1 EMISSIONS SCENARIOS

One of the RSWG's first tasks was to prepare some initial scenarios of possible future greenhouse gas emissions for the use of the three IPCC Working Groups. An experts' group was formed that looked at four hypothetical future patterns of greenhouse gas emissions and their effect on the atmosphere. The cumulative effect of these emissions was calculated using the concept of equivalent CO_2 concentrations (e.g., the contributions of all greenhouse gases to radiative forcing are converted into their equivalent in terms of CO_2 concentrations). Global economic growth rates were taken from World Bank projections, and population estimates were taken from United Nations (UN) studies, and assumed equal for all scenarios.

The first of the scenarios, called the "Business as Usual" or the 2030 High Emissions Scenario, assumes that few or no steps are taken to limit greenhouse gas emissions. Energy use and clearing of tropical forests continue and fossil fuels, in particular coal, remain the world's primary energy source. The Montreal Protocol comes into effect, but without strengthening, and with less than 100 percent compliance. Under this scenario, the equivalent of a doubling of pre-industrial CO_2 levels occurs, according to Working Group I, by around 2025.

The predicted anthropogenic contributions to greenhouse gas emissions in 2025 are shown in Table 1. The RSWG attempted to synthesize and compare the results of the AFOS/EIS Reference Scenario and the Task A "Business as Usual" (or "2030 High Emissions") Scenario (see Figure 2). The figure shows the equivalent CO_2 concentrations for the Task A "Business as Usual" Scenario and the AFOS/EIS Reference Scenario with its higher CO_2 emissions and the CFC phase-out agreed to by the Parties to the Montreal Protocol. The results indicate that the CO_2 equivalent concentrations and thus the effect on the global climate are similar for both scenarios.

The second of the scenarios, the 2060 Low Emissions Scenario, assumes that a number of environmental and economic concerns result in steps to reduce the growth of greenhouse gas emissions.

Energy efficiency measures, which might only be possible with government intervention, are implemented, emissions controls are adopted globally, and the share of the world's primary energy provided by natural gas increases. Full compliance with the Montreal Protocol is achieved and tropical deforestation is halted and reversed. Under this scenario, the cumulative effect of such measures is a CO_2 equivalent doubling around 2060.

The remaining two scenarios reflect futures where steps in addition to those in the 2060 Low Emissions Scenario are taken to reduce greenhouse gas emissions. These steps include rapid utilization of renewable energy sources, strengthening of the Montreal Protocol, and adoption of agricultural policies to reduce emissions from livestock systems, rice paddies, and fertilizers.

All of the above scenarios provide a conceptual basis for considering possible future patterns of emissions and the broad responses that might affect those patterns. However, they represent assumptions rather than cases derived from specific studies. In addition, no full assessment was made as yet of the total economic costs and benefits, technological feasibility, or market potential of the underlying policy assumptions.

FIGURE 2: EIS/AFOS Reference Scenario—Task A "Business as Usual" CO_2 Equivalent Concentrations

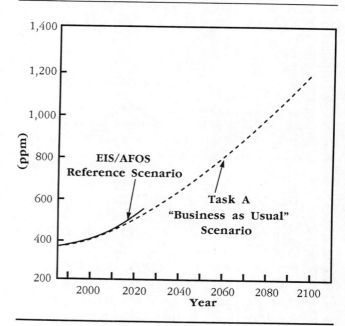

TABLE 1:   Anthropogenic Greenhouse Gas Emissions From Working Group III Scenarios

	AFOS/EIS REFERENCE SCENARIO MODIFIED TO INCLUDE CFC PHASE-OUT[a]		TASK A "BUSINESS AS USUAL" SCENARIO	
	1985	2025	1985	2025
CO_2 Emissions (BTC)				
Energy	5.1	12.4	5.1	9.9
Deforestation	1.7[b]	2.6	0.7[c]	1.4
Cement	0.1	0.2	0.1	0.2
Total	**6.9**	**15.2**	**5.9**	**11.5**
CH_4 Emissions ($TgCH_4$)[d]				
Coal Mining	44.0	126.0	35.0	85.0
Natural Gas	22.0	59.0	45.0	74.0
Rice	110.0	149.0	110.0	149.0
Enteric Ferm.	75.0	125.0	74.0	125.0
Animal Waste	37.0	59.0	–	–
Landfills	30.0	60.0	40.0	71.0
Biomass Burning	53.0	73.0	53.0	73.0
Total	**371.0**	**651.0**	**357.0**	**577.0**
N_2O (TgN)[d]	4.6	8.7	4.4	8.3
CO (TgC)[d]	473.0	820.0	443.0	682.0
NO_X (TgN)[d]	38.0	69.0	29.0	47.0
CFCs (Gg)				
CFC-11	278.0	11.0	278.0	245.0
CFC-12	362.0	10.0	362.0	303.0
HCFC-22	97.0	1572.0	97.0	1340.0
CFC-113	151.0	0.0	151.0	122.0
CFC-114	15.0	0.0	15.0	9.0
CFC-115	5.0	0.0	5.0	5.0
CCl_4	87.0	110.0	87.0	300.0
CH_3CCl_3	814.0	664.0	814.0	1841.0
Halon 1301	2.1	1.8	2.1	7.4

[a] The estimates for emissions of CFCs in 1985 and 2025 reflect the decisions taken at the meeting of the Parties to the Montreal Protocol in London in June 1990. At that meeting, the Parties agreed to accelerate the phase-out of the production and consumption of CFCs, halons, carbon tetrachloride and methyl chloroform.

[b] Midrange estimates for deforestation and biomass consistent with preferred value from Working Group I.

[c] Assuming low biomass per hectare and deforestation rates.

[d] Differences in the 1985 emissions figures are due to differences in definitions and qualifying the emissions from these particular sources.

2.2   REFERENCE SCENARIO

Table 2 shows the results of the EIS Reference Scenario (for CO_2 emissions from the energy sector only) divided by region. The table is incomplete and does not include CO_2 emissions from non-energy sources or other greenhouse gases and sinks. While it is not directly a measure of a region's climate forcing contribution, this table does portray a future where, in the absence of specific policy measures, global emissions of one major gas, CO_2, grow from 5.15 BTC in 1985, to 7.30 BTC in 2000 and 12.43 BTC in 2025. Primary energy demand more than doubles between 1985 and 2025, an average annual growth rate of 2.1 percent.

The annual rate of growth in CO_2 emissions

TABLE 2: Gross CO$_2$ Emissions from the Energy Sector* (From the Reference Scenario)

	CO$_2$ EMISSIONS IN BILLION TONNES CARBON					
	1985	%	2000	%	2025	%
Global Totals	**5.15**	**(100)**	**7.30**	**(100)**	**12.43**	**(100)**
Industrialized	3.83	(74)	4.95	(68)	6.95	(56)
North America	1.34	(26)	1.71	(23)	2.37	(19)
Western Europe	0.85	(16)	0.98	(13)	2.37	(19)
OECD Pacific	0.31	(6)	0.48	(7)	1.19	(10)
Centrally Planned Europe	1.33	(26)	1.78	(24)	2.77	(22)
Developing	1.33	(26)	2.35	(32)	5.48	(44)
Africa	0.17	(3)	0.28	(4)	0.80	(6)
Centrally Planned Asia	0.54	(10)	0.88	(12)	1.80	(14)
Latin America	0.22	(4)	0.31	(4)	0.65	(5)
Middle East	0.13	(3)	0.31	(4)	0.67	(5)
South and East Asia	0.27	(5)	0.56	(8)	1.55	(12)

	CO$_2$ EMISSIONS IN TONNES CARBON PER CAPTIA AND BY CARBON INTENSITY					
	1985		2000		2025	
	PC[a]	CI[b]	PC	CI	PC	CI
Global	**1.06**	**15.7**	**1.22**	**15.8**	**1.56**	**16.0**
Industrialized	3.12	16.3	3.65	16.1	4.65	16.0
North America	5.08	15.7	5.75	15.8	7.12	16.6
Western Europe	2.14	15.6	2.29	15.1	2.69	14.6
OECD Pacific	2.14	16.1	3.01	16.1	3.68	14.8
Non-OECD Europe	3.19	17.5	3.78	16.9	5.02	16.4
Developing	0.36	14.2	0.51	15.2	0.84	16.0
Africa	0.29	12.3	0.32	13.2	0.54	15.2
Centrally Planned Asia	0.47	17.3	0.68	18.8	1.15	19.6
Latin America	0.55	11.5	0.61	11.4	0.91	11.8
Middle East	1.20	16.7	1.79	16.1	2.41	15.5
South and East Asia	0.19	12.3	0.32	14.3	0.64	15.6

* This table presents regional CO$_2$ emissions and does not include CFCs, CH$_4$, O$_3$, N$_2$O, or sinks. Climate change critically depends on all GHG from all economic sectors. Totals and subtotals reflect rounding. *This table should be interpreted with care.*

[a] PC–Per capita carbon emissions in tonnes carbon per person.

[b] CI–Carbon intensity in kilograms carbon per gigajoule.

varies from 0.7 percent in Western Europe, 1.3 percent in North America and the Pacific OECD Countries, to 3.6 percent in developing countries. The share of emissions between regions varies over time.

Under this scenario, the per capita emissions in the industrialized countries increase from 3.1 tonnes carbon (TC) per capita in 1985 to 4.7 TC per capita in 2025. For the developing countries, the per capita emissions rise from 0.4 TC per capita in 1985 to 0.8 TC per capita in 2025.

The Reference Scenario sets out an example of the scope of the reductions in total global emissions that might be necessary to stabilize or reduce CO$_2$ emissions. The stabilization of global emissions at 1985 levels would require reductions of 29 percent by

2000 and 50 percent by 2025. A reduction of global emissions to 20 percent below 1985 levels would require reductions of 44 percent in 2000 and 67 percent by 2025.

The carbon intensity figures show, for each region, the amount of carbon emitted per unit of energy consumed. The contribution of energy consumption in a region to global warming is largely a function of its carbon intensity, total fuel use, and of the efficiency with which it consumes fossil fuels. Carbon intensity for industrialized countries changes from 16.3 tonnes carbon per gigajoule (TC-GJ) in 1985 to 15.5 in 2025. In the developing world the change is from 14.2 TC-GJ to 15.6.

3. RESPONSE STRATEGIES FOR ADDRESSING GLOBAL CLIMATE CHANGE

Because climate change could potentially result in significant impacts on the global environment and human activities, it is important to begin considering now what measures might be taken in response. Working Group I found that under a "Business as Usual" Scenario global average temperature could rise by 0.3 degrees centigrade per decade; it also found that under the Accelerated Control Policies Scenario (Scenario D) with extremely stringent emissions reductions the temperature rise could perhaps be reduced to 0.1 degree centigrade per decade. The RSWG identified a wide range of options for the international community to consider. These include measures both to limit net greenhouse gas emissions and to increase the ability of society and managed ecosystems to adapt to a changing climate.

Strategies that focus only on one group of emissions sources, one type of abatement option, or one particular greenhouse gas will not achieve this. Policy responses, should, therefore, be balanced against alternative abatement options among the energy, industry, forestry, and agricultural sectors, and adaptation options and other policy goals where applicable at both national and international levels. Ways should be sought to account for other countries, and for intergenerational issues, when making policy decisions.

The consideration of climate change response strategies, however, presents formidable difficulties for policymakes. On the one hand, the information available to make sound policy analyses is inadequate because of: (a) remaining scientific uncertainties regarding the magnitude, timing, rate, and regional consequences of potential climate change; (b) uncertainty with respect to how effective specific response options or groups of options would be in actually averting potential climate change; and (c) uncertainty with respect to the costs, effects on economic growth, and other economic and social implications of specific response options or groups of options. The potentially serious consequences of climate change on the global environment, however, give sufficient reasons to begin by adopting response strategies that can be justified immediately even in the face of such significant uncertainties.

Recognizing these factors, a large number of options were preliminarily assessed. It appears that some of these options may be economically and socially feasible for implementation in the near term while others, because they are not yet technically or economically viable, may be more appropriate for implementation in the longer term. In general, the RSWG found that the most effective response strategies, especially in the short term, are those which are:

- beneficial for reasons other than climate change and justifiable in their own right—for example, increased energy efficiency and lower greenhouse gas emission technologies, better management of forests and other natural resources, and reductions in emissions of CFCs and other ozone-depleting substances that are also radiatively important gases;
- economically efficient and cost effective, in particular those that use market-based mechanisms;
- able to serve multiple social, economic, and environmental purposes;
- flexible and phased, so that they can be easily modified to respond to increased understanding of scientific, technological, and economic aspects of climate change;
- compatible with economic growth and the concept of sustainable development;
- administratively practical and effective in terms of application, monitoring, and enforcement; and

reflecting obligations of both industrialized and developing countries in addressing this issue, while recognizing the special needs of developing countries, in particular in the areas of financing and technology.

The degree to which options are viable will also vary considerably depending on the region or country involved. For each country, the implications of specific options will depend on its social, environmental, and economic context. Only through careful analysis of all available options will it be possible to determine which are best suited to the circumstances of a particular country or region. Initially, the highest priority should be to review existing policies with a view to minimizing conflicts with the goals of climate change strategies. New policies will be required.

4. OPTIONS FOR LIMITING GREENHOUSE GAS EMISSIONS

The RSWG reviewed potential measures for mitigating climate change by limiting net emissions of greenhouse gases from the energy, industry, transportation, housing and building, forestry, agriculture, and other sectors. These measure include those that limit emissions from greenhouse gas sources (such as energy production and use), those that increase the use of natural sinks (such as immature forests and other biomass) for sequestering greenhouse gases, as well as those measures aimed at protecting reservoirs such as existing forests. While the RSWG was not mandated to consider the role of the oceans, Working Group I noted that oceans also play an equally important role as sinks and reservoirs for carbon dioxide. A discussion of both short- and long-term options for each major emissions sector is provided below.

It also should be recognized that the large projected increase in the world population, to as much as ten billion people during the next century, will be a major factor in causing the projected increase in global greenhouse gases. This is because larger populations will be accompanied by increased consumption of energy and of food, more land clearing, and other activities, all of which will cause an increase in net greenhouse gas emission. It is essential, therefore, that policies designed to deal effectively with the issue of potential global climate change include strategies and measures to reduce the rate of growth of the world population.

4.1 LIMITATION OF NET EMISSIONS FROM THE ENERGY SECTOR

The energy sector plays a vitally important role in economic well-being and development for all nations. At the same time, because energy production and use account for approximately one half of the radiative forcing from human activities, energy policies need to ensure that continued economic growth occurs in a manner that, globally, conserves the environment for future generations. However, there is no single, quick-fix technological option for limiting greenhouse gas emissions from energy sources. A comprehensive strategy is necessary that deals with improving efficiency on both the demand and supply sides as a priority and emphasizes technological research, development, and deployment.

The RSWG recognizes the particular difficulties that will be faced by countries, particularly developing countries whose economy is heavily dependent on the production and/or export of fossil fuels, as a consequence of actions taken by other countries to limit or reduce energy-related greenhouse gas emissions. These difficulties should be taken into account when elaborating international strategies.

Various potential options have been identified for reducing greenhouse gas emissions from energy systems. The most relevant categories of options appear to be:

- efficiency improvements and conservation in energy supply, conversion, and end use;
- fuel substitution by energy sources that have lower or no greenhouse gas emissions;
- reduction of greenhouse gas emissions by removal, recirculation, or fixation; and
- management and behavioral changes (e.g., increased work in homes through information technology) and structural changes (e.g., modal shift in transport).

From an analysis of the technologies in these categories, it appears that some technologies are

TABLE 3: Examples of Short-Term Options

I. IMPROVE EFFICIENCY IN THE PRODUCTION, CONVERSION, AND USE OF ENERGY

ELECTRICITY GENERATION	INDUSTRY SECTOR	TRANSPORT SECTOR	BUILDING SECTOR
· Improved efficiency in electricity generation: –repowering of existing facilities with high efficiency systems; –introduction of integrated gasification combined cycle systems; –introduction of atmospheric fluidized bed combustion; –introduction of pressurized fluidized bed combustion with combined cycle power systems; –improvements of boiler efficiency. · Improved system for cogeneration of electricity and steam. · Improved operation and maintenance. · Introduction of photovoltaics, especially for local electricity generation. · Introduction of fuel cells.	· Promotion of further efficiency improvements in production process. · Materials recycling (particularly energy-intensive materials). · Substitution with lower energy intensity materials. · Improved electro-mechanical drives and motors. · Thermal process optimization, including energy cascading and cogeneration. · Improved operation and maintenance.	· Improved fuel efficiency of road vehicles: –electronic engine management and transmission control systems; –advanced vehicle design: reduced size and weight, with use of lightweight composite materials and structural ceramics; improved aerodynamics, combustion chamber components, better lubricants and tire design, etc. –regular vehicle maintenance; –higher capacity trucks; –improved efficiency in transport facilities; –regenerating units. · Technology development in public transportation: –intra-city modal shift (e.g., car to bus or metro); –advanced train control system to increase traffic density on urban rail lines; –high-speed inter-city trains; –better intermodal integration. · Improved driver behavior, traffic management, and vehicle maintenance.	· Improved heating and cooling equipment and systems: –improvement of energy efficiency of air conditioning; –promotion of introduction of area heating and cooling including use of heat pumps; –improved burner efficiency; –use of heat pumps in buildings; –use of advanced electronic energy management control systems. · Improved space conditioning efficiency in house/building: –improved heat efficiency through highly efficient insulating materials; –better building design (orientation, window, building, envelope, etc.); –improved air-to-air heat exchangers. · Improved lighting efficiency. · Improved appliance efficiency. · Improved operation and maintenance. · Improved efficiency of cook stoves (in developing countries).

TABLE 3 (*continued*):   Examples of Short-Term Options

II. NON FOSSIL AND LOW EMISSION ENERGY SOURCES

ELECTRICITY GENERATION	OTHER SECTORS
· Construction of small-scale and large-scale hydro projects. · Expansion of conventional nuclear power plants. · Construction of gas-fired power plants. · Standardized design of nuclear power plants to improve economics and safety. · Development of geothermal energy projects. · Introduction of wind turbines. · Expansion of sustainable biomass combustion. · Replacement of scrubbers and other energy-consuming control technology with more energy efficient emission control.	· Substitution of natural gas and biomass for heating oil and coal. · Solar heating. · Technologies for producing and utilizing alternative fuels: —improved storage and combustion systems for natural gas; —introduction of flexible-fuel and alcohol fuel vehicles.

III. REMOVAL, RECIRCULATION, OR FIXATION

ENERGY/INDUSTRY	LANDFILLS
· Recovery and use of leaked or released CH_4 from fossil fuel storage, coal mining. · Improved maintenance of oil and natural gas and oil production and distribution systems to reduce CH_4 leakage. · Improved emission control of CO, SO_X, NO_X and VOCs to protect sinks of greenhouse gases.	· Recycle and incineration of waste materials to reduce CH_4 emissions. · Use or flaring of CH_4 emissions. · Improved maintenance of landfill to decrease CH_4 emissions.

---

available now or in the short term while others need further development to lower costs or to improve their environmental characteristics.

Tables 3 and 4 provide various examples of technological options within each of the broad categories defined above, and their possible application in the short, medium, and longer term. This distinction among time frames is used in order to reflect the remaining technological needs in each category and to assist in formulating technological strategies. Short-term technologies are those that apparently are or will be both technically and economically ready for introduction and/or demonstration by the year 2005. Mid-term technologies are those that while technically available now, are not yet economic and thus may not be implemented until the period from 2005 to 2030. Longer-term technologies are not yet available but may emerge after 2030 as a result of research and development. Such time frames could be influenced by such factors as the pace of the technological changes and economic conditions.

The technical, economic, and market potential of technological options will vary depending upon the sector in which they are to be applied. The technical potential of an energy technology is its capacity to reduce potential emissions, irrespective of the costs involved, and is largely a function of technical feasibility and resource availability. Economic potential refers to whether the application of the options is economically efficient and cost-effective—it may be significantly less than technical potential where there are positive resource costs. Market potential refers to whether the consumer or user is likely to adopt the option—it might be even less than economic potential due to market imperfections, attitudes to risk, and the presence of non-monetary costs.

There is, in general, extensive information available on the technical potential of the many technological options listed. For example:

• in the Transportation sector, vehicle efficiency improvements have very high technical

TABLE 4: Examples of Medium-/Long-Term Options

I. IMPROVE EFFICIENCY IN THE PRODUCTION, CONVERSION, AND USE OF ENERGY

ELECTRICITY GENERATION	INDUSTRY SECTOR	TRANSPORT SECTOR	BUILDING SECTOR
· Advanced technologies for storage of intermittent energy. · Advanced batteries. · Compressed air energy storage. · Superconducting energy storage.	· Increased use of less energy-intensive materials. · Advanced process technologies. · Use of biological phenomena in processes. · Localized process energy conversion. · Use of fuel cells for cogeneration.	· Improved fuel efficiency of road vehicles. · Improvements in aircraft and ship design: –advanced propulsion concepts; –ultra high-bypass aircraft engines; –contra rotating ship propulsion.	· Improved energy storage systems: –use of information technology to anticipate and satisfy energy needs; –use of hydrogen to store energy for use in buildings. · Improved building systems: –new building materials for better insulation at reduced cost; –windows which adjust opacity to maximize solar gain. · New food storage systems which eliminate refrigeration requirements.

II. NON FOSSIL AND LOW EMISSION ENERGY SOURCES

ELECTRICITY GENERATION	OTHER SECTORS
· Nuclear power plants: –passive safety features to improve reliability and acceptability. · Solar power technologies: –solar thermal; –solar photovoltaic (especially for local electricity generation). · Advanced fuel cell technologies.	· Other technologies for producing and utilizing alternative fuels: –improved storage and combustion systems for hydrogen. –control of gases boiled off from cryogenic fuels. –improvements in performance of metal hydrides. –high-yield processes to convert ligno-cellulosic biomass into alcohol fuels. –introduction of electric and hybrid vehicles. –reduced re-charging time for advanced batteries.

III. REMOVAL, RECIRCULATION, OR FIXATION

· Improved combustion conditions to reduce N_2O emissions.

· Treatment of exhaust gas to reduce N_2O emissions.

· CO_2 separation and geological and marine disposal.

potential (e.g., 50 percent improvement from the average vehicle on the road in some countries);

- in the Electricity Generation sector, efficiency improvements of 15 to 20 percent could be achieved for retrofits of coal plants and up to 65 percent of new generation versus average existing coal plants; fuel substitution could achieve 30 percent (for oil to natural gas) to 40 percent (for coal to natural gas) reduction in emissions of CO_2;
- in the Building sector, new homes could be roughly twice as energy efficient and new commercial buildings up to 75 percent more energy efficient than existing buildings; retrofitting existing homes could average 25 percent improvement and existing commercial buildings around 50 percent.
- in the Industry sector, the technical potential for efficiency improvements ranges from around 15 percent in some sub-sectors to over 40 percent in others (i.e., the best available technology versus the stock average).

The constraints to achieving the technical potential in these sectors can be generally categorized as:

- capital costs of more efficient technologies vis-à-vis the cost of energy;
- relative prices of fuels (for fuel substitution);
- lack of infrastructure;
- remaining performance drawbacks of alternative technologies;
- replacement rates;
- reaching the large number of individual decision makers involved.

Each of these constraints may be more or less significant depending on the sector in question. While not a constraint, behavioral changes (e.g., improved driver behavior, better vehicle maintenance, and turning off unused lights) can make significant contributions to emissions reduction in all sectors. Achieving such changes requires the engagement of both the energy supplier and the consumer. Likewise, improvements in operational practices on the part of industry and government (e.g., better traffic management or boiler operation) offer significant potential but require increased attention. Transport and housing policies (e.g., promotion of public transport, home insulation) could also reduce

greenhouse gas emissions. A more comprehensive assessment of the measures to overcome these constraints is contained in section 7 of this policymakers summary.

Factors external to the energy sector also significantly constrain potential. These include the difficulty of:

- making basic changes in the structure of economies (e.g., development of new transportation and housing infrastructure);
- making fundamental changes in attitudinal and social factors (e.g., preferences for smaller and higher-efficiency vehicles).

The challenge to policymakers is to enhance the market uptake of technological options and behavioral and operational changes as well as to address the broader issues outside the energy sector in order to capture more of the potential that exists.

Options and Strategies

Tables 3 and 4 summarize the technological, regulatory, and institutional approaches that could form elements of strategies to control greenhouse gases.

A list of options recommended by EIS as measures for addressing greenhouse gas emissions is given below. Countries are encouraged to evaluate the social, economic, and environmental consequences of these options:

- taking steps now* to attempt to limit, stabilize, or reduce the emission of energy-related greenhouse gases and prevent the destruction and improve the effectiveness of sinks (one option that governments may wish to consider is the setting of targets for CO_2 and other greenhouse gases);
- adopting a flexible progressive approach, based on the best available scientific, economic, and technological knowledge, to action needed to respond to climate change;
- drawing up specific policies and implementing wide-ranging comprehensive programmes that cover all energy-related greenhouse gases;

---

* There was significant concern expressed at the RSWG meeting about the immediacy implied by the word "now" in option one, when implementation could only be considered at a rate consistent with countries' level of knowledge and particular circumstances.

- starting with implementing strategies that have multiple social, economic, and environmental benefits, are cost effective, are compatible with sustainable development, and make use of market forces in the best way possible;
- intensifying international, multilateral, and bilateral cooperation in developing new energy strategies to cope with climate change. In this context, industrialized countries are encouraged to promote the development and transfer of energy-efficient and clean technologies to other countries;
- increasing public awareness of the need for external environmental costs to be reflected in energy prices, markets, and policy decisions to the extent that they can be determined;
- increasing public awareness of energy-efficient technologies and products and alternatives, through public education and information (e.g., labeling);
- strengthening research and development and international collaboration in energy technologies, and economic and energy policy analysis, which are relevant for climate change;
- encouraging the participation of industry, the general public, and NGOs in the development and implementation of strategies to limit greenhouse gas emissions.

Short-Term Strategy Options

Short-term strategies for all individual nations include:

- improving diffusion of energy-efficient and alternative energy technologies that are technically and commercially proven;
- improving energy efficiency of mass-produced goods, including motor vehicles and electrical appliances, and equipment and buildings (e.g., through improved standards);
- developing, diffusing, and transferring technologies to limit energy-related greenhouse gas emissions;
- reviewing energy-related price and tariff systems and policy decisions on energy planning to better reflect environmental costs.

Long-Term Strategy Options

Over the longer term, sustainable development will remain a central theme of policies and strategies. Specific approaches within a sustainable development policy framework will evolve as our understanding of climate change and its implications improves.

Long-term strategies for all individual nations include:

- accelerating work to improve the long-term potential of efficiency in the production and use of energy; encouraging a relatively greater reliance on no or lower greenhouse gas emissions energy sources and technologies; and enhancing natural and man-made means to sequester greenhouse gases;
- further reviewing, developing, and deploying policy instruments, which may include public information, standards, taxes and incentives, tradeable permits, and environmental impact assessments, which will induce sustainable energy choices by producers and consumers without jeopardizing energy security and economic growth;
- developing methodologies to evaluate the trade-off between limitation and adaptation strategies and establishing changes in infrastructure (e.g., pipelines, electrical grids, dams) needed to limit or adapt to climate change.

4.2 LIMITATION OF NET EMISSIONS FROM THE INDUSTRY SECTOR

The most significant source of greenhouse gases associated with industrial activity not related to energy use is the production and use of CFCs and other halocarbons. CFCs represent a very important source of greenhouse gas emissions and account for about 24 percent of the total contributions to the enhanced radiative forcing for the period of the 1980s. While the RSWG did not consider control strategies for these gases, since the issue is already addressed under the Montreal Protocol on Substances that Deplete the Ozone Layer, it noted that the review of the Montreal Protocol now under way should take into account the global warming potential of potential CFC substitutes.

The RSWG did develop future emission scenarios

for CFCs and HCFC-22 (HCFC-22 was used as a surrogate for a potential mix of HCFCs and HFCs substitutes). The potential impact of such substitutes on radiative forcing was assessed by Working Group I. For a given emission rate, HCFCs and HFCs are less effective greenhouse gases than the CFCs because of their shorter lifetimes. The growth rates assumed in the IPCC scenarios will result in the atmospheric concentrations of HFCs and HCFCs becoming comparable to the CFCs during the next several decades, assuming that the CFCs had continued to be used at current rates. Assuming the IPCC scenarios for HFCs and HCFCs, Working Group I calculated that these gases would contribute up to 10 percent of the total additional radiative forcing for the period 2000–50.

4.3 LIMITATION OF NET EMISSIONS FROM THE AGRICULTURE SECTOR

About 9 percent of anthropogenic greenhouse gas emissions can be attributed to the agricultural sector—in particular, livestock systems, rice cultivation, and the use of nitrogenous fertilizers. Limitation of emissions from this sector presents a challenge because the processes by which greenhouse gases—in particular, methane and nitrous oxide—are released in agricultural activities are not well understood. In addition, response options in the agricultural sector must be designed to ensure maintenance of food supply. There appear, however, to be a number of short-term response options, some economically beneficial in their own right, that could contribute to a limitation of net emissions from agricultural sources. Where appropriate, the removal of subsidies, incentives, and regulatory barriers that encourage greenhouse gas emissions from the agricultural sector would be both environmentally and economically beneficial. In addition, there are a number of promising technologies and practices that, in the longer term, could significantly reduce greenhouse gas emissions.

Short-Term Options

Livestock systems: Methane emissions could be reduced through improved management of livestock wastes, expansion of supplemental feeding practices, and increased use of production- and growth-enhancing agents, with safeguards for human health.

Fertilizer use: Nitrous oxide emissions may be reduced by using existing improved fertilizer formulations, judicious use of animal manures and compost, and improved application technology and practices.

Marginal lands: Areas marginally suitable for annual cropping systems may be shifted to perennial cover crops for fodder, pastoral land use, or forests if soils are suitable. Such actions would increase carbon uptake, both in the vegetation and soil, and would yield other benefits.

Sustainable agricultural practices: Where possible, minimum or no-till systems should be introduced for those countries currently using tillage as part of the annual cropping sequence, thus maintaining and increasing soil organic matter.

Longer-Term Options

Rice cultivation: A comprehensive approach, including management of water regimes, improvement of cultivars, efficient use of fertilizers, and other management practices, could lead to a 10 to 30 percent reduction in methane emissions from flooded rice cultivation, although substantial research is necessary to develop and demonstrate these practices. It is estimated that at least twenty years would be needed to introduce such practices. Adaptable alternative crops research is needed to provide a more diverse crop base for rice-growing regions.

Livestock: Through a number of technologies it appears that methane emissions may be reduced from livestock systems by up to 25–75 percent per unit of product in dairy and meat production, although many uncertainties exist.

Fertilizers: Fertilizer-derived emissions of nitrous oxide potentially can be reduced (although to what extent is uncertain) through changes in practices, such as using fertilizers with controlled nitrogen conversion rates, improving fertilizer-use efficiency, and adopting alternative agricultural systems where possible.

Desertification: Enhanced research on control measures.

4.4 LIMITATION OF NET EMISSIONS FROM FORESTRY AND OTHER ACTIVITIES

Forestry and related aspects of land use cannot be considered in isolation, and solutions must be based on an integrated approach that links forestry to other policies, such as those concerned with poverty and land resources, which should be supported by strong institutions in order to enhance overall forest management. The forest crisis is rooted in the agricultural sector and in people's needs for employment and income. Deforestation will be stopped only when the natural forest is economically more valuable for the people who live in and around the forests than alternative uses for the same land.

Forestry practices and other human activities associated with land use, such as biomass burning and landfills, account for about 18 percent of anthropogenic greenhouse gas emissions. A number of short- and long-term response options for limiting net emissions from these sectors have been identified.

Short-Term Options

1) Improvement of forest-management and reduction of deforestation and forest degradation, which should be supported by:
   - reduction of air pollution, which contributes to forest degradation;
   - elimination of inappropriate economic incentives and subsidies that contribute to forest loss, where appropriate;
   - integration of forest conservation requirements and sustainable development in all relevant sectors of national development planning and policy, taking account of the interests of local communities;
   - coordinated remote sensing, data collection, and analyses to provide the required data;
   - a meeting of interested countries from the developing and the industrialized worlds and of appropriate international agencies to identify possible key elements of a world forest conservation protocol in the context of a climate convention process that also addresses energy supply and use, and practical means of implementing it. Such a meeting should also develop a framework and methodology for analyzing the feasibility of the Noordwijk remit, including alternative targets, as well as the full range of costs and benefits;
   - strengthening Tropical Forestry Action Plan (TFAP) and, in the light of the independent review that is being undertaken, the International Tropical Timber Organization (ITTO), and other international organizations whose objective is to help developing countries in achieving conservation and sustainable development and management of forests;
   - an assessment of incentives and disincentives for sustainable forest management—for example, the feasibility of labeling;
   - introduction of sustainable forest harvesting and management;
   - development of enhanced regeneration methods;
   - development and implementation of (large-scale) national afforestation and forest conservation plans, where feasible.

2) Where appropriate, expand forest areas, especially by afforestation, agroforestry, and regreening of available surplus agricultural, urban, and marginal lands.

3) Where appropriate, strengthen and improve the use of forest products and wood through measures such as substituting a portion of fossil energy sources by wood or other sustainable managed biomass; partial replacement of high energy input materials by wood; further recycling of forest products; and improved efficiency of use of fuelwood.

4) Development of methane recovery systems for landfill and wastewater treatment facilities and their use, in particular, in industrialized countries.

Longer-Term Options

1) Maintain the health and the continuance of existing forests as major natural carbon reser-

voirs, especially through the development and implementation of:

- silvicultural adjustment and stress management strategies;
- special forest protection strategies (developed under climate change scenarios);
- environmentally sound treatment practices for peatlands;
- standardization of methods of forest inventory and bio-monitoring to facilitate global forest management.

2) Expand forest biomass, especially of intensively managed temperate forests, by silviculture measures and genetically improved trees.

3) With regard to waste management, use of gas collection and flaring to reduce methane emissions from landfills and development of biogas plants to reduce methane emissions from wastewater treatment. Demonstration, training, and technology transfer are necessary to realize these potentials, which may range from 30 to 90 percent for landfills and up to 100 percent for wastewater treatment.

5. FURTHER WORK ON GREENHOUSE GAS EMISSION LIMITATION GOALS

There has been considerable international discussion of targets for specific greenhouse gas emissions, in particular, CO_2, which is the most abundant of the greenhouse gases. The final declaration at the November 1989 Noordwijk Conference on Atmospheric Pollution and Climate Change encouraged the IPCC to include in its First Assessment Report an analysis of quantitative targets to limit or reduce CO_2 emissions, and urged all industrialized countries to investigate the feasibility of achieving such targets, including, for example, a 20 percent reduction of CO_2 emissions by the year 2005. The Conference also called for assessing the feasibility of increasing net global forest growth by 12 million hectares per year. During its Third Plenary, the IPCC accepted the mandate.

Although the feasibility of quantitative targets on greenhouse gas emissions fell within the RSWG's original mandate through its Energy and Industry

Subgroup (EIS), it was agreed that these new, specific tasks would require more time, data, and analyses in order to be dealt with properly. It was decided, therefore, that the results of the deliberation of the EIS on these remits could not be fully included in its report but only treated in an incomplete and preliminary way. A progress report is to be presented to the fourth IPCC Plenary following an international workshop to be hosted by the United Kingdom in June 1990. As for the Noordwijk remit on global forest growth, the RSWG through its Agriculture, Forestry, and Other Human Activities Subgroup (AFOS) noted that a framework and methodology for analyzing its feasibility should be developed.

While the technical potential of a number of options has been demonstrated, there is very little information available on the actual economic and social feasibility associated with implementation of such options. An adequate understanding of the benefits, in terms of changes in climate variables that are avoided, is also seriously lacking. It is imperative that further work on the cost and benefit implications of response strategies be undertaken. These issues have been identified as one of the most important areas for future research by the RSWG, concerned international organizations, and individual countries.

The material available to the EIS demonstrates the important role emissions of industrialized countries play in total global emissions in the near term. The material also indicates that the technical potential for reduction is large, and differs greatly between regions and countries. Therefore, in the near term, no significant progress in limiting global emissions will occur without actions by the industrialized countries. Some countries have already decided to stabilize or reduce their emissions.

6. MEASURES FOR ADAPTING TO GLOBAL CLIMATE CHANGE

In addition to the limitation options discussed above, the RSWG reviewed measures for adapting to potential climate change. The consideration of adaptation options is critical for a number of reasons. First, because it is believed that there is likely to be a lag time between emissions and subsequent climate change, the climate may already be committed to a certain degree of change. Implementation of adaptation measures may thus be necessary regardless of any limitation actions that may be taken. Second, natural climate variability itself necessitates adaptation.

Furthermore, should significant adverse climate change occur, it would be necessary to consider limitation and adaptation strategies as part of an integrated package in which policies adopted in the two areas complement each other so as to minimize costs. Limitation and adaptation options should be developed and analyzed recognizing the relationship between the timing and costs of limitation and adaptation. For example, the more net emissions are reduced and the rate of climate change potentially slowed, the easier it would be to adapt. A truly comprehensive approach should recognize that controlling the different gases might have different effects on the adaptive capacity of natural resources.

The RWSG explored two broad categories of adaptation options:

- *Coastal zone management*, or options that maximize the ability of coastal regions to adapt to the projected sea level rise and to reduce vulnerability to storms; and
- *Resource use and management*, or options that address the potential impacts of global climate change on food security, water availability, natural and managed ecosystems, land, and biodiversity.

6.1 COASTAL ZONE MANAGEMENT

Under the 2030 High Emissions Scenario, global climate change is predicted to raise global mean sea level 65 cm (with an uncertainty range of 30 to 110 cm) by the year 2100. If sea level rises by 1 meter, hundreds of thousands of square kilometers of coastal wetlands and other lowlands could be inundated, while ocean beaches could erode as much as a few hundred meters over the next century. Flooding would threaten lives, agriculture, livestock, and structures, while salt water would advance inland into aquifers, estuaries, and soils, thus threatening water supplies and agriculture in some areas. Loss

of coastal ecosystems would threaten fishery resources.

Some nations would be particularly vulnerable to such changes. Eight to ten million people live within one meter of high tide in each of the unprotected river deltas of Bangladesh, Egypt, and Vietnam. Half a million people live in coral atoll nations that lie almost entirely within three meters of sea level, such as the Maldives, the Marshall Islands, Tuvalu, Kiribati, and Tokelau. Other states with coastal areas, archipelagos, and island nations in the Pacific and Indian Oceans and the Caribbean could lose much of their beaches and arable lands, which would cause severe economic and social disruption.

Available responses to sea level rise fall broadly into three categories:

- *Retreat:* Under this option no actions would be taken to protect the land from the sea—the focus would instead be on providing for people and ecosystems to shift landward in an optimal fashion. This choice could be motivated by either excessive costs of protection or by a desire to maintain ecosystems.
- *Accommodation:* Under this strategy, while no attempt would be made to protect the land at risk, measures would be taken to allow for continued habitation of the area. Specific responses under this option would include erecting flood shelters, elevating buildings on pilings, converting agriculture to fish farming, or growing flood- or salt-tolerant species.
- *Protection:* A protection strategy uses site-specific features such as seawalls, dikes, dunes, and vegetation to protect the land from the sea so that existing land uses can be retained.

There are various environmental, economic, social, cultural, legal, institutional, and technological implications for each of these options. Retreat could lead to a loss of property, potentially costly resettlement of populations, and, in some notable cases, refugee problems. Accommodation could result in declining property values and costs for modifying infrastructure. Protecting existing development from a one-meter sea level rise would require about 360,000 kilometers of coastal defenses at a total cost of U.S.$500 billion, over the next one hundred years. The annual cost of protection represents, on average, 0.04 percent of total gross national product (GNP), and ranges from zero to 20 percent for individual countries. The estimate is not discounted and does not reflect present coastal defense needs or impacts of salt water intrusion or flooding of unprotected lands. Further, the protection could have negative impacts on fisheries, wildlife, and recreation. The loss of traditional environments could potentially disrupt family life and create social instability.

Actions to Prepare for Possible Sea Level Rise

A number of response options are available which not only enhance the ability of coastal nations to adapt to sea level rise, but are also beneficial in their own right. Implementation of such options would be most effective if undertaken in the short term, not because there is an impending catastrophe, but because there are opportunities to avoid adverse impacts by acting now—opportunities that may not be as effective if the process is delayed. These options include:

National Coastal Planning

- *Development and implementation in the short term of comprehensive national coastal zone management plans*, which (a) deal with both sea level rise and other impacts of global climate change and (b) ensure that risks to populations are minimized while recognizing the need to protect and maintain important coastal ecosystems.
- *Identification of coastal areas at risk.* National efforts are needed to (a) identify functions and resources at risk from a one-meter rise in sea level and (b) assess the implications of adaptive response measures on them.
- *Provisions to ensure that coastal development does not increase vulnerability to sea level rise.* Actions in particular need of review include river levees and dams, conversions of mangroves and other wetlands for agriculture and human habitation, harvesting of coral, and increased settlement in low-lying areas. In addition, while structural measures to prepare for sea level rise are not yet warranted, the design and location of coastal infrastructure and coastal defenses should include consideration of sea level rise and other coastal impacts of climate change. It is sometimes less expensive to design a structure today, incorporating these factors, than to rebuild it later.
- *Review and strengthening of emergency preparedness and coastal zone response mechanisms.* Efforts are needed to develop emergency preparedness plans for reducing vulnerability to coastal storms through better evacuation planning and the development of coastal defense mechanisms that recognize the impact of sea level rise.

International Cooperation

- *Maintenance of a continuing international focus on the impacts of sea level rise.* Existing international organizations should be augmented with new mechanisms to focus attention and awareness on sea level change and to encourage the nations of the world to develop appropriate responses.
- *Provision of technical assistance and cooperation to developing nations.* Institutions offering financial support should take into account the need for technical assistance and cooperation in developing coastal management plans, assessing coastal resources at risk, and increasing a nation's ability—through education, training, and technology transfer—to address sea level rise.
- *Support by international organizations for national efforts to limit population growth in coastal areas.* In the final analysis, rapid population growth is the underlying problem with the greatest impact on both the efficacy of coastal zone management and the success of adaptive response options.

Research, Data, and Information

- *Strengthening of research on the impacts of global climate change on sea level rise.* International and national climate research programmes need to be directed at understanding and predicting changes in sea level, extreme events, precipitation, and other impacts of global climate change on coastal areas.
- *Development and implementation of a global ocean-observing network* — for example, through the efforts of the IOC, WMO, and UNEP—to establish a coordinated international ocean-observing network that will allow for accurate assessment and continuous mon-

itoring of changes in the world's oceans and coastal areas, particularly sea level changes and coastal erosion.

- *Dissemination of data and information on sea level change and adaptive options.* An international mechanism could be identified with the participation of the parties concerned for collecting and exchanging data and information on climate change and its impact on sea level and the coastal zone and on various adaptive options. Sharing this information with developing countries is critically important for preparation of coastal management plans.

A programme could begin now to enable developing countries to implement coastal zone management plans by the year 2000. The programme would provide for training of country experts, data collection, and technical assistance and cooperation. Estimated funding to provide the necessary support over the next five years is U.S.$10,000,000. It is suggested that international organizations such as UNEP and WMO consider coordinating this programme in consultation with interested nations.

6.2   RESOURCE USE AND MANAGEMENT

The reports of Working Groups I and II indicate significant and unavoidable impacts, both positive and negative, upon the very resources that humans and other species rely on to live. These resources include water, agriculture, livestock, fisheries, land, forests, and wildlife. The RSWG addressed these resource issues in the context of considering options for ensuring food security; conserving biological diversity; maintaining water supplies; and using land rationally for managed and unmanaged ecosystems.

The potential impacts of climate change on natural resources and human activities are poorly understood. First, credible regional estimates of changes in critical climatic factors—such as temperature, soil moisture, annual and seasonal variability, and frequencies of droughts, floods, and storms—are simply not available. For many of these critical climate factors even the direction of change is uncertain. Second, methods for translating these changes into effects on the quantity and quality of

resources are generally lacking. While it is clear that some of the impacts of climate change on resources could be negative and others positive, a more specific quantification of those impacts is not possible at this time. Nevertheless, these uncertainties do not preclude taking appropriate actions, especially if they are worthwhile for other, non-climate related, reasons. However, it can be said that: (a) those resources that are managed by humans (e.g., agriculture, forestry) are more suited to successful adaptation than unmanaged ecosystems; and (b) the faster the rate of change, the greater the impact. In that regard, it is very important to realize that some species will not be able to survive rapid climate changes.

Through the ages societies and living things have developed the capability to adapt to the climate's natural variability and to extreme events. Several climatic zones span the globe, and resource use and management is an ongoing challenge in each of these zones. Therefore, society could borrow from this existing large reservoir of experience and knowledge in developing policies to adapt to possible climate change. In addition, expected future economic and technological progress would provide the financial and technical resources required to better adapt to a changing climate. Nevertheless, significant costs and legal, institutional, and cultural adjustments may be necessary to implement adaptation measures.

In recognition of the uncertainties regarding the impacts of climate change on resource use and management, the following sections provide general, rather than specific, options in three categories. The appropriateness of these options for individual countries may vary depending on the specific social, environmental, and economic context.

Short-Term Research Related Options

There are a number of actions that would augment our knowledge base for making reasoned judgments about response strategies. These include:

- developing inventories, data bases, monitoring systems, and catalogues of the current state of resources and resource use and management practices;
- improving our scientific understanding of and predictive tools for critical climatic factors,

their impacts on natural resources, and their socioeconomic consequences;

- undertaking studies and assessments to gauge the resilience and adaptability of resources and their vulnerability to climate change;
- encouraging research and development by both public and private enterprises directed toward more efficient resource use and biotechnological innovation (with adequate safeguards for health, safety, and the environment), including allowing innovators to benefit from their work;
- continuing existing research and development of methods to cope with the potentially worst consequences of climate change, such as developing more drought- or salinity-resistant cultivars or using classical and modern breeding techniques to help keep farming and forestry options open, and research on agrometeorology or agroclimatology;
- increasing research on the preservation of biological resources *in situ* and *ex situ*, including investigations into the size and location of protected natural areas and conservation corridors.

Short-Term Policy Options

Some response strategies are available that are probably economically justified under present-day conditions and that could be undertaken for sound resource management reasons, even in the absence of climate change. In general, these relate to improving the efficiency of natural resource use, fuller utilization of the "harvested" component of resources, and waste reduction. Measures that could be implemented in the short term include:

- increased emphasis on the development and adoption of technologies that may increase the productivity or efficiency (per unit of land or water) of crops, forests, livestock, fisheries, and human settlements, consistent with the principles of sustainable development. Such efficiencies reduce the demand for land for human activities and could also help reduce emissions of greenhouse gases. Examples of specific options include more efficient milk and meat production, improved food storage and distribution, and better water management practices;

- increased promotion and strengthening of resource conservation and sustainable resource use—especially in highly vulnerable areas. Various initiatives could be explored for conserving the most sensitive and valuable resources, including strengthening conservation measures, managing development of highly vulnerable resources, and promoting reforestation and afforestation;
- acceleration of economic development efforts in developing countries. Because these countries often have largely resource-based economies, efforts at improving agriculture and natural resource use would be particularly beneficial. Such efforts would also promote capital formation, which would generally make adaptation to climate change and sustainable development more feasible;
- developing methods whereby local populations and resource users gain a stake in conservation and sustainable resource use—for example, by investing resource users with clear property rights and long-term tenure, and allowing voluntary water transfer or other market mechanisms;
- decentralizing, as practicable, decision making on resource use and management.

Longer-Term Options

There are also a number of other possible responses that are costly or otherwise appear to be more appropriate for consideration in the longer term, once uncertainties regarding climate change impacts are reduced. Options in this category include:

- building large capital structures (such as dams) to provide for enhanced availability of water and other resources;
- strengthening and enlarging protected natural areas and examining the feasibility of establishing conservation corridors to enhance the adaptation prospects for unmanaged ecosystems;
- as appropriate, reviewing and eliminating direct and indirect subsidies and incentives for inefficient resource use, and other institutional barriers to efficient resource use.

7. MECHANISMS FOR IMPLEMENTING RESPONSE STRATEGIES

The RSWG also considered several priority areas that must be addressed in order to adequately implement limitation or adaptation responses. These "implementation mechanisms" represent the primary vehicles through which national, regional, and international responses to climate can be brought into force. The specific implementation mechanisms considered were:

- public education and information;
- technology development and transfer;
- economic (market) mechanisms;
- financial mechanisms; and
- legal and institutional mechanisms, including possible elements of a framework convention on climate change.

The results of the RSWG's deliberations on these issues are provided below.

7.1 PUBLIC EDUCATION AND INFORMATION

A well-informed global population is essential for addressing and coping with an issue as complex as climate change. Because climate change would affect, either directly or indirectly, almost every sector of society, broad global understanding of the issue will facilitate the adoption and implementation of such response options as deemed necessary and appropriate. The dissemination of information also represents a powerful economic instrument for ensuring that markets accurately take into account potential consequences and/or opportunities of climate change.

The core aims of public education and information programmes are to:

- promote awareness and knowledge of climate change issues;
- provide guidance for positive practices to limit and/or adapt to climate change;
- encourage wide participation of all sectors of the population of all countries, both developed and developing, in addressing climate change issues and developing appropriate responses; and
- especially emphasize key target groups, such as children and youth, as well as individuals at household levels, policymakers and leaders, media, educational institutions, scientists, business and agricultural sectors.

Given the importance of a well-informed population, the RSWG developed suggestions and approaches for improving international awareness of the potential causes and impacts of climate change. In this process it was recognized that, while broad-based understanding is essential, no single mechanism can work for every group or in every culture or country. The social, economic, and cultural diversity of nations will likely require educational approaches and information tailored to the specific requirements and resources of particular locales, countries, or regions. The importance of education and information for developing countries cannot be overemphasized.

A number of national and international actions should be taken to disseminate broadly information on climate change. These include the:

- establishment of national committees or clearinghouses to collect, develop, and disseminate objective materials on climate change issues. This could help provide focal points for information on issues such as energy efficiency, energy savings, forestry, agriculture, etc.;
- use by international organizations (UNESCO, UNEP, WMO, etc.) and non-governmental organizations of IPCC and other relevant reports in developing and providing to all countries an adequate understanding for future actions;
- use of an existing international institution, or development of a new institution, if necessary, to serve as a clearinghouse for informational and educational materials;
- upon completion of the IPCC reports, or earlier, arranging a series of short seminars targeted to inform high-priority decision makers, world leaders, and others of causes and effects of climate change.

7.2 TECHNOLOGY DEVELOPMENT AND TRANSFER

The development and transfer of technologies are vital to any effort to address global climate change. The development of new technologies may provide the means by which societies can meet their energy, food, and other needs in the face of changes in global climate, while at the same time minimizing emissions of greenhouse gases. Prompt transfer of technologies, especially to developing countries, is likewise an important aspect of any effort to limit or adapt to climate change.

Technology Research and Development

Technological development, including improvement and reassessment of existing technologies, is needed to limit or reduce anthropogenic greenhouse gas emissions; absorb such gases by protecting and increasing sinks; adapt human activities and resource use and management to the impacts of climate change; and detect, monitor, and predict climate change and its impacts. Technological development could be pursued in a wide range of activities such as energy, industry, agriculture, transport, water supply, coastal protection, management of natural resources, and housing and building construction.

Adequate and trained human resources are a prerequisite for development and transfer of technologies, and technological actions, founded on a sound scientific basis, must be consistent with the concept of sustainable development.

Criteria for selecting technologies include such factors as the existence of economic and social benefits in addition to environmental benefits, economic efficiency taking into account all the external costs, suitability to local needs, ease of administration, information needs, and acceptability to the public.

Appropriate pricing policies where applicable, information exchange on the state of development of technologies, and the support of governments are important measures that can promote technology development. Also of importance are international collaborative efforts, especially between the industrialized and the developing countries in the bilateral and multilateral context.

Technology Transfer

There is a need for the rapid transfer to the developing countries, on a preferential basis, of technologies for addressing climate change. Developing countries are of the view that transfer of technologies on a noncommercial basis is necessary and that specific bilateral and multilateral arrangements should be established to promote this. Some other countries where technologies are not owned by the government believe that transfer of technologies would be a function of commercial negotiations. The issue of intellectual property rights also presents a case where international opinion is mixed.

A number of impediments also exist that hinder the effective transfer of technologies to developing countries. These include lack of financial resources, necessary institutions, and trained human resources. Existing institutions could be strengthened, or new mechanisms established, where appropriate, to finance technology transfers, train human resources, and evaluate, introduce, and operate existing or new technologies. Legal barriers and restrictive trade practices are also impeding factors.

It has not been possible to bridge the difference on views on some of the questions mentioned above. It is extremely important to reach early international agreement on these issues in order to promote effective flow of technologies to monitor, limit, or adapt to climate change. One area where international agreement may be possible is the promotion of CFC substitutes and provision of assistance and cooperation to the developing countries in the acquisition and manufacture of such substitutes.

Several countries have suggested that the issue of technology transfer to Eastern European countries be addressed.

7.3 ECONOMIC MECHANISMS

It is important that any potential measures to limit or adapt to global climate change be as economically efficient and cost-effective as possible, while taking into account important social implications. In general, environmental objectives can be achieved either through regulations requiring the use of a specific technology or attainment of specific goals, or economic instruments such as emissions fees, subsidies, tradeable permits, or sanctions.

Economic instruments, through their encouragement of flexible selection of abatement measures, frequently offer the possibility of achieving environmental improvements at lower cost than regulatory mechanisms. Unlike many regulations, they tend to encourage innovation and the development of improved technologies and practices for reducing emissions. Economic mechanisms also have the potential to provide the signals necessary for more environmentally sensitive operation of markets. It is unlikely, however, that economic instruments will be applicable to all circumstances.

Three factors are considered as potential barriers to the operation of markets and/or the achievement of environmental objectives through market mechanisms. These are: *information problems*, which can often cause markets to produce less effective or unfavorable environmental outcomes; *existing measures and institutions*, which can encourage individuals to behave in environmentally damaging ways; and *balancing competing objectives* (social, environmental, and economic). An initial response strategy may therefore be to address information problems directly and to review existing measures that may be barriers. For example, prior to possible adoption of a system of emission charges, countries should examine existing subsidies and tax incentives on energy and other relevant greenhouse gas producing sectors.

A general advantage of market based *economic instruments* is that they encourage limitations or reductions in emissions by those who can achieve them at least cost. They also provide an ongoing incentive for industry and individual consumers to apply the most efficient limitation/reduction measures through, for example, more efficient and cleaner technologies. Such incentives may be lacking in the case of regulations.

Regulations are the customary means of controlling pollution in both market and centrally planned economies. An advantage of regulations is that, in certain circumstances, they create more certainty as to desired outcomes, whereas major disadvantages are that they may discourage innovation, introduce inflexibilities in meeting objectives, can discourage resource use efficiency, and offer few or no incentives to reduce emissions below specified levels.

It is evident that the question of adoption of any form of economic instrument, whether domestically or internationally, raises many complex and difficult issues. Careful and substantive analysis of all implications of such instruments that have been identified for consideration include:

- *A system of tradeable emissions permits:* An emission permit system is based on the concept that the economic costs of attaining a given environmental goal can be minimized by allowing for the trading of emissions rights. Once an overall limit on emissions has been set, emissions entitlements amounting to that limit could be provided to emitting sources and free trading of such entitlements allowed. This would reduce the costs of meeting a given emission target because: (a) as in trade, comparative advantages between trading entities would be maximized; and (b) economic incentives would be created for the development of improved greenhouse gas limitation technologies, sink enhancement, and resource use efficiency (energy conservation). Concerns with this approach include the limited experience with this instrument, the potential scope and size of trading markets, and the need for the development of an administrative structure not currently in place.

- *A system of emission charges:* Emission charges are levied on specified emissions depending on their level of contribution to climate change. Such changes may provide a means of encouraging emitters to limit or reduce emissions and provide an incentive for diverse parties to implement efficient means of limiting or reducing emissions. Another advantage of charges is that they generate revenue that could provide a funding base for further pollution abatement, research, and administration, or allow other taxes to be lowered. Concerns with this approach include the difficulty of deciding on the basis and size of the tax, and the lack of certainty that the tax will achieve the agreed emission reduction target.

- *Subsidies:* Subsidies are aimed at encouraging environmentally sound actions by lowering their costs. Subsidies could be used, *inter alia*, to encourage the use of energy-efficient equipment and non-fossil energy sources, and the development and greater use of environmentally sound technologies. Concerns with subsidies include the possible size of the required financial commitment of governments, the need for careful design, the need for review, and the international trade aspects of such measures.

- *Sanctions*. A final type of economic instrument is the use of economic sanctions for the enforcement of international agreements. This would require an international convention to establish a system of agreed trade or financial sanctions to be imposed on countries not adhering to agreed regimes. Many contributors expressed considerable reservations about applying this approach to greenhouse gas emissions because of the complexity of the situation. The concerns include a belief that sanctions could appear to be arbitrary, could create confusion and resentment, and could be used as a pretext to impose new non-tariff trade barriers.

It has also been suggested that environmental protection could be advanced and the economic costs of meeting greenhouse gas limitation targets, if any, minimized by addressing, to the extent feasible, all greenhouse gas sources and sinks comprehensively. This approach could employ an "index" relating net emissions of various greenhouse gases by further development of the index formulated by Working Group I.

Each of the approaches outlined above, however, poses potentially significant challenges in terms of implementation and acceptability. There is an incomplete understanding of the economic and social consequences of these various approaches. It is evident that further work is required in all countries, and in ongoing IPCC work, to fully evaluate the practicality of such measures and costs and benefits associated with different mechanisms, especially with their use internationally. It has, however, been pointed out that an international system of tradeable permits, or, alternatively, an international system of emissions charges, could offer the potential of serving as a cost-efficient main instrument for achieving a defined target for the reduction of greenhouse gas emissions.

Finally, it was stressed that in order to share equitably the economic burdens, implementation of any of the international economic instruments discussed above should take into account the circumstances that most emissions affecting the atmosphere at present originated in the industrialized countries where the scope for change is the greatest, and that, under present conditions, emissions from developing countries are growing and may need to grow in order to meet their development requirements and thus, over time, are likely to represent an increasingly significant percentage of global emissions. It is appreciated that each instrument assessed has a role in meeting greenhouse gas emission objectives, but the suitability of particular instruments is dependent on the particular circumstances and at this stage no measure can be considered universally superior to any other available mechanisms.

7.4 FINANCIAL MECHANISMS

Industrialized and developing countries consider it important that assurances of financial mechanisms are needed for undertaking adequate measures to limit and/or adapt to climate change.

Guiding Principles

The following principles should guide the financial approach:

a) Industrialized countries and developing countries have a common responsibility in dealing with problems arising from climate change, and effective responses require a global effort.

b) Industrialized countries should take the lead and have specific responsibilities on two levels:

(i) major part of emissions affecting the atmosphere at present originates in industrialized countries where the scope for change is greatest. Industrialized countries should adopt domestic measures to limit climate change by adapting their own economies in line with future agreements to limit emissions;

(ii) to cooperate with developing countries in international action, without standing in the way of the latter's development, by contributing additional financial resources, by appropriate transfer of technology, by engaging in close cooperation concerning scientific observation, by analysis and research, and finally by means of technical cooperation geared to forestalling and managing environmental problems.

c) Emissions from developing countries are growing and may need to grow in order to meet their development requirements and thus, over time, are likely to represent in increasingly significant percentage of global emissions. Developing countries should, within the limits feasible, take measures to suitably adapt their economies. Financial resources channeled to developing countries would be most effective if focused on those activities that contribute both to limiting greenhouse gas emissions and promoting economic development. Areas for cooperation and assistance could include:

• efficient use of energy resources and the increased use of fossil fuels with lower greenhouse gas emission rates or non-fossil sources;

• rational forest management practices and agricultural techniques that reduce greenhouse gas emissions;

• facilitating technology transfer and technology development;

• measures that enhance the capacity of developing countries to develop programmes to address climate change, including research and development activities and public awareness and education;

• participation by developing countries in international forums on global climate change, such as the IPCC.

It was also recognized that cooperation and assistance for adaptive measures would be required, noting that for some regions and countries, adaptation rather than limitation activities are potentially most important.

A number of possible sources for generating financial resources were considered. These include general taxation, specific taxation on greenhouse gas emissions, and emissions trading. For the significant complexities and implications of such taxes, reference is made to the economic measures paper (section 7.3). Creative suggestions include using undisbursed official resources, which might result from savings on government energy bills and lower levels of military expenditures, a fixed percentage tax on travel tickets, and levies on countries that have been unable to meet their obligations. The question has also been raised of whether such financial cooperation and assistance should be given only to those countries that abstain from activities producing greenhouse gases. A positive international economic environment, including further reduction of trade barriers and implementation of more equitable trade practices, would help to generate resources that can be applied toward pressing needs.

With respect to institutional mechanisms for providing financial cooperation and assistance to developing countries, a two-track approach was considered.

i) One track built on work under way or planned in existing institutions. In this regard, the World Bank, a number of regional banks, other multilateral organizations, and bilateral agencies have initiated efforts to incorporate global climate change issues into their programmes. Bilateral donors could further integrate and reinforce the environmental components of their assistance programmes and develop co-financing arrangements with multilateral institutions while ensuring that this does not impose inappropriate environmental conditions.

ii) Parallel to this track the possibility of new mechanisms and facilities was considered. Some developing and industrialized countries suggested that new mechanisms directly related to a future climate convention and protocols, such as a new international fund, were required. It was added that such new instruments could be located within the World Bank (with new rules) or elsewhere. It was also noted that the Global Environmental Facility proposed by the World Bank in collaboration with UNEP and UNDP was welcomed by industrialized and developing countries at the World Bank Development Committee meeting in May 1990.

It was noted that the issue of generating financial resources was distinct from that of allocating those resources.

Areas identified for future work include studies, with donor assistance, for developing countries on their current and projected net emissions levels and assistance and cooperation needs for limiting such emissions. Further consideration is also needed of the important role which the private sector might play, through technology transfer, foreign direct investment, and other means to assist and cooperate with developing countries to respond to climate change.

7.5 LEGAL AND INSTITUTIONAL MECHANISMS

A number of institutions and international legal mechanisms exist that have a bearing on the climate change issue, in particular those dealing with the environment, science and technology, energy, natural resources, and financial assistance. One of these existing international legal mechanisms, the Vienna Convention on the Protection of the Ozone Layer and its associated Montreal Protocol on Substances that Deplete the Ozone Layer, deals specifically with reducing emissions of important greenhouse gases that also deplete the ozone layer. However, there is a general view that, while existing legal instruments and institutions related to climate change should be fully utilized and further strengthened, they are insufficient alone to meet the challenge.

A consensus emerged at the 44th session of the UN General Assembly on the need to prepare as a matter of urgency a framework convention on climate change, laying down, as a minimum, general principles and obligations. It should, in the view of RSWG, be framed in such a way as to gain the adherence of the largest possible number and most suitably balanced range of countries while permitting timely action to be taken. It may contain provisions for separate annexes/protocol(s) to deal with specific obligations. As part of the commitment of the parties to action on greenhouse gas emissions and adverse effects of climate change, the convention should also address the particular financial and other needs of the developing countries (notably those most vulnerable to climate change agriculturally or otherwise), the question of access to and transfer of technology, the need for research and monitoring, and institutional requirements.

Decisions will have to be taken on a number of key issues. These include:

- the political imperative of striking the correct balances (a) between the arguments for a far-reaching, action-oriented convention and the need for urgent adoption of a convention so as to begin tackling the problem of climate change; and (b) among the risks of inaction, the costs of action, and current levels of scientific uncertainty;
- the extent to which specific obligations, particularly on the control of emissions of greenhouse gases, should be included in the convention itself, possibly as annexes, or be the subject of a separate protocol(s);
- the timing of negotiation of protocol(s) in relation to the negotiations on the convention;
- the introduction, as appropriate, of sound scientific bases for establishing emission targets (such as total emission levels, per capita emissions, emissions per GNP, emissions per energy use, climatic conditions, past performance, geographic characteristics, fossil fuel resource base, carbon intensity per unit of energy, energy intensity per GNP, socioeconomic costs and benefits, or other equitable considerations);
- the extent to which specific goals with respect to global levels of emissions or atmospheric concentrations of greenhouse gases should be addressed;
- whether obligations should be equitably differentiated according to countries' respective responsibilities for causing and combating climate change and their level of development;
- the need for additional resources for developing countries and the manner in which this should be addressed, particularly in terms of the nature, size, and conditions of the funding, even if detailed arrangements form the subject of a separate protocol;
- the basis on which the promotion of the development and transfer of technology and provision of technical assistance and cooperation to developing countries should take place, taking into account considerations such as terms of transfer (preferential or non-preferential, commercial or non-commercial), assured access, intellectual property rights, the environmental

soundness of such technology, and the financial implications;

- the nature of any new institutions to be created by the convention (such as Conference of the Parties, an Executive Organ, as well as other bodies), together with their functions, composition and decision-making powers, e.g., whether or not they should exercise supervision and control over the obligations undertaken.

The international negotiation on a framework convention should start as quickly as possible after the completion of the IPCC interim report. The full and effective participation of developing countries in this process is essential. Many, essentially developing, countries stressed that the negotiation must be conducted in the forum, manner, and with the timing to be decided by the UN General Assembly. This understanding also applies to any associated protocols. In the view of many countries and international and non-governmental organizations, the process should be conducted with a view of concluding it not later than the 1992 UN Conference on Environment and Development.

The foregoing does not necessarily constitute an exclusive list of issues that will arise in the negotiations. However, a readiness to address these fundamental problems will be a prerequisite for ensuring the success of the negotiations and the support of a sufficiently wide and representative spread of nations. The legal measures topic paper developed by the Working Group is given in Annex I.

ANNEX I
LEGAL AND INSTITUTIONAL MECHANISMS

EXECUTIVE SUMMARY

1) The coordinators' report has as its primary objective the compilation of elements that might be included in a future framework Convention on Climate Change, and a discussion of the issues that are likely to arise in the context of developing those elements.

2) There is a general view that while existing legal instruments and institutions with a bearing on climate should be fully utilized and further strengthened, they are insufficient alone to meet the challenge. A very broad international consensus has therefore emerged in the IPCC, confirmed notably at the 44th United Nations General Assembly, on the need for a framework Convention on Climate Change. Such a Convention should generally follow the format of the Vienna Convention for the Protection of the Ozone Layer, in laying down, as a minimum, general principles and obligations. It should further be framed in such a way as to gain the adherence of the largest possible number and most suitably balanced spread of countries while permitting timely action to be taken; it should contain provision for separate annexes/protocols to deal with specific obligations. As part of the commitment of the parties to action on greenhouse gas emissions and the adverse effects of global warming, the Convention would also address the particular

financial needs of the developing countries, the question of the access to and transfer of technology, and institutional requirements.

3) The paper points out a number of issues to be decided in the negotiation of a Convention. In general these are:

- the political imperative of striking the correct balances: on the one hand, between the arguments for a far-reaching, action-oriented Convention and the need for urgent adoption of such a Convention so as to begin tackling the problem of climate change; and, on the other hand, between the cost of inaction and the lack of scientific certainty;
- the extent to which specific obligations, particularly on the control of emissions of carbon dioxide and other greenhouse gases, should be included in the Convention itself or be the subject of separate protocol(s):
- the timing of negotiations of such protocol(s) in relation to the negotiations on the Convention.

4) In particular, within the Convention the following specific issues will need to be addressed:

(a) *Financial needs of developing countries.* The need for additional resources for developing countries and the manner in which this should be addressed, particularly in terms of the nature, size, and conditions of the funding, even if detailed arrangements form the subject of a separate protocol, will have to be considered by the negotiating parties.

(b) *Development and transfer of technology.* The basis on which the promotion of the development and transfer of technology and provision of technical assistance to developing countries should take place will need to be elaborated, taking into account considerations such as terms of transfer, assured access, intellectual property rights, and the environmental soundness of such technology.

(c) *Institutions.* Views differ substantially on the role and powers of the institutions to be created by the Convention, particularly in exercising supervision and control over the obligations undertaken.

5) The inclusion of any particular element in the paper does not imply consensus with respect to that element, or the agreement of any particular government to include that element in a Convention.

6) The coordinators have not sought to make a value judgment in listing and summarizing in the attached paper the elements proposed for inclusion in a framework Convention: their text seeks merely to assist the future negotiators in their task. They note, however, that a readiness to address the foregoing fundamental problems in a realistic manner will be a prerequisite for ensuring the success of the negotiations and the support of a sufficiently wide and representative spread of nations.

POSSIBLE ELEMENTS FOR INCLUSION IN A FRAMEWORK CONVENTION ON CLIMATE CHANGE

PREAMBLE

In keeping with common treaty practice including the format of the Vienna Convention, the Climate Change Convention would contain a preamble which might seek to address some or all of the following items:

- a description of the problem and reasons for action (need for timely and effective response without awaiting absolute scientific certainty);
- reference to relevant international legal instruments (such as the Vienna Convention and Montreal Protocol) and declarations (such as UNGA Resolution 43/53 and Principle 21 of the Stockholm Declaration);
- recognition that climate change is a common concern of mankind, affects humanity as a whole, and should be approached within a global framework, without prejudice to the sovereignty of states over the airspace superadjacent to their territory as recognized under international law;
- recognition of the need for an environment of a quality that permits a life of dignity and well-being for present and future generations;
- reference to the balance between the sovereign right of states to exploit natural resources and the concomitant duty to protect and conserve climate for the benefit of mankind, in a manner not to diminish either;

- endorsement and elaboration of the concept of sustainable development;
- recognition of the need to improve scientific knowledge (e.g., through systematic observation) and to study the social and economic impacts of climate change, respecting national sovereignty;
- recognition of the importance of the development and transfer of technology and of the circumstances and needs, particularly financial, of developing countries; need for regulatory, supportive, and adjustment measures to take into account different levels of development and thus differing needs of countries;
- recognition of the responsibility of all countries to make efforts at the national, regional, and global levels to limit or reduce greenhouse gas emissions and prevent activities that could adversely affect climate, while bearing in mind that:
  - most emissions affecting the atmosphere at present originate in industrialized countries where the scope for change is greatest;
  - implementation may take place in different time frames for different categories of countries and may be qualified by the means at the disposal of individual countries and their scientific and technical capabilities;
  - emissions from developing countries are growing and may need to grow in order to meet their development requirements and thus, over time, are likely to represent an in-

creasingly significant percentage of global emissions;

- recognition of the need to develop strategies to absorb greenhouse gases, i.e., protect and increase greenhouse gas sinks; to limit or reduce anthropogenic greenhouse gas emissions; and to adapt human activities to the impacts of climate change.

Other key issues that will have to be addressed during the development of the preambular language include:

- should mankind's interest in a viable environment be characterized as a fundamental right?
- is there an entitlement not to be subjected, directly or indirectly, to the adverse effects of climate change?
- should there be a reference to the precautionary principle?
- in view of the interrelationship among all greenhouse gases, their sources and sinks, should they be treated collectively?
- should countries be permitted to meet their aggregate global climate objectives through joint arrangements?
- should reference be made to weather modification agreements such as the ENMOD treaty as relevant legal instruments?
- is there a common interest of mankind in the development and application of technologies to protect and preserve climate?
- does the concept of sustainable development exclude or include the imposition of new conditionality in the provision of financial assistance to developing countries, and does it imply a link between the protection and preservation of the environment, including climate change, and economic development, so that both are to be secured in a coherent and consistent manner?
- should the preamble address the particular problems of countries with an agricultural system vulnerable to climate change and with limited access to capital and technologies, recognizing the link with sustainable development?
- is there a minimum standard of living that is a prerequisite to adopting response strategies to address climate change?

DEFINITIONS

As is the practice, definitions will need to be elaborated in a specific article on definitions. The terms that will need to be defined will depend on the purpose of the Convention and thus the language used by the negotiating parties.

GENERAL OBLIGATIONS

Following the format of such treaties as the Vienna Convention, an article would set out the general obligations agreed to by the parties to the Convention. Such obligations may relate to, for example:

- the adoption of appropriate measures to protect against the adverse effects of climate change, to limit, reduce, adapt to, and, as far as possible, prevent climate change in accordance with the means at the disposal of individual countries and their scientific and technical capabilities; and to avoid creating other environmental problems in taking such measures;
- the protection, stabilization, and improvement of the composition of the atmosphere in order to conserve climate for the benefit of present and future generations;
- taking steps having the effect of limiting climate change but which are already justified on other grounds;
- the use of climate for peaceful purposes only, in a spirit of good neighborliness;
- cooperation by means of research, systematic observation, and information exchange in order to understand better and assess the effects of human activities on the climate and the potential adverse environmental and socio-economic impacts that could result from climate change, respecting national sovereignty;
- the encouragement of the development and transfer of relevant technologies, as well as the provision of technical and financial assistance, taking into account the particular needs of developing countries to enable them to fulfill their obligations;

- cooperation in the formulation and harmonization of policies and strategies directed at limiting, reducing, adapting to, and, as far as possible, preventing climate change;
- cooperation in the adoption of appropriate legal or administrative measures to address climate change;
- provision for bilateral, multilateral, and regional agreements or arrangements not incompatible with the Convention and any annex/protocol, including opportunities for groups of countries to fulfill the requirements on a regional or subregional basis;
- cooperation with competent international organizations effectively to meet the objectives of the Convention;
- the encouragement of and cooperation in the promotion of public education and awareness of the environmental and socio-economic impacts of greenhouse gas emissions and of climate change;
- the strengthening or modification if necessary of existing legal and institutional instruments and arrangements relating to climate change;
- a provision on funding mechanisms.

Other key issues will have to be addressed in the process of elaborating this article, such as the questions below:

- should there be a provision setting any specific goals with respect to levels of emissions (global or national) or atmospheric concentrations of greenhouse gases while ensuring stable development of the world economy, particularly stabilization by industrialized countries, as a first step, and later reduction of CO_2 emissions and emissions of other greenhouse gases not controlled by the Montreal Protocol? Such a provision would not exclude the application of more stringent national or regional emission goals than those that may be provided for in the Convention and/or any annex/protocol.
- in light of the preambular language, should there be a provision recognizing that implementation of obligations may take place in different time frames for different categories of country and/or may be qualified by the means at the disposal of individual countries and their scientific and technical capabilities?

- should there be a commitment to formulate appropriate measures such as annexes, protocols, or other legal instruments and, if so, should such formulation be on a sound scientific basis or on the basis of the best available scientific knowledge?
- in addressing the transfer of technology particularly to developing countries, what should be the terms of such transfers (i.e., commercial versus non-commercial, preferential versus non-preferential, the relationship between transfers and the protection of intellectual property rights)?
- should funding mechanisms be limited to making full use of existing mechanisms or also entail new and additional resources and mechanisms?
- should provision be made for environmental impact assessments of planned activities that are likely to cause significant climate change as well as for prior notice of such activities?
- what should be the basis of emission goals—e.g., total emission levels, per capita emissions, emissions per GNP, emissions per energy use, climatic conditions, past performance, geographic characteristics, fossil fuel resource base, carbon intensity per unit of energy, energy intensity per GNP, socio-economic costs and benefits, or other equitable considerations?
- should the particular problem of sea level rise be specifically addressed?
- is there a link between nuclear stockpiles and climate change?

INSTITUTIONS

It has been the general practice under international environmental agreements to establish various institutional mechanisms. The parties to a Climate Change Convention might, therefore, wish to make provision for a Conference of the Parties, an Executive Organ, and a Secretariat. The Conference of the Parties may, among other things: keep under continuous review the implementation of the Convention and take appropriate decisions to this end; review current scientific information; and promote harmonization of policies and strategies directed at limiting, reducing, adapting to, and, as far as possible, preventing climate change.

Questions that will arise in developing provisions for appropriate institutional mechanisms include:

- should any of the Convention's institutions (e.g., the Conference of the Parties and/or the Executive Organ) have the ability to take decisions, *inter alia*, on response strategies or functions in respect of surveillance, verification, and compliance that would be binding on all the parties and, if so, should such an institution represent all of the parties or be composed of a limited number of parties, e.g., based on equitable geographic representation?
- what should be the role of the Secretariat?
- what should be the decision-making procedures, including voting requirements (e.g., consensus, majority)?
- if a trust fund or other financial mechanism were established under the Convention, how should it be administered?
- should scientific and/or other bodies be established on a permanent or ad hoc basis, to provide advice and make recommendations to the Conference of the Parties concerning research activities and measures to deal with climate change?
- should the composition of the above bodies reflect equitable climatic or geographic representation?
- should there be a provision for working groups, e.g., on scientific matters as well as on socio-economic impacts and response strategies?
- is there a need for innovative approaches to institutional mechanisms in the light of the nature of the climate change issue?
- what should be the role of non-governmental organizations?

RESEARCH, SYSTEMATIC OBSERVATIONS, AND ANALYSIS

It would appear to follow general practice to include provision for cooperation in research and systematic monitoring. In terms of research, each party might be called upon to undertake, initiate, and/or cooperate in, directly or through international bodies, the conduct of research on and analysis of:

- physical and chemical processes that may affect climate;
- substances, practices, processes, and activities that could modify the climate;
- techniques for monitoring and measuring greenhouse gas emission rates and their uptake by sinks;
- improved climate models, particularly for regional climates;
- environmental, social, and economic effects that could result from modifications of climate;
- alternative substances, technologies, and practices;
- environmental, social, and economic effects of response strategies;
- human activities affecting climate;
- coastal areas with particular reference to sea level rise;
- water resources; and
- energy efficiency.

The parties might also be called upon to cooperate in establishing and improving, directly or through competent international bodies, and taking fully into account national legislation and relevant ongoing activities at the national, regional, and international levels, joint or complementary programmes for systematic monitoring and analysis of climate, including a possible worldwide system; and cooperate in ensuring the collection, validation, and transmission of research, observational data, and analysis through appropriate data centers.

Other issues that will arise in developing this provision include:

- should consideration be given to the establishment of panels of experts or of an independent scientific board responsible for the coordination of data collection from the above areas of research and analysis and for periodic assessment of the data?
- should provision be made for on-site inspection?
- should there be provision for open and nondiscriminatory access to meteorological data developed by all countries?
- should a specific research fund be established?

INFORMATION EXCHANGE AND REPORTING

Precedents would suggest the inclusion of a provision for the transmission of information through the Secretariat to the Conference of the Parties on measures adopted by them in implementation of the Convention and of protocols to which they are party. In an annex to the Vienna Convention, the types of information exchanged are specified and include scientific, technical, socio-economic, commercial, and legal information.

For the purposes of elaborating this provision, issues having to be addressed by the negotiating parties include the following:

- is there a need for the elaboration of a comprehensive international research programme in order to facilitate cooperation in the exchange of scientific, technological, and other information on climate change?
- should parties be obliged to report on measures they have adopted for the implementation of the Convention, with the possible inclusion of regular reporting on a comparable basis of their emissions of greenhouse gases?
- should each party additionally be called upon to develop a national inventory of emissions, strategies, and available technologies for addressing climate change? If so, the Convention might also call for the exchange of information on such inventories, strategies, and technologies.

DEVELOPMENT AND TRANSFER OF TECHNOLOGY

While the issue of technology has been addressed in the section on General Obligations, it might be considered desirable to include separate provisions on technology transfer and technical cooperation. Such provisions could call upon the parties to promote the development and transfer of technology and technical cooperation, taking into account particularly the needs of developing countries, to enable them to take measures to protect against the adverse effects of climate change, to limit, reduce, and, as far as possible, prevent climate change, or to adapt to it.

Another issue that will arise is: should special terms be attached to climate-related transfers of technology (such as a preferential and/or non-commercial basis and assured access to, and transfer of, environmentally sound technologies on favorable terms to developing countries), taking into consideration the protection of intellectual property rights?

SETTLEMENT OF DISPUTES

It would be usual international practice to include a provision on the settlement of disputes that may arise concerning the interpretation or application of the Convention and/or any annex/protocol. Provisions similar to those in the Vienna Convention for the Protection of the Ozone Layer might be employed, i.e., voluntary resort to arbitration or the International Court of Justice (with a binding award), or, if neither of those options is elected, mandatory resort to conciliation (with a recommendatory award).

OTHER PROVISIONS

It would be the usual international practice to include clauses on the following topics:

- amendment of the Convention;
- status, adoption, and amendment of annexes;
- adoption and entry into force of, and amendments to, protocols;
- signature;
- ratification;
- accession;
- right to vote;
- relationship between the Convention and any protocol(s);

- entry into force;
- reservations;
- withdrawal;
- depositary;
- authentic texts.

ANNEXES AND PROTOCOLS

The negotiating parties may wish the Convention to provide for the possibility of annexes and/or protocols. Annexes might be concluded as integral parts of the Convention, while protocols might be concluded subsequently (as in the case of the Montreal Protocol to the Vienna Convention on Protection of the Ozone Layer). While it is recognized that the Convention is to be all-encompassing, the negotiating parties will have to decide whether greenhouse gases, their sources and sinks, are to be dealt with, individually, in groups, or, comprehensively, in annexes or protocols to the Convention. The following, among others, might also be considered as possible subjects for annexes or protocols to the Convention:

- agricultural practices;
- forest management;
- funding mechanisms;
- research and systematic observations;
- energy conservation and alternative sources of energy;
- liability and compensation;
- international emissions trading;
- international taxation system;
- development and transfer of climate change-related technologies.

Issues that will arise in connection with the development of annexes and protocols include:

- timing, i.e., negotiating parties advocating a more action-oriented Convention may seek to include specific obligations in annexes as opposed to subsequent protocols and/or negotiate one or more protocols in parallel with the Convention negotiations;
- sequence, i.e., if there is to be a series of protocols, in what order should they be taken up?

I I

IPCC RESPONSE STRATEGIES WORKING GROUP REPORTS

Contents
IPCC Response Strategies
Working Group Reports

1. INTRODUCTION

1.1 ESTABLISHMENT OF THE RESPONSE STRATEGIES WORKING GROUP

The Intergovernmental Panel on Climate Change (IPCC) was established under the auspices of the World Meteorological Organization (WMO) and the United Nations Environment Program (UNEP) to address the need for an international organization that could deal with the issue of climate change. As summarized in WMO Executive Council resolution 4 (EC-XL) of 1987, the organization's objectives are to address climate change by:

(i) Assessing the scientific information that is related to the various components of the climate change issue, such as emissions of major greenhouse gases and modification of the Earth's radiation balance resulting therefrom, and that are needed to enable the environmental and socio-economic consequences of climate change to be evaluated; and

(ii) Formulating realistic response strategies for the management of the climate change issue.

At the IPCC's first meeting in Geneva in November 1988, the Panel agreed that its work included three main tasks:

(i) Assessment of available scientific information on climate change;

(ii) Assessment of environmental and socio-economic impacts of climate change; and

(iii) Formulation of response strategies.

To accomplish these tasks in the most efficient and expeditious manner possible, the IPCC decided to establish three Working Groups to deal with each of the tasks identified above. The IPCC agreed that the three working groups, on science, impacts, and response strategies, would be chaired, respectively, by the United Kingdom, Soviet Union, and United States.

The latter of these Working Groups, the Response Strategies Working Group (RSWG), held its first meeting in Washington in January 1989 under the chairmanship of Dr. Frederick M. Bernthal of the United States. RSWG Vice-Chairs were also named from Canada, China, Malta, the Netherlands, and Zimbabwe. At that first meeting, the RSWG established a Steering Committee and four Subgroups to carry out a work plan for formulating response strategies (see Figure 1.1). The RSWG Steering Committee was given responsibility for coordinating the Working Group's activities in general and for specifically addressing two crosscutting tasks: (1) the development of greenhouse gas emissions scenarios; and (2) the development of a strategy for considering implementation mechanisms. The four RSWG Subgroups were tasked with developing a range of climate change response strategies in the areas of: (1) Energy and Industry; (2) Agriculture and Forestry; (3) Coastal Zone Management; and (4) Resource Use and Management. It was agreed that the first two subgroups would consider measures for limiting net greenhouse gas emissions from the energy, industry, agriculture, and forestry sectors, and that the latter two subgroups would deal with measures for adapting to the impacts of climate change on coastal regions and natural resources.

FIGURE 1.1:   **Organization of Working Group III**

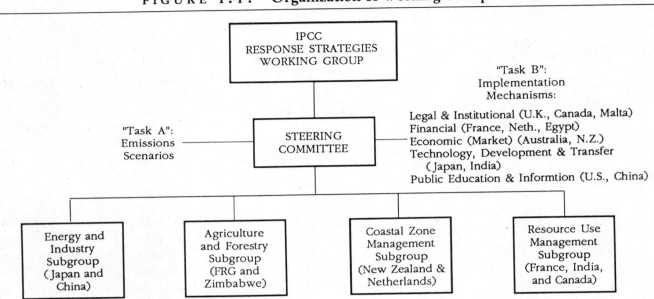

1.2   RSWG STEERING COMMITTEE

The RSWG Steering Committee was established to provide for overall coordination and direction of the RSWG's work. It was also agreed that the Steering Committee would undertake cross-cutting tasks relevant to the work of all the RSWG Subgroups or to the activities of the other two IPCC Working Groups.

1.2.1   EMISSIONS SCENARIOS

At its first meeting the RSWG requested that, as its first task, or "Task A," the Steering Committee conduct an analysis of possible future scenarios of global emissions of greenhouse gases. The purpose of these scenarios was to provide the four RSWG subgroups and the IPCC Science and Impacts Working Groups with a preliminary basis for conducting long-range analyses. By April 1989 a United States–Netherlands team of experts developed three possible scenarios of future emissions corresponding to: (1) the equivalent of a CO_2 doubling from pre-industrial levels by about the year 2030; (2) a CO_2 equivalent doubling by approximately 2060; and (3) a doubling by about 2090 with stabilization thereafter. The group subsequently developed two additional emissions scenarios corresponding to emissions projections in which atmospheric concentrations of greenhouse gases are stabilized at a level less than a CO_2 equivalent doubling. In addition, the Steering Committee's emissions projections have been complemented by more recent work developed by the Energy and Industry and Agriculture and Forestry Subgroups based on individual country studies of likely long-term greenhouse gas emissions trends.

1.2.2   IMPLEMENTATION MECHANISMS

The Steering Committee's second task, or "Task B," was to develop a plan for identifying "implementation mechanisms," or, in other words, the specific means through which response strategies can be brought into force in an effective manner. The Steering Committee agreed that it would consider five categories of implementation mechanism and named two or three countries to act as co-coordinators for each topic area:

- *Public education and information,* which comprises those mechanisms designed to stimulate global awareness of the climate change issue and possible response strategies (co-coordinators: United States and China).

- *Economic (market) measures,* or those mechanisms that ensure that response strategies are designed in the most cost-effective and economically viable manner possible (co-coordinators: Australia and New Zealand).
- *Technology development and transfer,* which relates both to mechanisms for promoting the development of new technologies to limit or adapt to climate change and to those that encourage the transfer of climate change related technologies internationally (co-coordinators: Japan and India).
- *Financial measures,* or those mechanisms that assist in the ability of countries, in particular developing countries, to address climate change (co-coordinators: Netherlands, France, and Egypt).
- *Legal and institutional measures,* which deal with assessing legal and institutional mechanisms for addressing climate change, including the possible development of a framework convention on climate change (co-coordinators: United Kingdom, Canada, and Malta).

The RSWG held a special workshop on these implementation mechanisms in October 1989 in Geneva. The workshop was attended by forty-three countries and eight international organizations and provided an opportunity for a broad exchange of views on these important mechanisms for addressing climate change.

1.3 RSWG SUBGROUPS

The RSWG agreed that consideration of specific response strategies would be conducted by four subgroups in the areas of: (1) Energy and Industry; (2) Agriculture, Forestry, and Other Human Activities; (3) Coastal Zone Management; and (4) Resource Use and Management. The first two subgroups were designed to address "limitation" issues, e.g., measures to limit net greenhouse gas emissions from the energy, industry, agriculture, and forestry sectors. The latter two subgroups were set up to consider measures for adapting to the impacts of climate change, e.g., the impacts of sea level rise on coastal regions or of changing temperature and precipitation patterns on natural resources.

1.3.1 ENERGY AND INDUSTRY SUBGROUP

The Energy and Industry Subgroup (EIS), co-chaired by Japan and China, was tasked with defining policy options for climate change response strategies related to greenhouse gas emissions produced by energy production, conversion, and use, as well as emissions from industrial sources not related to energy use. The EIS considered energy uses in the industrial, transportation, and residential sectors that produce carbon dioxide, methane, nitrous oxide, and other gases, and defined technological and policy options to reduce emissions of these gases. The EIS also developed estimates of future greenhouse gas emissions from the energy and industry sector.

1.3.2 AGRICULTURE, FORESTRY, AND OTHER HUMAN ACTIVITIES SUBGROUP

The Agriculture, Forestry, and Other Human Activities Subgroup (AFOS), co-chaired by the Federal Republic of Germany and Zimbabwe, was given the mandate of dealing with issues related to the limitation of greenhouse gas emissions from the agriculture, forestry, and other sectors not related to the production or use of energy or industrial activities. The AFOS reviewed in particular methane emissions from livestock, rice, biomass, and waste sources, carbon dioxide emissions from deforestation or CO_2 uptake from reforestation, and nitrous oxide emissions from the use of fertilizers. The AFOS also developed estimates of future greenhouse gas emissions from the agriculture, forestry, and other sectors.

1.3.3 COASTAL ZONE MANAGEMENT SUBGROUP

The Coastal Zone Management Subgroup (CZMS), co-chaired by New Zealand and the Netherlands, was tasked with considering response strategies for dealing with the impacts of sea level rise and the increased incidence of storms and other extreme events on coastal regions. The CZMS reviewed information from governments, institutions, and non-governmental organizations on technologies, practices, and other relevant activities for the coastal zone and held workshops on technologies and practices in both the Southern and Northern hemispheres. Based on this work, the CZMS developed a series of options for dealing with potential climate change impacts on coastal regions.

1.3.4   RESOURCE USE AND MANAGEMENT SUBGROUP

The Resource Use and Management Subgroup (RUMS), co-chaired by Canada, France, and India, was tasked with considering measures for adapting to the impacts of climate change on agriculture, fisheries, animal husbandry, water resources, forests, wildlife and biological diversity, and other natural resources. The RUMS considered possible strategies for either reducing the potential negative impacts or taking advantage of possible positive impacts of climate change on food security, water availability, and natural ecosystems in general.

1.4   OTHER FACTORS

In conducting its activities, the RSWG recognized that the task of developing response strategies was both complex and difficult, particularly because its work would depend on analyses being developed simultaneously in the IPCC Science and Impacts Working Groups. The RSWG was also faced with the need to complete an interim assessment report by the summer of 1990 to form part of the IPCC's first assessment report. Given these constraints, the RSWG agreed that it should concentrate on a short-term (18-month) work plan that would focus on the following elements:

- development and distribution of preliminary emissions scenarios;
- refinement of a strategy for considering implementation mechanisms;
- carrying out of short-term work plans of the

four RSWG subgroups for integration into an overall RSWG report; and
- development of longer-term work plans.

The report of the Energy and Industry Subgroup (EIS) was so voluminous that, for space reasons, only the Executive Summary is included in this volume.

This RSWG report represents the analysis it was considered feasible to complete in the time available from the first RSWG meeting in January 1989 to the adoption of this report by the RSWG in June 1990. This report identifies a wide range of possible response strategies for limiting or adapting to climate change and reviews available mechanisms for implementing these strategies. It is recognized, however, that there is considerable work to be done in further defining and assessing the response strategies. The RSWG has thus developed a work plan for the next 18-month period and thereafter, with an emphasis on areas where further information is needed to develop response strategies, so that future efforts can be directed in the most effective manner possible.

It must also be emphasized that the RSWG's task is to identify and evaluate response strategies, not to determine which actions should be undertaken by the international community to deal with climate change. The response strategies that have been identified therefore represent options rather than recommendations. While the RSWG has sought to provide useful guidance for policymakers, the determination of what actions should be undertaken is a subject for formal international negotiations.

Finally, the RSWG, and the IPCC as a whole, have had to deal with the difficulties presented by the relatively limited participation by the international community in some aspects of the Working Group's activities. The participation of centrally planned and developing countries, in particular, has not been as extensive in some of the RSWG's technical work as would be ideal for an exercise of this nature. The RSWG has made great efforts to increase the participation of all countries, in particular developing countries, in its work program. This remains an issue that needs continued attention.

—FREDERICK M. BERNTHAL
Chairman
Response Strategies Working Group

2

Emissions Scenarios

COORDINATORS
D. Tirpak (U.S.A.)
P. Vellinga (Netherlands)

CONTENTS

EMISSIONS SCENARIOS

EXECUTIVE SUMMARY

TASK A

The Response Strategies Working Group (RSWG) of the Intergovernmental Panel on Climate Change (IPCC) formed an expert group to develop scenarios of future emissions of greenhouse gases. These scenarios are to serve as initial reference and guidance for the work of the subgroups of RSWG and as a first basis for analyses by the Science Working Group and the Impacts Working Group.

The scenarios depict five different ways that future emissions of greenhouse gases might evolve over the next century and serve to illustrate the types of changes that would be needed to stabilize emissions while continuing to allow growth and improvement in the standard of living. There are limitations regarding our ability to estimate future rates of population growth, economic growth, and technological innovation and these lend uncertainty to projections of greenhouse gases over long time horizons. Based on this analysis, carbon dioxide (CO_2) emissions and atmospheric concentrations in

the years 2025 and 2075 could take the values in Executive Summary Table 2.1.

Executive Summary Table 2.2 contains the future emissions of other greenhouse gases for the five scenarios. Executive Summary Figure 2.1 shows the impact of these emissions on atmospheric concentrations of CO_2 and on the greenhouse effect in total. The impacts range from rapid increases in atmospheric concentrations of greenhouse gases throughout the next century in one scenario to declines in the rate of growth in atmospheric concentrations eventually leading to stabilization by early next century in another scenario. Significant policy changes would be required to achieve the latter scenario, although the specific costs of such policies have not been estimated as part of this analysis.

The first of the scenarios, called the *2030 High Emissions Scenario*, depicts a world in which few or no steps are taken to reduce emissions in response to concerns about greenhouse warming. Continued population and economic growth produces increases in the use of energy and in the rate of clearing of tropical forests. The Montreal Protocol comes

EXECUTIVE SUMMARY TABLE 2.1: Annual CO_2 Emissions and Atmospheric Concentrations

	2025		2075	
SCENARIO	EMISSIONS (PG C)	CONC. (PPM)	EMISSIONS (PG C)	CONC. (PPM)
2030 High Emissions	11.5	437	18.7	679
2060 Low Emissions	6.4	398	8.8	492
Control Policies	6.3	398	5.1	469
Accelerated Policies	5.1	393	3.0	413
Alternative Accelerated Policies	3.8	381	3.5	407

	1985	2000	2025	2050	2075	2100
2030 HIGH EMISSIONS SCENARIO						
CO_2 (Petagrams C)	6.0	7.7	11.5	15.2	18.7	22.4
N_2O (Teragrams N)	12.5	14.2	16.4	17.3	17.3	17.6
CH_4 (Teragrams)	540.5	613.9	760.8	899.1	992.0	1062.9
CFC-11 (Gigagrams)	278.3	305.1	244.7	251.5	253.0	253.0
CFC-12 (Gigagrams)	361.9	376.0	302.7	314.1	316.1	316.1
HCFC-22 (Gigagrams)	96.9	522.9	1340.4	2681.3	2961.0	2961.0
2060 LOW EMISSIONS SCENARIO						
CO_2 (Petagrams C)	5.9	5.5	6.4	7.5	8.8	10.3
N_2O (Teragrams N)	12.5	13.1	13.9	14.1	14.3	14.6
CH_4 (Teragrams)	540.5	576.8	665.4	723.8	732.4	735.6
CFC-11 (Gigagrams)	278.3	302.1	226.7	223.2	223.2	223.2
CFC-12 (Gigagrams)	361.9	372.2	279.1	277.9	277.9	277.9
HCFC-22 (Gigagrams)	96.9	525.7	1357.2	2707.2	2988.1	2988.1
CONTROL POLICIES SCENARIO						
CO_2 (Petagrams C)	5.9	5.6	6.3	7.1	5.1	3.5
N_2O (Teragrams N)	12.5	12.9	13.2	13.0	12.5	12.2
CH_4 (Teragrams)	540.5	557.9	607.7	621.6	562.0	504.9
CFC-11 (Gigagrams)	278.3	197.1	10.6	0.0	0.0	0.0
CFC-12 (Gigagrams)	361.9	262.2	10.2	0.0	0.0	0.0
HCFC-22 (Gigagrams)	96.9	638.5	1571.9	2927.6	3208.5	3208.5
ACCELERATED POLICIES SCENARIO						
CO_2 (Petagrams C)	6.0	5.6	5.1	2.9	3.0	2.7
N_2O (Teragrams N)	12.5	12.9	13.1	12.7	12.5	12.3
CH_4 (Teragrams)	540.7	565.8	583.5	553.0	530.1	502.1
CFC-11 (Gigagrams)	278.3	197.1	10.6	0.0	0.0	0.0
CFC-12 (Gigagrams)	361.9	262.2	10.2	0.0	0.0	0.0
HCFC-22 (Gigagrams)	96.9	638.5	1571.9	2927.6	3208.5	3208.5
ALTERNATIVE ACCELERATED POLICIES SCENARIO						
CO_2 (Petagrams C)	6.0	4.6	3.8	3.7	3.5	2.6

EXECUTIVE SUMMARY FIGURE 2.1: CO$_2$ and Equivalent CO$_2$ Concentrations

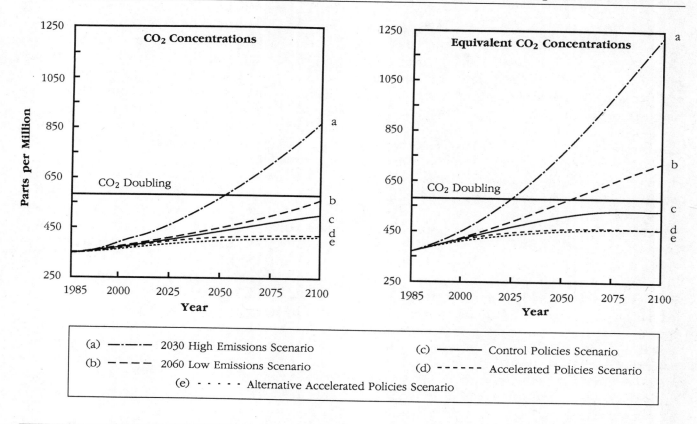

into effect but without strengthening and with less than 100 percent compliance. Fossil fuels continue to dominate energy supply, with coal taking a much larger share of energy supply in the future. Emissions of greenhouse gases such as CO$_2$, CH$_4$, and N$_2$O increase continuously throughout the next century with emissions of CO$_2$ doubling within forty years. Emissions of many of the chlorofluorocarbons stabilize and decline due to compliance to the Montreal Protocol but emissions of substitutes such as HCFC-22 increase. These increases in emissions yield increases in atmospheric concentrations of greenhouse gases with an equivalent greenhouse effect of a doubling of CO$_2$ concentrations from pre-industrial levels by 2030 and continued increase throughout the rest of the century.

The second of the scenarios, called the *2060 Low Emissions Scenario*, portrays a world in which a number of environmental and economic concerns result in steps to reduce the growth of greenhouse gas emissions. In this scenario, energy efficiency

improves more rapidly due to such factors as efficiency standards and technology transfer. Emission controls are adopted globally, reducing emissions of CO and NO$_x$. The share of primary energy provided by natural gas increases. Full compliance to the Montreal Protocol is realized. Tropical deforestation is halted and a global reforestation effort begins. These steps reduce growth in emissions by 50 to 75 percent and significantly slow down the growth in atmospheric concentrations of greenhouse gases. CO$_2$ emissions do not double until 2100, but the equivalent greenhouse effect of a doubling of CO$_2$ concentrations over pre-industrial levels is achieved by 2060 and continues to grow, albeit at a slower rate than in the first scenario.

The third of the scenarios, called the *Control Policies Scenario*, reflects a future where concern over global climate change and other environmental issues, such as stratospheric ozone depletion, motivate steps over and above those taken in the 2060 Low Emissions Scenario. Technological development, commercialization, and government efforts result in rapid penetrations of renewable energy sources in the last half of the next century. The Montreal Protocol is strengthened to include a full phase-out of CFCs and freezes on methyl-chloroform and carbon tetrachloride. Agricultural policies yield reduction in emissions of greenhouse gases from enteric fermentation in domestic animals, from rice paddies, and from fertilizer. As a result, emissions of CO_2, N_2O, and CH_4 grow slowly through the middle of the next century, then start to decline. Emissions of CO and NO_x decline sharply along with emissions of CFCs. These emission trends yield increases in atmospheric concentrations of greenhouse gases equivalent to slightly less than a doubling of CO_2 from pre-industrial levels by 2090 with concentrations stable after 2090.

The fourth and fifth scenarios, called the *Accelerated Policies Scenarios*, are similar to the Control Policies Scenario but feature much more rapid development and penetration of renewable energy sources encouraged in part by global adoption of carbon fees. Biomass energy represents 10 to 25 percent of primary energy supply by 2025, depending on economic growth assumptions. The results of these two scenarios differ only in emissions of CO_2 and primarily in the short run. In the first of these scenarios, carbon emissions from energy continue to increase through 2000 while total emissions of carbon decline, due to sequestering of carbon through reforestation. After 2000, carbon emissions

from all sources decline through the end of the century to levels less than half those in 1985. In the alternative scenario, CO_2 emissions start declining immediately but reach the same levels by the end of the next century. These emission scenarios yield very similar atmospheric concentrations of greenhouse gases. In both scenarios, atmospheric concentrations of greenhouse gases continue to increase but stabilize by the middle of the next century at levels 25 percent greater than current levels but well below an equivalent doubling of CO_2 over pre-industrial levels.

In summary, the 2030 High Emissions and 2060 Low Emissions scenarios may be viewed as two different paths that global greenhouse gas emissions could follow over the next several decades. The latter case assumes sizeable improvements in energy efficiency, which may only be possible with government action. The Control Policies and Accelerated Policies scenarios require deliberate actions by governments (e.g., phasing out of CFCs, increasing fossil energy prices or using other measures to ensure penetration by renewables). In general, these scenarios do not achieve the goals of the Toronto Conference "The Changing Atmosphere—1988," which is a 20 percent reduction in CO_2 emissions by early in the next century. This goal is achieved in the Alternative Accelerated Policy scenario submitted by the Netherlands. The economic implications of all of the scenarios have not been analyzed.

2.1 INTRODUCTION

At its first meeting in Washington, D.C., in January 1989, Working Group III, or the Response Strategies Working Group (RSWG), decided to develop three global emission scenarios that would serve as (1) an initial reference and guidance for the work of the subgroups of the RSWG (Industry and Energy, Forestry and Agriculture, etc.), and (2) a first basis for the work of Working Group I (the Science Working Group) and Working Group II (the Impacts Working Group).

An expert group developed these three scenarios and a draft report and presented them to the RSWG in May 1989. Later, at the general IPCC meeting in June 1989 in Nairobi, the group decided to add a fourth scenario that would lead to stabilization of greenhouse gas concentrations at CO_2 equivalent levels well below the CO_2 doubling level. In addition, Working Group I requested some changes in the initial three scenarios and the expert group decided to provide an alternative fourth scenario that was quite similar to the newly adopted fourth scenario, but which featured lower estimates of carbon dioxide emissions during the next few decades.

This report and the accompanying Appendix represent the outcome of two experts meetings, one held at the National Institute for Public Health and Environmental Protection (RIVM) in Bithoven, the Netherlands, on April 7–8, 1989, and a subsequent meeting held in Washington, D.C., in December 1989. The experts meeting in April was attended by representatives from the U.S. Environ-mental Protection Agency (U.S. EPA), the U.S. Department of Energy, the Netherlands' Ministry of Housing, Physical Planning and Environment and its research institute RIVM, and United Kingdom observers from Working Group I.

2.2 DEFINITION OF SCENARIOS AND METHODOLOGY

2.2.1 DEFINITION AND USE OF SCENARIOS

The scenarios presented in this report represent very different possible futures. RSWG has constructed these scenarios through a process of identifying and estimating natural and anthropogenic sources of emissions, identifying the key factors that are likely to influence future emissions from these sources, making different sets of assumptions about how these factors may change in the future, and then estimating the impact of these combined changes on emissions. The resulting scenarios meet the needs of Working Group I by providing emissions estimates that behave according to specifications, consistent with a world that is evolving in a specified and reasonable manner.

The expert group used two alternative models to construct these scenarios: the Atmospheric Stabilization Framework (ASF) developed by the U.S. EPA, and the Integrated Model for the Assessment of the Greenhouse Effect (IMAGE) developed by RIVM. These models provide both a structure for

understanding how different economic and physical factors influence emissions of greenhouse gases, as well as a means of applying different estimates of how these factors may change and calculating the effect that changes will have on future emissions.

2.2.2 SPECIFICATIONS OF SCENARIOS

The scenarios of future greenhouse gas emissions represent vastly different views of the future and a wide range of changes in atmospheric concentrations of greenhouse gases. One scenario depicts a continued rapid buildup in atmospheric concentrations of greenhouse gases through the end of the next century, while another scenario represents a world where concentrations quickly stabilize.

To account for the wide range of impacts on greenhouse warming from the different greenhouse gases the expert group used the concept of *equivalent CO₂ concentration* to define the scenarios. Equivalent CO_2 concentration, or the concentration of CO_2, is defined as the concentration of CO_2 that, by itself, would produce the increase in direct radiative forcing produced by all of the greenhouse gases. Equivalent CO_2 concentration is derived by first estimating the increase, over pre-industrial levels, in direct radiative forcing from all of the greenhouse gases and then calculating the concentration of CO_2 that would produce the same increase, assuming atmospheric concentrations of all other greenhouse gases stayed at pre-industrial levels.

The equivalent CO_2 concentration is greater than the atmospheric concentration of CO_2 as long as the concentrations of other gases such as methane (CH_4) and nitrous oxide (N_2O) are equal to or greater than pre-industrial levels. For example, assume that atmospheric concentrations of the most important greenhouse gases were as follows: CO_2—444 parts per million (ppm), N_2O—341 parts per billion (ppb), CH_4—2510 ppb, CFC-11—537 parts per trillion (ppt), CFC-12—1077 ppt, and HCFC-22—558 ppt. Direct radiative forcing from all of the greenhouse gases then would be about 4.0 watts/meter^2 above pre-industrial levels, which is less than the 4.3 watts/meter^2 that would be produced by a doubling of CO_2. The equivalent CO_2 concentration would be 550 ppm, which represents slightly less than a doubling of CO_2 over pre-industrial levels.

The four emission scenarios are as follows:

1) *2030 High Emissions:* Equivalent CO_2 concentrations reach a value double that of pre-industrial atmospheric concentrations of CO_2 by 2030.

2) *2060 Low Emissions:* Equivalent CO_2 concentrations reach a value double that of pre-industrial atmospheric concentrations of CO_2 by 2060.

3) *Control Policies:* Equivalent CO_2 concentrations reach a value double that of pre-industrial atmospheric concentrations of CO_2 by 2090 and stabilize thereafter.

4) *Accelerated Policies:* Equivalent CO_2 concentrations stabilize at a level less than a doubling of pre-industrial atmospheric concentrations of CO_2.

No specifications were made as to the relative contribution of different greenhouse gases to the equivalent CO_2 concentrations or, in the first two scenarios, to the pattern of concentrations after they doubled. The expert group also developed an *Alternative Accelerated Policies Scenario.* While the fourth scenario assumes that economic, political, and technological constraints would prevent any significant reduction in emissions in the short run, the alternative fourth scenario assumes a political climate that stresses the urgency of rapidly slowing down the rate of climate change, and assumes earlier reductions in CO_2 emissions.

2.2.3 METHODOLOGY

The methodology used to create the four emission scenarios incorporated two broad steps. First, the group designed and developed two *detailed scenarios* of future greenhouse gas emissions for each of the four requested emission scenarios, with the main difference between the two scenarios being the rate of economic growth. Each of the eight detailed scenarios represented a much different view of how the world might evolve and produce levels of greenhouse gases that meet the scenario specifications. The second step involved combining, for each of the

four requested emission scenarios, the two detailed (lower economic growth and higher economic growth) scenarios to create an *average* scenario. For the alternative fourth scenario (Alternative Accelerated Policies) the detailed scenario assumed that immediate action would be taken to stabilize concentrations. Selected results from the eight detailed scenarios, the four average scenarios, and the Alternative Accelerated Policies Scenario are presented in this report.

The design and development of the *detailed* emission scenarios consisted of four steps as follows:

1) Identify emission sources and estimate current (1985) emissions from these sources.

2) Identify alternative assumptions for key parameters that influence greenhouse gas emissions (e.g., economic growth and energy efficiency).

3) For each detailed scenario, combine assumptions for different parameters and implement the greenhouse models (ASF and/or IMAGE) using these assumptions.

4) Validate that the models produce results that meet the scenario specifications for each set of assumptions.

For the eight detailed scenarios, the ASF was used to combine the input assumptions and produce estimates of future emissions. The atmospheric and ocean models incorporated within the ASF were used to estimate future atmospheric concentrations of the greenhouse gases and the equivalent CO_2 concentrations in order to validate that the assumptions provided results consistent with the scenario specifications. The Dutch then reviewed the emission specifications and used IMAGE to further validate that the emission estimates provided the specified results.

2.3 ASSUMPTIONS

To develop scenarios of *future* emissions, certain assumptions must first be made about *current* sources of emissions; then additional assumptions must be made concerning the path of economic and technical change as well as the behavior of other factors that could influence greenhouse gas emissions in the future. The analytical frameworks described above (ASF and IMAGE) serve primarily to organize these assumptions and account for their consequences.

2.3.1 CURRENT EMISSION SOURCES

A great deal of uncertainty surrounds the estimates of many of the greenhouse gas emissions. This uncertainty reflects a number of factors, such as poor or inconsistent measurements of emissions, as well as a lack of information on activities that cause emissions, such as the area of tropical forest cleared and the amount of biomass burned as a result of the clearing. The level of uncertainty varies considerably by gas and by emission source. Emissions of CO_2 from the combustion of fossil fuels are estimated within an error of plus or minus 10 percent (Marland and Rotty, 1984), while high and low estimates of emissions of N_2O from fertilizer (including leaching) can vary by a factor of 4 (Bolle et al., 1986).

Estimates of current emissions of CO_2 from fossil fuel combustion and cement production are from Marland et al. (1988), and emissions of CO_2 from tropical deforestation are from Houghton (1988). For 1985, these sources estimate emissions of 5.2 petagrams of carbon (Pg C) from fossil fuel combustion, 0.1 Pg C from cement manufacturing, and 0.7 Pg C from tropical deforestation.

The estimates of current emissions for CO, CH_4, and NO_x reflect the recommendations of Working Group I made in October 1989. Two emission sources for CO have been excluded: oxidation of man-made hydrocarbons and oxidation of natural hydrocarbons, since these two sources are estimated endogenously in the atmospheric model within the ASF. Also, changes in anthropogenic emissions of non-methane hydrocarbons are not estimated. Similarly, NO_x from stratospheric subsidence is excluded. Table 2.1 shows the emission assumptions for CO, Table 2.2, the assumptions for CH_4 (from Cicerone and Oremland, 1988), Table 2.3, the assumptions for NO_x, and Table 2.4, the assumptions for N_2O. The estimates for N_2O emissions reflect the need to balance recent data on emis-

sions of N_2O from stationary sources with esti-mates of the atmospheric life of N_2O and current growth rates in atmospheric concentrations. Using 0.25 percent as the estimated annual rate of growth in atmospheric concentrations of N_2O and an atmospheric life of 160 years yields 12.5 teragrams of nitrogen ($Tg\ N$) as the estimate of N_2O emissions in 1985. Of this total, an estimated 8.0 $Tg\ N$ is from natural sources, for example, from nitrification and denitrification in soils and from oceans and fresh waters. Other emission sources include 1.0 $Tg\ N$ from combustion of fossil fuels and 1.6 $Tg\ N$ from fertilizer, which includes the leaching of fertilizer through the groundwater.

2.3.2 DEMOGRAPHIC, ECONOMIC, AND TECHNOLOGICAL CHANGE

Key factors expected to influence future changes in emissions of greenhouse gases include population growth, economic growth, the costs of technology used to convert energy from one form to another, end-use efficiency, deforestation rates, CFC emissions, and agricultural emissions. The expert group made different assumptions about how these factors may change in the future, then combined these different assumptions to construct the detailed emission scenarios.

The combinations of assumptions used to construct the scenarios are displayed in Table 2.5 and

TABLE 2.1:   Current Annual CO Emissions

SOURCE	Tg CO	Tg C
Fossil Fuels + Industry	440	189
Oxidation of Man-Made Hydrocarbons*	100	43
Oxidation of Natural Hydrocarbons*	110	47
Wood Fuel	110	47
Biomass Burning	550	236
Oceans	40	17
Vegetation Emissions	130	56
Total	1480	634
Total (Excluding Hydrocarbons)	1270	544

*Not included in ASF

TABLE 2.2:   Current Annual CH₄ Emissions

SOURCE	Tg CH_4
Enteric Fermentation	80
Wetlands	115
Rice Paddies	110
Biomass Burning	55
Termites	40
Landfills	40
Oceans	10
Freshwaters	5
Methane Hydrate Destabilization	5
Coal Mining	35
Gas Drilling, Venting, and Transmission	45
Total	540

Source: Cicerone and Oremland, 1988

TABLE 2.3:   Current Annual NOₓ Emissions

SOURCE	Tg N
Fossil Fuel Combustion	21
Biomass Burning	9
Soil Emission	12
Lightning	8
Stratospheric Subsidence*	1
Total	48

*Not included in ASF

TABLE 2.4:   Current Annual N₂O Emissions

SOURCE	Tg N
Fossil Fuel Combustion	1.0
Fertilizer (Including Leaching)	1.6
Gain of Cultivated Land	0.4
Natural Emissions from Soils	6.0
Emissions from Oceans and Freshwater	2.0
Tropical Deforestation	0.5
Savanna Burning and Wildfires	0.3
Fuelwood and Industrial Biomass	0.2
Burning of Agricultural Wastes	0.4
Total	12.5

TABLE 2.5: Scenario Assumptions

	2030 High Emissions		2060 Low Emissions		Control Policies		Accelerated Policies	
	Higher Growth	Lower Growth	Higher Growth	Lower Growth	Higher Growth	Lower Growth	Higher Growth	Lower Growth
Population[a]	World Bank	World Bank	World Bank	World Bank	World Bank	World Bank	World Bank	World Bank
GNP[b]	Higher	Lower	Higher	Lower	Higher	Lower	Higher	Lower
Energy Supply[c]	Carbon-intensive	Carbon-intensive	Gas-intensive	Gas-intensive	Non-Fossil-intensive	Non-Fossil-intensive	Early Non-Fossil-intensive	Early Non-Fossil-intensive
Energy Demand[d]	Moderate Efficiency	Moderate Efficiency	High Efficiency	High Efficiency	High Efficiency	High Efficiency	High Efficiency	High Efficiency
Control Technology[e]	Modest Controls	Modest Controls	Stringent Controls	Stringent Controls	Stringent Controls	Stringent Controls	Stringent Controls	Stringent Controls
CFCs[f]	Protocol/Low Compliance	Protocol/Low Compliance	Protocol/Full Compliance	Protocol/Full Compliance	Phase Out	Phase Out	Phase Out	Phase Out
Deforestation[g]	Moderate	Rapid	Reforest	Reforest	Reforest	Reforest	Reforest	Reforest
Agriculture[h]	Current Factors	Current Factors	Current Factors	Current Factors	Declining Factors	Declining Factors	Declining Factors	Declining Factors

a Population estimates, taken from the World Bank (Zachariah and Vu, 1988) were not varied between scenarios. Global population reaches 9.5 billion in 2050 and 10.4 billion in 2100. In developing countries, population growth rates decline markedly after 2000, achieving a net reproduction rate of unity in every country by 2040.

b GNP growth rates were based on estimates from the World Bank (1987). For the Higher Growth cases, the average annual rate of GNP growth for the world decreases from 3.6 percent per year for the period 1985–2000, to about 3.3 percent per year for the period 2000–2025, to about 2.6 percent per year for the period 2025–2100. For the Lower Growth cases, the average annual rate of GNP growth for the world decreases from about 2.2 percent per year for the period 1985–2000, to about 2.1 percent per year for the period 2000–2025, to about 1.3 percent per year for the period 2025–2100.

c For the Carbon-intensive cases, energy supply is dominated by fossil fuel technologies. For the Gas-intensive cases, fossil fuels continue to play a major role, but the natural gas share of the energy supply market is increased. For the Non-Fossil-intensive cases, non-fossil technologies become economic in the latter part of the next century and supply most primary energy needs after 2075. In the Early Non-Fossil cases, non-fossil technologies play a much larger role starting in the early part of the next century.

d In the Moderate Efficiency cases, the annual rate of improvement in energy intensity (or primary energy use per dollar of GNP) decreases from an initial value between 1.0 and 1.5 percent to an average value of 0.7 to 1.2 percent for the years 2075 to 2100. The average annual rate of improvement in energy intensity for the years 1985 to 2100 ranges between 0.8 and 1.3 percent. In the High Efficiency cases, the annual rate of improvement in energy intensity decreases from an initial value between 1.5 and 2.5 percent to an average value of 1.1 to 1.8 percent for the years 2075 to 2100. The average annual rate of improvement in energy intensity for the years 1985 to 2100 ranges between 1.2 and 1.9 percent. In the Stringent

e In the Modest Controls cases, current emission control technologies are assumed. In the Stringent

Controls cases, the following controls are assumed: more stringent NO_x and CO controls on mobile and stationary sources, including all gas vehicles using three-way catalysts (in OECD countries by 2000 and in the rest of the world by 2025); from 2000 to 2025 conventional coal burners used for electricity generation are retrofit with low NO_x burners, with 85 percent retrofit in the developed countries and 40 percent in developing countries; starting in 2000 all new combustors used for electricity generation and all new industrial boilers require selective catalytic reduction in the developed countries and low NO_x burners in the developing countries, and after 2025 all new combustors of these types require selective catalytic reduction; other new industrial non-boiler combustors such as kilns and dryers require low NO_x burners after 2000.

f In the Protocol/Low Compliance cases, the Montreal Protocol is assumed to come into force and apply through 2100, with 100 percent participation by the United States and developed countries and 85 percent participation by developing countries. The assumptions are the same in the Protocol/Full Compliance case except that all countries participate. In the Phase-Out cases, CFCs are completely phased out by 2000 and production of CCl_4 and CH_3CCl_3 are frozen with 100 percent participation.

g Tropical deforestation increases gradually in the Moderate case, from 11 million ha/yr in 1985 to 15 million ha/yr by about 2100. Tropical deforestation increases exponentially in the Rapid case, reaching 34 million ha/yr in about 2050, with almost complete tropical forest deforestation by about 2075. In the Reforest cases, deforestation stops by 2025, and about 1,000 million ha are reforested by 2100.

h Levels of activity for the major agricultural activities were not varied between scenarios, since population levels were not varied. Emission factors, however, were changed between scenarios. Current estimates of average emission factors were used in the Current Factors cases, while reduced factors (assuming technological and management improvements) were used in the Declining Factors cases.

explained briefly below. As described earlier, two detailed scenarios were developed for each of the four requested emission scenarios, with the assumptions on the future rate of economic growth being the main difference between the two detailed scenarios. In Table 2.5 "Higher Growth" refers to higher economic growth detailed scenario and "Lower Growth" refers to lower economic growth detailed scenario. Additional information on the assumptions and results of these scenarios are included in the Appendix.

2.3.2.1 Population

The population estimates, taken from the World Bank (Zachariah and Vu, 1988), project a global population of 9.5 billion by 2050 and 10.4 billion by 2100. Regional population growth assumptions are shown in Figure 2.1. The bulk of the increase during the first half of the next century occurs in the developing countries, while population in the de-

veloped countries remains fairly stable during the entire period.

2.3.2.2 Economic Growth

Economic growth rates (Table 2.6) were derived from estimates from the World Bank (1987); adjustments to these estimates are described in the Appendix. The overall rate of economic development is one of the most important determinants of future greenhouse gas emissions. Robust economic growth is certainly the goal of most governments, and other things being equal, higher levels of economic activity would be associated with greater greenhouse gas emissions. In the scenarios defined here, for example, some regions reach a per capita income level of $5,000 (typical of Western Europe during the 1970s) by 2020 under the Higher Growth assumptions, but not until 2075 under the Lower Growth assumptions. For other regions, per capita income levels remain below $5,000 throughout the

TABLE 2.6: Economic Growth Rate Assumptions (Annual Percentage Growth)

| | HIGHER GROWTH | | | | |
	1985–2000	2000–2025	2025–2050	2050–2075	2075–2100
Region					
United States	3.0	2.5	2.0	1.5	1.0
Western Europe and Canada	3.0	2.5	2.0	1.5	1.0
Japan, Australia, New Zealand	3.5	2.5	2.0	1.5	1.0
USSR and Eastern Europe	4.6	4.1	3.1	2.6	2.1
Centrally Planned Asia	5.5	5.0	4.5	4.0	3.5
Middle East	4.1	4.6	3.6	3.1	2.6
Africa	4.5	4.0	3.5	3.0	2.5
Latin America	4.7	4.2	3.7	3.2	2.7
Rest of South and East Asia	5.3	4.8	4.3	3.8	3.3
	LOWER GROWTH				
	1985–2000	2000–2025	2025–2050	2050–2075	2075–2100
Region					
United States	2.0	1.5	1.0	1.0	1.0
Western Europe and Canada	2.0	1.5	1.0	1.0	1.0
Japan, Australia, New Zealand	2.5	1.5	1.0	1.0	1.0
USSR and Eastern Europe	2.6	2.1	1.6	1.6	1.6
Centrally Planned Asia	3.5	3.0	2.5	2.5	2.5
Middle East	3.3	2.8	2.1	2.1	2.1
Africa	3.0	2.6	2.1	2.1	2.1
Latin America	2.7	2.2	1.7	1.7	1.7
Rest of South and East Asia	3.3	2.8	2.3	2.3	2.3

FIGURE 2.1: Global Population Levels by Region, 1985–2100

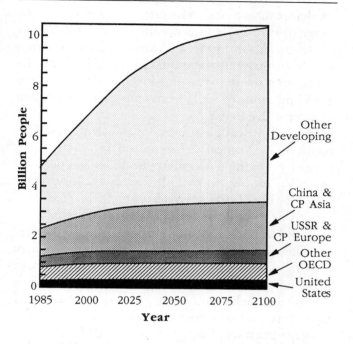

period under the Lower Growth assumptions. Scenarios of sustainable development in which all developing countries achieve a higher income level by the middle of the next century could be explored in future analyses.

2.3.2.3 *Energy Supply*

Four alternative assumptions on energy supply provide for a wide range of possible futures. The first set of assumptions, labeled Carbon-intensive, represents a pessimistic view of future non-fossil developments, where the cost of non-fossil energy supply technologies remains high compared to fossil supplies, and/or the technical potential for these energy sources remains low. Fossil energy resources, primarily coal, are adequate to satisfy demand for energy, and synthetic fuels from coal become competitive early in the twenty-first century. Under these assumptions, costs of commercial solar electricity remain at current levels of around $0.11/kwh (1988 U.S. dollars), use of nuclear energy is limited because of concerns over safety, and energy from commercial biomass remains low be-

cause of a combination of high development costs, lack of infrastructure, and competition with other uses for the land.

The Gas-intensive assumptions are similar to the Carbon-intensive assumptions except that increased use of natural gas is encouraged in order to meet a number of environmental goals, including reducing CO_2 emissions. Carbon fees of 20 to 25 percent of delivered prices help reduce demand for oil and coal and help provide price subsidies for natural gas of up to 15 percent. Research and development help accelerate reductions in gas exploration and production costs on the order of 0.5 percent annually.

The third set of assumptions, labeled Non-Fossil-intensive, represents a world where technological developments result in competitive and abundant non-fossil energy sources, but not until late in the next century. Costs of commercial solar electricity fall to $0.06/kwh (1988 U.S. dollars) after 2050. Similarly, costs of nuclear energy fall to $0.055/kwh, and political and technological constraints inhibiting its use are relaxed. Finally, costs of producing large quantities of biomass energy are significantly reduced.

The fourth set, labeled Early Non-Fossil-intensive, is similar to the previous set of assumptions but with several key differences. Here, constraints that impede the penetration and use of non-fossil supplies are relaxed, and cost reductions are achieved, *early* in the twenty-first century, so that non-fossil energy sources can make a significant impact by 2025. Fees based on the carbon content of fossil fuels, set at a rate equivalent to $50/ton on coal (1988 U.S. dollars), are phased in on a global scale by 2025.

The use of the assumptions for the different emission scenarios is straightforward. The Carbon-intensive energy supply assumptions are used to construct the 2030 High Emissions scenarios. The Gas-intensive assumptions are used for the 2060 Low Emissions scenarios. The Non-Fossil-intensive assumptions are used with the Control Policies scenarios, and the Early Non-Fossil-intensive assumptions are used in the Accelerated Policies scenarios.

2.3.2.4 *Energy Demand*

Two alternative assumptions are made concerning future demand for energy under each set of economic growth assumptions. These alternatives reflect both the variance in historical trends as well as a wide range of views as to what is possible in the future. The first set of these assumptions, labeled Moderate Efficiency, yields annual improvements in energy intensity (measured as primary energy use per dollar GNP) of around 0.9 percent under the lower economic growth assumptions and 1.3 percent under the higher economic growth assumptions. The High Efficiency assumptions yield annual improvements in energy intensity of around 1.2 percent under lower economic growth and 1.8 percent under the higher economic growth (with annual improvements from 1985 to 2025 averaging 2.3 percent).

2.3.2.5 *CFCs and Halons*

Three sets of assumptions are made concerning future emissions of perhalogenated hydrocarbons and hydrohalocarbons (including CFCs). The first of the assumptions, labeled Protocol/Low Compliance, represents a very pessimistic view of efforts to further control emissions of CFCs. They include no strengthening of the Montreal Protocol and only 85 percent participation by developing countries. The second set of assumptions, labeled Protocol/Full Compliance, still represents a pessimistic view, in that there is no strengthening of the Montreal Protocol, but all countries participate and there is 100 percent compliance to the existing terms. The third set of assumptions (Phase Out) represents an optimistic view of efforts to strengthen the Montreal Protocol; these assumptions include a complete phase-out of CFCs by the year 2000 and limits on emissions of carbon tetrachloride and methylchloroform, as well as 100 percent participation in the strengthened agreement. In all of the above sets of assumptions, the substitution of HCFCs (HCFC-22 is used as proxy) for controlled gases occurs at a rate of .35 to 1.

The use of these assumptions to construct the emission scenarios is also straightforward. For the 2030 High Emissions scenarios, the Protocol/Low Compliance assumptions are used. The Protocol/Full Compliance assumptions are used with the 2060 Low Emissions scenarios. And the Phase Out assumptions are used with both the Control Policies and Accelerated Policies scenarios.

2.3.2.6 *Deforestation*

Three alternate sets of assumptions concerning rates of deforestation and reforestation provide a wide range of futures, from a continued increase in rates of deforestation to rapid cessation of deforestation and vigorous reforestation efforts. All three deforestation scenarios assume low initial biomass and 1985 emissions of CO_2 from tropical deforestation at around 0.7 Pg C.

In the first set of assumptions, labeled Rapid Deforestation, tropical deforestation increases exponentially, reaching 34 million hectares (ha) per year by 2050, leading to complete global deforestation by 2075. No efforts are made to establish tree plantations or to reforest areas.

In the second set of assumptions, labeled Moderate Deforestation, tropical deforestation increases gradually, reaching 15 million ha per year by about 2100, compared with 11 million ha/yr in 1985. The rate of establishment of tree plantations is assumed to be zero and no reforestation occurs.

In the third set of assumptions, labeled Reforestation, tropical deforestation stops by 2025, while about 1,000 million ha is reforested by 2100. Only land that once supported forests and is not intensively cultivated is assumed to be available for reforestation. Of the reforested land, 380 million ha are assumed to be in plantations; the rest absorbs carbon at a slower rate but eventually reaches a much higher level of biomass.

The Rapid Deforestation assumptions were used for the 2030 High Emissions/Lower Growth scenario. The Moderate Deforestation assumptions were used for the 2030 High Emissions/Higher Growth scenario. The Reforestation assumptions were used for all other scenarios.

2.3.2.7 *Agriculture*

The assumptions affecting emissions from agriculture fall into two groups: (1) those that have an impact on consumption and production of agricultural products and (2) those that might change the emissions attributed to a given level of agricultural activity (emission coefficients). Since population growth was the same in all scenarios, future activity levels for the major agricultural variables affecting trace gas emissions (e.g., land use for rice production, meat consumption, dairy consumption, fertilizer use, etc.) did not vary among scenarios.

Two separate sets of assumptions are made concerning future emissions from these activities. The first set, labeled Current Factors, assumes that emissions coefficients from agricultural activities remain constant over time. As examples, the percent of N applied as fertilizer that evolves as N_2O would remain constant in the future, and CH_4 emissions from a single hectare of land cultivated for rice production would also remain constant. The second set of assumptions (Declining Factors) has the emission coefficients declining 0.5 percent annually as a result of efforts to control emissions, for example, by changing types of fertilizer and methods of application, by altering rice cultivation practices and rice cultivars, and by adopting meat and dairy production techniques such as methane-inhibiting ionophores. Examples of Current Factor coefficients, used in the 2030 High Emissions and 2060 Low Emissions scenarios, are shown in Table 2.7. Declining Factors were used in all other Scenarios.

TABLE 2.7: Emission Coefficients for the Agricultural Sector

	N_2O FROM FERTILIZER	
FERTILIZER TYPE	% EVOLVED DIRECTLY	% EVOLVED FROM LEACHING
Ammonium Nitrate and Ammonium Salts	0.1	2.0
Nitrate	0.05	2.0
Urea	0.5	2.0
Other Nitrogenous and Other Complex	1.0	2.0
Anhydrous Ammonia	0.5	2.0

	METHANE FROM AGRICULTURAL ACTIVITIES
Rice Production	75 grams CH_4/meter²/harvest
Enteric Fermentation	(Kilograms CH_4/Head)*
Cattle	41
Sheep	6
Dairy Cows	64
Pigs	15

* Global Averages

2.4 RESULTS

Combining the diverse sets of assumptions described in the previous section in the modeling frameworks provides a wide range of different views of how the world could evolve. As described earlier, two alternative detailed scenarios provide results that are consistent with each of the four requested scenarios of greenhouse gas concentration buildup. The detailed scenarios were also combined to create an Average Scenario for each requested emission profile. An additional scenario provides an alternative view of how stabilization of greenhouse gas concentrations can be achieved. Table 2.8 shows the expected emissions of each gas from each source for each of the four average (2030 High Emissions, 2060 Low Emissions, Control Policies, Accelerated Policies) scenarios and the Alternative Accelerated Policies Scenario.

Both the ASF and IMAGE were used to estimate changes in atmospheric concentrations of the greenhouse gases in order to confirm that they met the specifications of the four emission scenarios. The results of the ASF and IMAGE were within 5 percent, with variations due primarily to the use of different approaches to estimate atmospheric concentrations of the greenhouse gases and a different conceptual approach to the modeling of land-use changes. Figure 2.2 displays CO_2 concentrations and equivalent CO_2 concentrations estimated by the Atmospheric Stabilization Framework for the four averaged scenarios and the Alternative Accelerated Policies Scenario along with the level of atmospheric CO_2 concentrations representing a doubling of pre-industrial concentrations. These divergent outcomes are described in more detail below for the energy, forestry, and agricultural sectors.

2.4.1 ENERGY SECTOR

The two most important determinants of greenhouse gas emissions are the level of energy demand and the combination of sources that are used to supply that energy. The basic differences between the scenarios in global primary energy supplies are illustrated in Figures 2.3 and 2.4. The regional differences in primary energy consumption are illustrated in Figures 2.5 and 2.6.

T A B L E 2 . 8 : Greenhouse Gas Emissions: 2030 High Emissions (Average) Scenario						
Gas/Source	1985	2000	2025	2050	2075	2100
CO_2 (Petagrams C/Yr)						
Commercial Energy	5.1	6.5	9.9	13.5	17.7	21.7
Tropical Deforestation	0.7	1.0	1.4	1.4	0.7	0.4
Cement	0.1	0.2	0.2	0.3	0.3	0.3
Total	6.0	7.7	11.5	15.2	18.7	22.4
N_2O (Teragrams N/Yr)						
Commercial Energy	1.1	1.4	2.1	2.8	3.5	4.1
Fertilizer	1.5	2.5	3.5	3.7	3.9	3.9
Gain of Cultivated Land	0.4	0.6	0.8	0.8	0.4	0.2
Natural Land Emissions	6.0	6.0	6.0	6.0	6.0	6.0
Oceans/Freshwater	2.0	2.0	2.0	2.0	2.0	2.0
Biomass Burning	1.4	1.8	2.0	2.0	1.5	1.3
Total	12.5	14.2	16.4	17.3	17.3	17.6
CH_4 (Teragrams CH_4/Yr)						
Commercial Energy	2.0	2.3	3.2	4.1	4.9	6.3
Fuel Production	80.0	98.1	158.7	212.5	268.9	326.9
Enteric Fermentation	74.5	94.5	124.6	156.0	163.6	166.9
Rice	110.0	125.7	148.7	167.9	168.8	157.1
Oceans/Freshwater	15.0	15.0	15.0	15.0	15.0	15.0
Landfills	40.0	50.1	71.1	103.5	149.2	179.6
Wetlands	115.0	115.0	115.0	115.0	115.0	115.0
Biomass Burning	55.1	64.4	75.6	76.1	57.7	47.2
Wild Ruminants and Termites	44.0	44.0	44.0	44.0	44.0	44.0
Methane Hydrate	5.0	5.0	5.0	5.0	5.0	5.0
Total	540.5	613.9	760.8	899.1	992.0	1062.9
NO_X (Teragrams N/Yr)						
Commercial Energy	23.2	28.2	39.1	51.4	63.2	75.4
Biomass Burning	6.1	7.0	8.3	8.3	6.4	5.2
Natural Land	12.0	12.0	12.0	12.0	12.0	12.0
Lightning	9.0	9.0	9.0	9.0	9.0	9.0
Total	50.2	56.2	68.3	80.7	90.5	101.5
CO (Teragrams C/Yr)						
Commercial Energy	182.2	198.3	289.0	386.8	468.8	627.8
Tropical Deforestation	135.0	189.6	255.2	255.2	135.0	72.1
Oceans	17.0	17.0	17.0	17.0	17.0	17.0
Agricultural Burning	146.0	153.7	162.8	165.7	160.7	153.0
Wood Use	47.0	47.0	47.1	47.3	47.6	48.0
Wildfires	10.0	10.0	10.0	10.0	10.0	10.0
Total	537.2	615.6	781.2	881.9	839.1	927.9
CFCs (Gigagrams/Yr)						
CFC-11	278.3	305.1	244.7	251.5	253.0	253.0
CFC-12	361.9	376.0	302.7	314.1	316.1	316.1
HCFC-22	96.9	522.9	1340.4	2681.3	2961.0	2961.0
Halon-1211	1.4	8.8	18.6	18.9	19.1	19.1
Halon-1301	2.1	5.1	7.4	7.5	7.7	7.8

Note: Totals reflect rounding.

Gas/Source	1985	2000	2025	2050	2075	2100
CO_2 (Petagrams C/Yr)						
Commercial Energy	5.1	5.6	6.6	7.6	8.7	10.3
Tropical Deforestation	0.7	−0.2	−0.5	−0.3	−0.2	−0.2
Cement	0.1	0.2	0.2	0.2	0.2	0.2
Total	5.9	5.5	6.4	7.5	8.8	10.3
N_2O (Teragrams N/Yr)						
Commercial Energy	1.1	1.2	1.3	1.4	1.5	1.8
Fertilizer	1.5	2.5	3.5	3.7	3.9	3.9
Gain of Cultivated Land	0.4	0.2	0.0	0.0	0.0	0.0
Natural Land Emissions	6.0	6.0	6.0	6.0	6.0	6.0
Oceans/Freshwater	2.0	2.0	2.0	2.0	2.0	2.0
Biomass Burning	1.1	1.2	1.0	1.0	1.0	0.9
Total	12.5	13.1	13.9	14.1	14.3	14.6
CH_4 (Teragrams CH_4/Yr)						
Commercial Energy	2.0	2.1	1.8	1.7	1.8	2.1
Fuel Production	80.0	88.8	118.2	132.2	136.6	151.5
Enteric Fermentation	74.5	94.5	124.6	156.0	163.6	166.9
Rice	110.0	125.7	148.7	167.9	168.8	157.1
Oceans/Freshwater	15.0	15.0	15.0	15.0	15.0	15.0
Landfills	40.0	43.0	55.4	50.0	47.1	45.6
Wetlands	115.0	115.0	115.0	115.0	115.0	115.0
Biomass Burning	55.1	43.9	38.0	37.0	35.5	33.4
Wild Ruminants and Termites	44.0	44.0	44.0	44.0	44.0	44.0
Methane Hydrate	5.0	5.0	5.0	5.0	5.0	5.0
Total	540.5	576.8	665.4	723.8	732.4	735.6
NO_X (Teragrams N/Yr)						
Commercial Energy	23.2	24.4	21.0	16.2	16.5	19.2
Biomass Burning	6.1	4.8	4.2	4.0	3.9	3.7
Natural Land	12.0	12.0	12.0	12.0	12.0	12.0
Lightning	9.0	9.0	9.0	9.0	9.0	9.0
Total	50.2	50.2	46.1	41.2	41.4	43.8
CO (Teragrams C/Yr)						
Commercial Energy	182.2	149.2	99.2	54.2	57.7	71.3
Tropical Deforestation	135.0	53.6	7.4	1.8	1.8	0.0
Oceans	17.0	17.0	17.0	17.0	17.0	17.0
Agricultural Burning	146.0	153.7	162.8	165.7	160.7	153.0
Wood Use	47.0	44.1	39.7	35.8	32.5	29.6
Wildfires	10.0	10.0	10.0	10.0	10.0	10.0
Total	537.2	427.6	336.1	284.6	279.8	280.9
CFCs (Gigagrams/Yr)						
CFC-11	278.3	302.1	226.7	223.2	223.2	223.2
CFC-12	361.9	372.2	279.1	277.9	277.9	277.9
HCFC-22	96.9	525.7	1357.2	2707.2	2988.1	2988.1
Halon-1211	1.4	8.8	18.2	17.8	17.5	17.5
Halon-1301	2.1	5.1	7.3	7.2	7.3	7.3

Gas/Source	1985	2000	2025	2050	2075	2100
CO_2 (Petagrams C/Yr)						
Commercial Energy	5.1	5.6	6.5	7.2	5.0	3.5
Tropical Deforestation	0.7	−0.2	−0.5	−0.3	−0.2	−0.2
Cement	0.1	0.2	0.2	0.2	0.2	0.2
Total	5.9	5.6	6.3	7.1	5.1	3.5
N_2O (Teragrams N/Yr)						
Commercial Energy	1.1	1.2	1.3	1.4	1.1	1.1
Fertilizer	1.5	2.3	2.8	2.6	2.4	2.2
Gain of Cultivated Land	0.4	0.2	0.0	0.0	0.0	0.0
Natural Land Emissions	6.0	6.0	6.0	6.0	6.0	6.0
Oceans/Freshwater	2.0	2.0	2.0	2.0	2.0	2.0
Biomass Burning	1.4	1.2	1.0	1.0	1.0	0.9
Total	12.5	12.9	13.2	13.0	12.5	12.2
CH_4 (Teragrams CH_4/Yr)						
Commercial Energy	2.0	2.0	1.7	1.7	1.5	1.7
Fuel Production	80.0	85.2	108.3	117.4	83.6	59.0
Enteric Fermentation	74.5	88.3	103.7	115.3	107.8	98.0
Rice	110.0	116.6	121.6	121.2	107.5	88.3
Oceans/Freshwater	15.0	15.0	15.0	15.0	15.0	15.0
Landfills	40.0	43.0	55.4	50.0	47.1	45.6
Wetlands	115.0	115.0	115.0	115.0	115.0	115.0
Biomass Burning	55.1	43.9	38.0	37.0	35.5	33.4
Wild Ruminants and Termites	44.0	44.0	44.0	44.0	44.0	44.0
Methane Hydrate	5.0	5.0	5.0	5.0	5.0	5.0
Total	540.5	557.9	607.7	621.6	562.0	504.9
NO_X (Teragrams N/Yr)						
Commercial Energy	23.2	24.5	21.2	16.9	14.4	14.7
Biomass Burning	6.1	4.8	4.2	4.0	3.9	3.7
Natural Land	12.0	12.0	12.0	12.0	12.0	12.0
Lightning	9.0	9.0	9.0	9.0	9.0	9.0
Total	50.2	50.3	46.4	41.9	39.3	39.4
CO (Teragrams C/Yr)						
Commercial Energy	182.2	149.0	99.3	55.0	52.2	60.6
Tropical Deforestation	135.0	53.6	7.4	1.8	1.8	0.0
Oceans	17.0	17.0	17.0	17.0	17.0	17.0
Agricultural Burning	146.0	153.7	162.8	165.7	160.7	153.0
Wood Use	47.0	44.1	39.7	35.8	32.5	29.6
Wildfires	10.0	10.0	10.0	10.0	10.0	10.0
Total	537.2	427.4	336.1	285.4	274.3	270.2
CFCs (Gigagrams/Yr)						
CFC-11	278.3	197.1	10.6	0.0	0.0	0.0
CFC-12	361.9	262.2	10.2	0.0	0.0	0.0
HCFC-22	96.9	638.5	1571.9	2927.6	3208.5	3208.5
Halon-1211	1.4	7.7	3.0	0.0	0.0	0.0
Halon-1301	2.1	4.1	1.8	0.0	0.0	0.0

TABLE 2.8 (*continued*): Greenhouse Gas Emissions: Accelerated Policies (Average) and Alternative Accelerated Policies Scenario

Gas/Source	1985	2000	2025	2050	2075	2100
CO_2 (Petagrams C/Yr)						
Accelerated Policies						
Commercial Energy	5.1	5.7	5.4	3.0	2.9	2.7
Tropical Deforestation	0.7	−0.2	−0.5	−0.3	−0.2	−0.2
Cement	0.1	0.2	0.2	0.2	0.2	0.2
Total	6.0	5.6	5.1	2.9	3.0	2.7
Alternative Accelerated Policies						
Commercial Energy	5.1	4.7	4.1	3.8	3.4	2.6
Tropical Deforestation	0.7	−0.2	−0.5	−0.3	−0.2	−0.2
Cement	0.1	0.2	0.2	0.2	0.2	0.2
Total	6.0	4.6	3.8	3.7	3.5	2.6
N_2O (Teragrams N/Yr)						
Commercial Energy	1.0	1.2	1.2	1.0	1.0	1.1
Fertilizer	1.6	2.3	2.8	2.6	2.4	2.2
Gain of Cultivated Land	0.4	0.2	0.0	0.0	0.0	0.0
Natural Land Emissions	6.0	6.0	6.0	6.0	6.0	6.0
Oceans/Freshwater	2.0	2.0	2.0	2.0	2.0	2.0
Biomass Burning	1.1	1.2	1.0	1.0	1.0	0.9
Total	12.5	12.9	13.1	12.7	12.5	12.3
CH_4 (Teragrams CH_4/Yr)						
Commercial Energy	2.0	1.9	1.9	1.6	1.7	1.9
Fuel Production	80.0	93.3	84.0	48.7	50.3	52.4
Enteric Fermentation	75.2	88.3	103.7	115.3	107.8	98.0
Rice	109.4	116.6	121.6	121.2	107.9	91.2
Oceans/Freshwater	15.0	15.0	15.0	15.0	15.0	15.0
Landfills	40.0	43.0	55.4	50.0	47.1	45.6
Wetlands	115.0	115.0	115.0	115.0	115.0	115.0
Biomass Burning	55.1	43.8	37.9	37.2	36.3	34.0
Wild Ruminants and Termites	44.0	44.0	44.0	44.0	44.0	44.0
Methane Hydrate	5.0	5.0	5.0	5.0	5.0	5.0
Total	540.7	565.8	583.5	553.0	530.1	502.1
NO_X (Teragrams N/Yr)						
Commercial Energy	24.2	24.7	24.7	18.7	18.0	18.8
Biomass Burning	6.1	4.8	4.2	4.1	3.9	3.8
Natural Land	12.0	12.0	12.0	12.0	12.0	12.0
Lightning	9.0	9.0	9.0	9.0	9.0	9.0
Total	51.2	50.5	49.9	43.8	42.9	43.5
CO (Teragrams C/Yr)						
Commercial Energy	185.8	133.3	104.2	55.4	58.4	68.0
Tropical Deforestation	135.0	53.6	7.4	1.8	1.8	0.0
Oceans	17.0	17.0	17.0	17.0	17.0	17.0
Agricultural Burning	146.0	153.1	162.6	166.9	165.3	156.4
Wood Use	47.0	44.1	39.7	35.8	32.5	29.6
Wildfires	10.0	10.0	10.0	10.0	10.0	10.0
Total	540.8	411.1	340.8	287.0	285.0	281.0
CFCs (Gigagrams/Yr)						
CFC-11	278.3	197.1	10.6	0.0	0.0	0.0
CFC-12	361.9	262.2	10.2	0.0	0.0	0.0
HCFC-22	96.9	638.5	1571.9	2927.6	3208.5	3208.5
Halon-1211	1.4	7.7	3.0	0.0	0.0	0.0
Halon-1301	2.1	4.1	1.8	0.0	0.0	0.0

FIGURE 2.2: CO_2 and Equivalent CO_2 Concentrations: Average

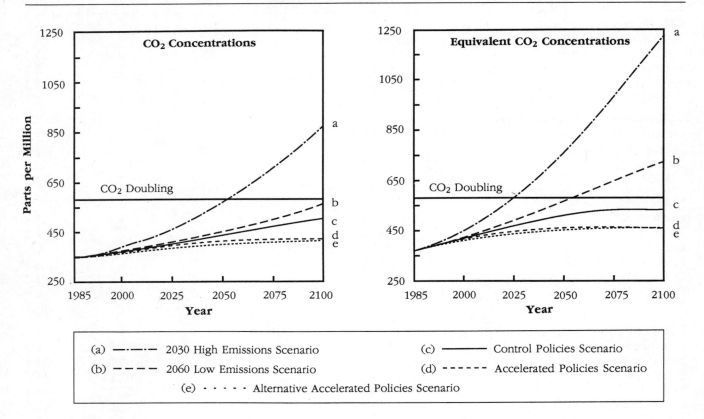

(a) —·—·— 2030 High Emissions Scenario   (c) ——— Control Policies Scenario

(b) — — — 2060 Low Emissions Scenario   (d) - - - - - Accelerated Policies Scenario

(e) · · · · · Alternative Accelerated Policies Scenario

2.4.1.1 *2030 High Emissions Scenarios*

In the 2030 High Emissions scenarios, fossil fuel consumption increases rapidly through the entire time horizon (1985 to 2100). With higher economic growth, primary energy supply increases from a 1985 level of 290 EJ to 660 EJ in 2025 (an annual growth rate of 2.1 percent) and reaches 1680 EJ in 2100 (an annual growth rate of 1.3 percent). Oil and natural gas supplies increase through 2025, then start to decline because of resource constraints. Coal becomes the dominant fuel; its share of primary energy supply rises from below 30 percent in 1985 to over 65 percent by the end of the time period. By 2100, half of all coal mined is used in synthetic fuel production. With lower economic growth, primary energy demand and supply increase at about half the rate seen with higher economic growth. Primary energy supply increases to 470 EJ by 2025 and to 730 EJ by 2100, with average annual rates of growth of 1.2 percent and 0.6 per-

cent, respectively. Oil and natural gas supply follows a pattern similar to that described above, and the share of primary energy supplied by coal increases to 55 percent in 2100.

Regional energy consumption patterns also differ between the two 2030 High Emissions detailed scenarios. Under the Higher Growth assumptions, all regions show an increase in primary energy consumption, but the increase in developing countries far exceeds that in the developed countries.

This result reflects the higher rates of economic growth assumed in developing countries as well as assumptions concerning energy use during this rapid development. The share of primary energy consumed in the developing countries increases from 22 percent in 1985 to over 60 percent in 2100. A similar pattern of regional growth is seen in the Lower Growth scenario, but the portion of energy consumed in the developing countries increases to only 45 percent in 2100.

2.4.1.2 *2060 Low Emissions Scenarios*

Reflecting the assumptions made concerning greater improvements in energy efficiency, the 2060 Low Emissions scenarios display smaller increases in primary energy supply and fossil fuel use than seen in the 2030 High Emissions scenarios. In addition, natural gas plays a larger role through 2025, but that role is reduced thereafter because of resource constraints. Primary energy supply reaches only 450 EJ in 2025 and 850 EJ in 2100 under the Higher Growth assumptions and 374 EJ in 2025 and 430 EJ in 2100 under the Lower Growth assumptions. The share of primary energy supply provided by natural gas grows from 19 percent in 1985 to 23 percent in 2100 under the Higher Growth assumptions and to 25 percent under the Lower Growth assumptions. The role of coal continues to expand; coal's share of primary energy increases from 28 percent in 1985 to 54 percent by 2100 (Higher Growth assumptions) and to 41 percent by 2100 (Lower Growth assumptions).

The regional patterns are similar to those shown in the scenarios that assume less efficiency improvement. The share of global primary energy consumed in the developing world increases from 22 percent in 1985 to over 50 percent by 2100 under the Higher Growth assumptions and to 45 percent by 2100 under the Lower Growth assumptions. Again, these results reflect higher rate of economic growth assumed for developing countries.

2.4.1.3 *Control Policies Scenarios*

The estimates of energy supply for the Control Policies scenarios are similar to those found in the 2060 Low Emissions scenarios, except that non-fossil energy supplies start to play a larger role after 2025, providing over 79 percent of primary energy by 2100 under Higher Growth assumptions and over 62 percent of primary energy by 2100 under Lower Growth assumptions. Most of the increase in non-fossil energy supplies occurs after 2050, since these supplies are not assumed to become competitive and widely available until the middle to end of the next century.

2.4.1.4 *Accelerated Policies Scenarios*

The Accelerated Policies scenarios provide a much different picture of future energy supply and demand. Non-fossil energy supplies, especially commercial biomass energy, play a much larger role and make an impact much earlier. Energy consumption in the scenarios is similar to the energy consumption in the Control Policies scenarios, except that in the Higher Growth scenario, energy use is slightly higher at mid-century because of the increased availability of biomass supplies and their lower costs. The share of primary energy provided by non-fossil sources increases from 13 percent in 1985 to 49 percent in 2025 to 78 percent by 2100 with high economic growth and to 33 percent in 2025 and 80 percent in 2100 with low economic growth.

Figure 2.7 shows the CO_2 emissions from fossil fuels under all of the scenarios. CO_2 emissions, which were 5.2 Pg C in 1985, reach levels in 2100 ranging from 30.8 Pg C in the 2030 High Emissions/Higher Growth scenario to 1.9 Pg C in the Accelerated Policies/Lower Growth scenario. Emissions of CO_2 from combustion of fossil fuels in the Alternative Accelerated Policies scenario decline immediately and continue to decline through 2100. This result differs from the other Stabilization scenarios (Higher Growth, Lower Growth, and Average), in which global emissions of CO_2 from fossil fuels grow through the year 2000 and then start to decline.

2.4.2 FORESTRY SECTOR

As shown in Figure 2.8, net emissions of CO_2 from tropical deforestation and reforestation varied considerably between the 2030 High Emissions scenario and all other scenarios.

2.4.2.1 *2030 High Emissions Scenarios*

In the 2030 High Emissions/Lower Growth scenario, future CO_2 emissions closely follow the rapidly increasing rate of land clearing assumed in the scenario. Emissions peak at over 2 Pg C/yr around the year 2050, with a total flux from the biosphere to the atmosphere of about 140 Pg C between 1985 and 2100.

In the 2030 High Emissions/Higher Growth scenario, emissions of CO_2 grow at a slower rate because a slower rate of land clearing is assumed. CO_2 emissions are spread out over a longer period, reaching a peak of slightly over 1 Pg C near the end of the century, with total flux of carbon over the time period at 120 Pg C.

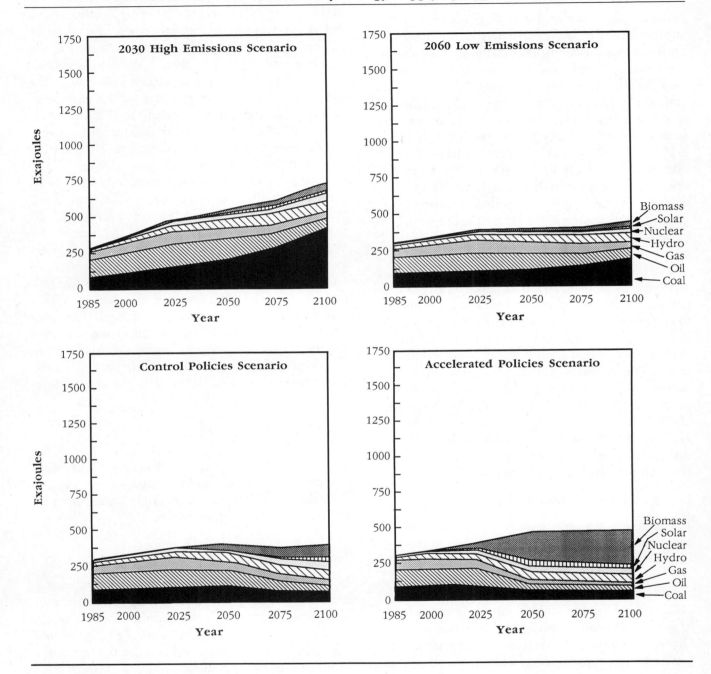

FIGURE 2.3: Global Primary Energy Supply by Type: Lower Growth

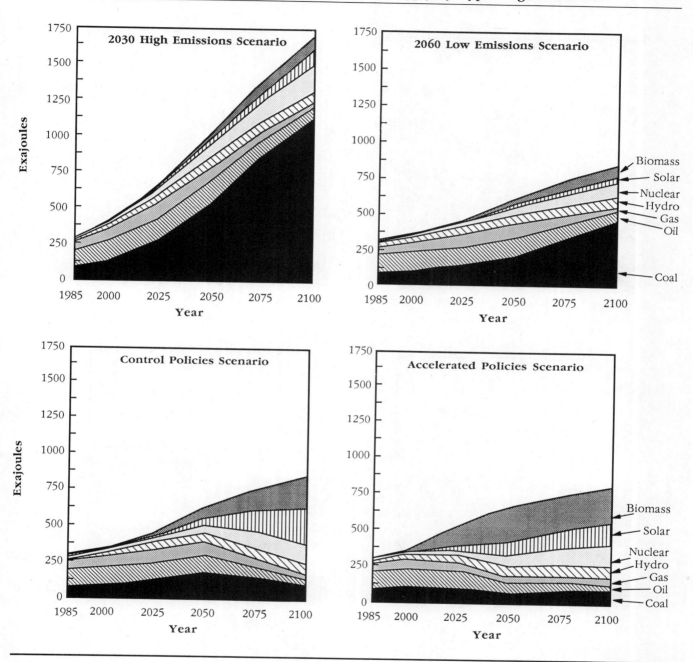

FIGURE 2.4: Global Primary Energy Supply by Type: Higher Growth

33

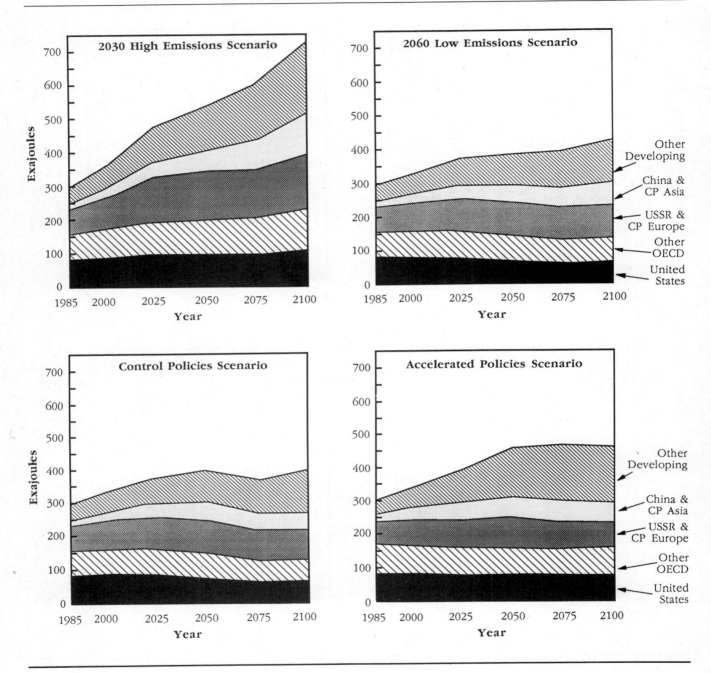

FIGURE 2.5 : Primary Energy Consumption by Region: Lower Growth

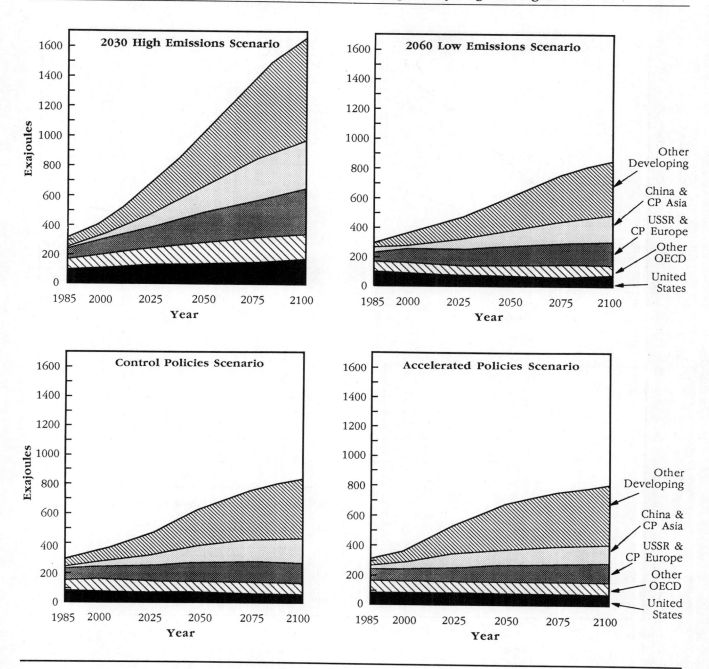

FIGURE 2.6: Primary Energy Consumption by Region: Higher Growth

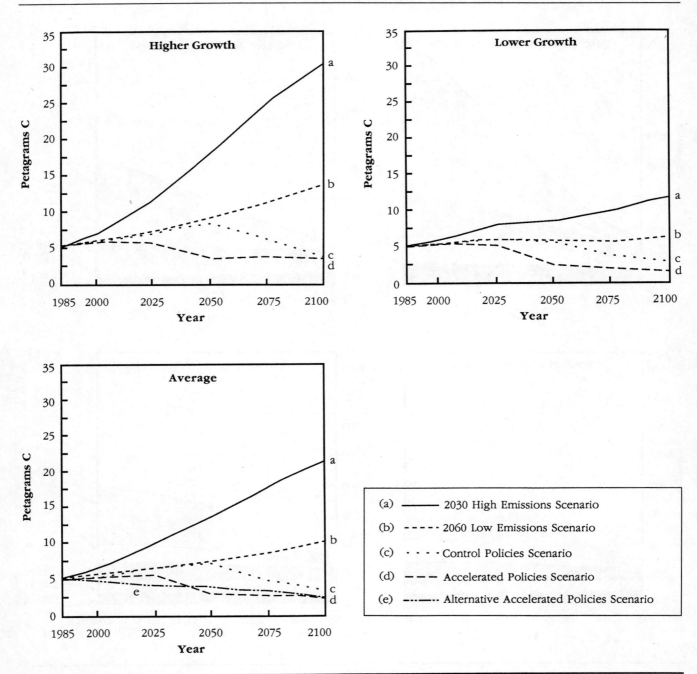

FIGURE 2.7: CO₂ Emissions from Fossil Fuels

Higher Growth

Lower Growth

Average

(a)	——	2030 High Emissions Scenario
(b)	-----	2060 Low Emissions Scenario
(c)	· · · ·	Control Policies Scenario
(d)	-- -- --	Accelerated Policies Scenario
(e)	-··-··-	Alternative Accelerated Policies Scenario

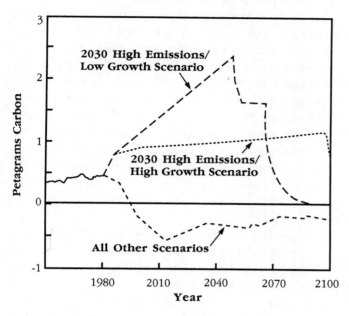

FIGURE 2.8: Net CO$_2$ Emissions from Deforestation (Petagrams Carbon)

production increases by about 120 percent. Satisfying the demands of increasing populations with a finite amount of land requires more intensive cultivation, increasing fertilizer use by 160 percent.

2.4.3.1 *2030 High Emissions and 2060 Low Emissions Scenarios*

Emissions of CH$_4$ and N$_2$O in the 2030 High Emissions and 2060 Low Emissions scenarios increase proportionately with agricultural activities. CH$_4$ from rice production increases 45 percent, and from enteric fermentation, 125 percent. N$_2$O from fertilizer use increases 155 percent by 2100.

2.4.3.2 *Control Policies and Accelerated Policies Scenarios*

By 2100, emissions of CH$_4$ and N$_2$O are roughly 45 percent lower in the Control Policies and Accelerated Policies scenarios owing to the assumed decline in emission coefficients.

2.4.2.2 *2060 Low Emissions, Control Policies, and Accelerated Policies Scenarios*

All of the other scenarios use the Reforestation assumptions, which yield a total release of carbon from deforestation of 10 Pg C, but a total accumulation of carbon from reforestation of 35 Pg C between 1985 and 2100. Therefore, net accumulation over the period is 25 Pg C, with annual accumulation peaking at about 0.7 Pg C before 2025.

Because of differences in modeling approaches and in input assumptions it is difficult to compare the results of the ASF to the results from IMAGE from land-use change. The results from IMAGE are more pessimistic and tend to yield greater net emissions of CO$_2$.

2.4.3 AGRICULTURAL SECTOR

The land area used for rice production increases by nearly 45 percent between 1985 and 2100. As a result of increased population and income, demand for animal protein rises: between 1985 and 2100, dairy production increases by 135 percent and meat

2.5 SENSITIVITY ANALYSIS

2.5.1 ECONOMIC GOALS AND EMISSIONS

At the May meeting of RSWG in Geneva, the Japanese delegation suggested that the implications of different CO$_2$ pathways be shown by estimating and graphing three parameters for each scenario: GNP, Energy/GNP, CO$_2$/Energy. These values for all nine scenarios are presented in Figures 2.9, 2.10, and 2.11; Figure 2.9 shows global GNP, Figure 2.10, energy/GNP, and Figure 2.11, CO$_2$/unit energy. Although the scenarios are believed to be technically feasible, the economic implications of these changes should be explored in the national analyses being performed by participating countries in RSWG. Also, it should be noted that the assumed levels of long run economic growth may not represent accurate measures of human welfare, especially in the developed countries.

The scenarios illustrate the possible potential growth in emissions from developing countries and highlight the importance of addressing the causes for this growth in global strategies to limit emis-

FIGURE 2.9: Alternative GNP Paths for Industrialized and Developing Nations (Trillions U.S.$)

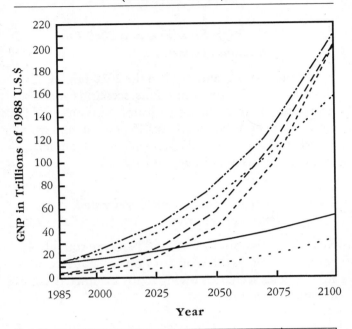

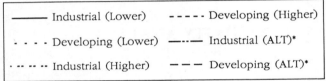

— Industrial (Lower)   - - - - Developing (Higher)

· · · · Developing (Lower)   —·—· Industrial (ALT)*

· — ·· Industrial (Higher)   — — Developing (ALT)*

* Path labeled "ALT" represents the Alternative Accelerated Policies Scenario.

sions. The higher rates of economic growth assumed for the developing countries account for much of the emission growth, and efforts to limit emissions and maintain economic growth could require both technology transfer and financial assistance from developed countries.

2.5.2 THERE ARE MANY WAYS TO GET THERE

The alternative societal, economic, policy, and technological paths incorporated in the scenarios do not cover the infinite variety of ways the future may unfold. At every step in the development of scenarios, trade-offs had to be made. It is possible to construct emission scenarios that would result in similar equivalent CO_2 concentrations assuming either high or low rates of economic growth, a high percentage of CO_2 emissions, or a high percentage of emissions of other greenhouse gases, that energy intensity reductions (due to technological change) increase end-use energy efficiency, that non-fossil energy supplies (e.g., renewable or nuclear) are substituted for fossil fuels, and that lifestyle changes alter the composition of final demand. Alternative scenarios could have been constructed with greater emphasis on the role of such factors as the amounts and modes of travel, the size of buildings, and the composition of industrial and agricultural output, for example.

While the assumed rate of economic growth is one of the most powerful determinants of the rate of greenhouse gas emissions, it is by no means the only determinant. Other important factors can potentially offset the association between higher economic growth and higher rates of greenhouse gas emissions. For example, a higher rate of economic growth makes it possible for capital stocks (e.g., power plants, factories, housing) to be refurbished or replaced more quickly, and makes the accelerated penetration of advanced technologies possible. Similarly, under slow economic growth an increase in deforestation may be more likely because of a lack of other economic opportunities. The alternative scenarios were constructed with these factors in mind.

Other greenhouse gas emissions could prove more important than indicated in the scenarios here. For example, higher anthropogenic emissions rates for CO and N_2O, still within the bounds of current scientific understanding, could have been assumed. Similarly, less optimistic assumptions about the ability to reduce emissions of CO, CH_4, and NO_x would have resulted in higher emissions of these gases, which would require greater reductions of greenhouse gases emissions from other sources in order to meet an equivalent CO_2 concentration target.

Less optimistic assumptions about the feasibility and impact of reforestation policies could also increase the need to reduce emissions of greenhouse gases from other sources in order to maintain equivalent CO_2 concentrations below stated goals. Similarly, if the competition for land—for agricultural use, biomass production, reforestation, and forest conservation—were to drive up costs faster than assumed in these scenarios, biomass production would decline and CO_2 emissions would be higher. This would require stronger emission reduction efforts in other areas, for example, in increased energy efficiency.

2.5.3 REFERENCE TO OTHER STUDIES AND PROPOSALS

The expert group responded only to the criteria established by Working Group I and RSWG to construct the reference scenarios. Other studies have suggested alternatives. Conclusions reached by the Bellagio Workshop (Jaeger, 1988) suggested that an ecological goal of a 0.1 degree C temperature increase per decade might be desirable. The Bruntland Commission emphasized the need for sustainable development and mentioned the desirability of a 50 percent reduction in primary energy consumption in industrialized countries by 2030, and the Toronto Conference, "The Changing Atmosphere—1988," suggested additional short-term emission reduction goals. The expert group did not attempt to analyze these alternatives. It should be noted that for the Accelerated Policies scenarios, primary energy use in industrialized countries stays relatively flat through 2025, as do global CO_2 emissions from the commercial energy sector. In the future it may be desirable to compare the results of these global analyses with results from studies using national models that can evaluate future emissions scenarios in greater detail.

2.5.4 TIME PROFILE OF EMISSIONS

The expert group recognized that emissions and concentrations in the future are not only a function of factors such as economic growth and technological change, but also of when measures to reduce emissions are adopted. To explore this issue, a preliminary sensitivity analysis was undertaken using IMAGE. In this analysis stable atmospheric concentrations equivalent to a doubling of CO_2 by the year 2090 were used as a target. The simulations utilized the emissions estimates from the 2030 High Emissions (Average) scenario, but modified the emissions at different points in time (2000, 2010, 2020, and 2030) in order to achieve the goal of keeping greenhouse gas concentrations from doubling. In all cases, CO_2 emissions have to decline in order to achieve stable concentrations at the $2xCO_2$-equivalent level in 2090, but Figures 2.12 and 2.13 clearly show that emissions would have to be curtailed sharply if the emissions reduction measures are delayed until a later date. For example, if emissions reduction measures are begun in 2000, a gradual 17 percent reduction in CO_2 emissions (from 2000 levels) is sufficient to meet the targeted concentrations. If measures are not implemented until 2020, however, the model calculates that a reduction of 60 percent (from 2020 levels) is necessary to achieve the targets.

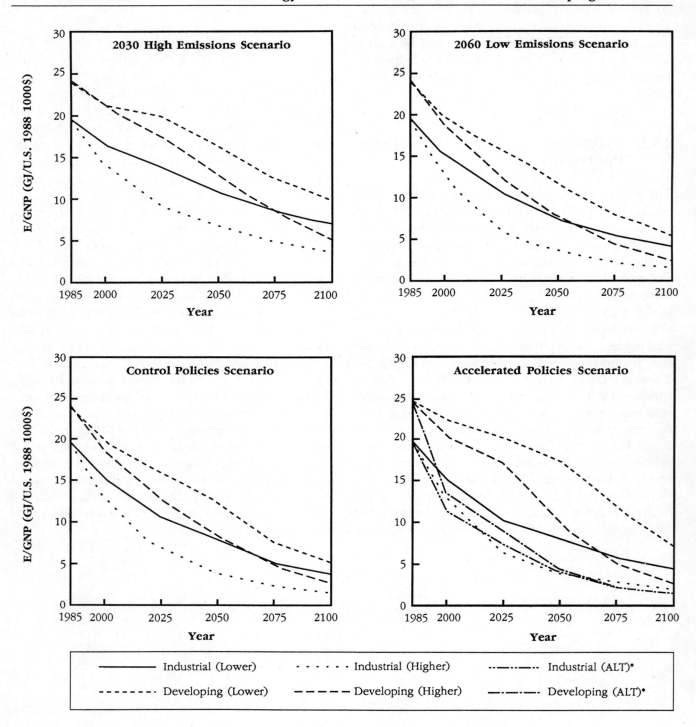

FIGURE 2.10: Alternative Energy Intensities for Industrialized and Developing Nations

* Path labeled "ALT" represents the Alternative Policies Scenario.

40

ilarly, if the competition for land—for agricultural use, biomass production, reforestation, and forest conservation—were to drive up costs faster than assumed in these scenarios, biomass production would decline and CO_2 emissions would be higher. This would require stronger emission reduction efforts in other areas, for example, in increased energy efficiency.

2.5.3 REFERENCE TO OTHER STUDIES AND PROPOSALS

The expert group responded only to the criteria established by Working Group I and RSWG to construct the reference scenarios. Other studies have suggested alternatives. Conclusions reached by the Bellagio Workshop (Jaeger, 1988) suggested that an ecological goal of a 0.1 degree C temperature increase per decade might be desirable. The Bruntland Commission emphasized the need for sustainable development and mentioned the desirability of a 50 percent reduction in primary energy consumption in industrialized countries by 2030, and the Toronto Conference, "The Changing Atmosphere—1988," suggested additional short-term emission reduction goals. The expert group did not attempt to analyze these alternatives. It should be noted that for the Accelerated Policies scenarios, primary energy use in industrialized countries stays relatively flat through 2025, as do global CO_2 emissions from the commercial energy sector. In the future it may be desirable to compare the results of these global analyses with results from studies using national models that can evaluate future emissions scenarios in greater detail.

2.5.4 TIME PROFILE OF EMISSIONS

The expert group recognized that emissions and concentrations in the future are not only a function of factors such as economic growth and technological change, but also of when measures to reduce emissions are adopted. To explore this issue, a preliminary sensitivity analysis was undertaken using IMAGE. In this analysis stable atmospheric concentrations equivalent to a doubling of CO_2 by the year 2090 were used as a target. The simulations utilized the emissions estimates from the 2030 High Emissions (Average) scenario, but modified the emissions at different points in time (2000, 2010, 2020, and 2030) in order to achieve the goal of keeping greenhouse gas concentrations from doubling. In all cases, CO_2 emissions have to decline in order to achieve stable concentrations at the $2xCO_2$-equivalent level in 2090, but Figures 2.12 and 2.13 clearly show that emissions would have to be curtailed sharply if the emissions reduction measures are delayed until a later date. For example, if emissions reduction measures are begun in 2000, a gradual 17 percent reduction in CO_2 emissions (from 2000 levels) is sufficient to meet the targeted concentrations. If measures are not implemented until 2020, however, the model calculates that a reduction of 60 percent (from 2020 levels) is necessary to achieve the targets.

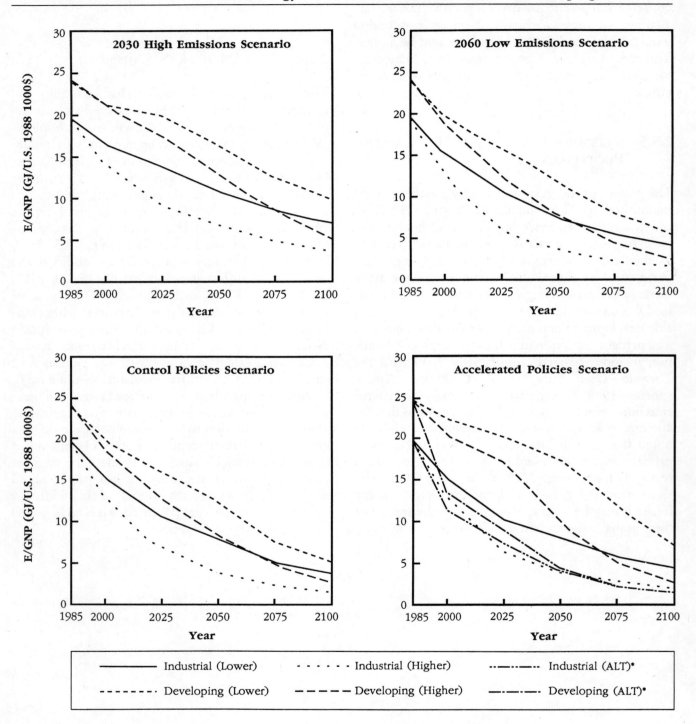

FIGURE 2.10: Alternative Energy Intensities for Industrialized and Developing Nations

* Path labeled "ALT" represents the Alternative Policies Scenario.

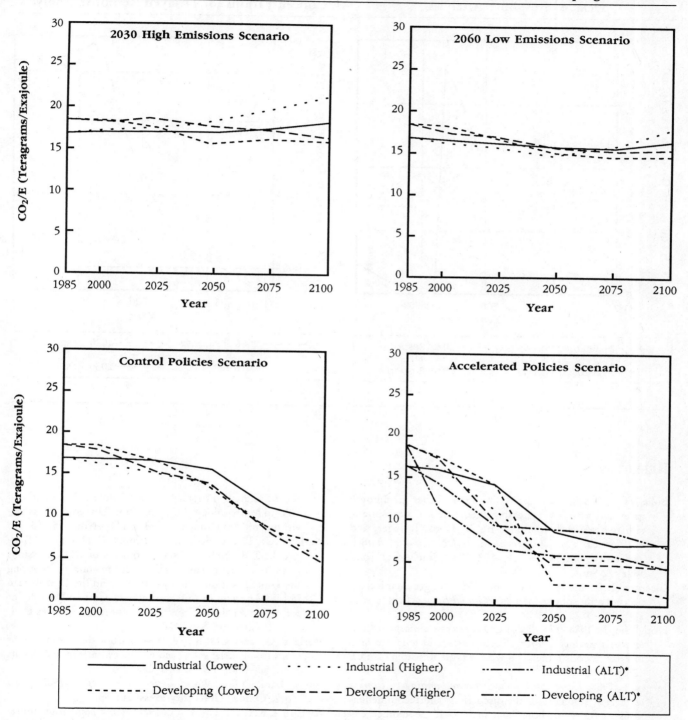

FIGURE 2.11: Alternative Carbon Intensities for Industrialized and Developing Nations

* Path labeled "ALT" represents the Alternative Accelerated Policies Scenario.

FIGURE 2.12:   CO$_2$ Emissions for
Delayed Response Analysis

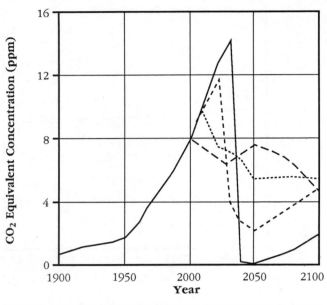

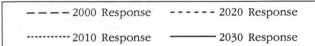

FIGURE 2.13:   CO$_2$ Equivalent
Concentrations for Delayed Response Analysis

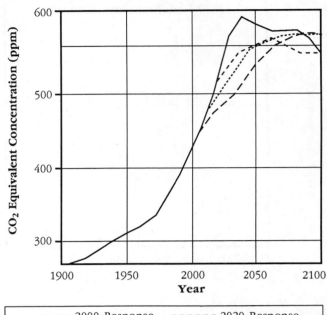

REFERENCES

Bolle, H., W. Seiler, and B. Bolin. 1986. Other greenhouse gases and aerosols: Assessing their role for atmospheric radiative transfer. In Bolin, B., B. Doos, J. Jager, and R. Warrick, eds. *The greenhouse effect, climate change, and ecosystems.* Scope 29, 157–203. Chichester: John Wiley & Sons.

Cicerone, R.J., and R.S. Oremland. 1988. Biogeochemical aspects of atmospheric methane. *Global Biogeochem. Cycles* 2(4): 299–327.

Houghton, R. 1988. *The flux of CO$_2$ between atmosphere and land as a result of deforestation and reforestation from 1850 to 2100.* Woods Hole: The Woods Hole Research Center. 12 pp.

Jaeger, J. 1988. *Developing policies for responding to climate change.* A summary of the discussions and recommendations of the workshops held in Villach (September 28–October 2, 1987) and Bellagio (November 9–13, 1987), under the auspices of the Beijer Institute, Stockholm

World Climate Program Impact Studies, WMO/UNEP (World Meteorological Organization/United Nations Environment Programme). WMO/TD—No. 225. 53 pp.

Marland, G., T.A. Boden, R.C. Griffin, S.F. Hurang, P. Kancircuk, T.R. Nelson. 1988. Estimates of CO$_2$ emissions from fossil fuel burning and cement manufacturing using the United Nations energy statistics and the U.S. Bureau of Mines cement manufacturing data. Carbon Dioxide Information Analysis Center, Environmental Sciences Division, Publication No. 3176.

Marland, G., and R.M. Rotty. 1984. Carbon dioxide emissions from fossil fuels: A procedure for estimation and results for 1950–1982. *Tellus,* 36B: 232–261.

World Bank. 1987. *World development report 1987.* New York: Oxford University Press. 285 pp.

Zachariah, K.C., and M.T. Vu. 1988. *World population projections, 1987–1988 Edition.* Baltimore: World Bank, Johns Hopkins University Press. 440 pp.

SUBGROUP REPORTS

3

Energy and Industry

CO-CHAIRS
K. Yokobori (Japan)
Shao-Xiong Xie (China)

CONTENTS

The following chapter is the Executive Summary, not the full report, of the Energy & Industry Subgroup.

ENERGY AND INDUSTRY

3.1 INTRODUCTION

3.1.1 THE ESTABLISHMENT OF THE ENERGY AND INDUSTRY SUBGROUP

The Energy and Industry Subgroup (EIS) was established at the first session of the Response Strategies Working Group (RSWG) and held its own first session on that occasion. Membership is open to all countries participating in the IPCC. As of March 1990, 26 countries have participated in the work of EIS together with eight international organizations and 29 non-governmental organizations (as observers). Representatives of Japan and the People's Republic of China were appointed as Co-Chairmen, and a representative of Canada as rapporteur. Membership grew in pace with the growing awareness of the significance of the issues involved.

3.1.2 TERMS OF REFERENCE

The EIS was charged with defining "policy options for national, regional and international responses to the possibility of climate change from greenhouse gas emissions produced by energy production, conversion and use." In doing so, the EIS was to consider greenhouse gases (GHGs), primarily carbon dioxide (CO_2), methane (CH_4), and nitrous oxide (N_2O) and "define technology and policy options to attempt to reduce emissions of these gases to a level consistent with, or below, emission scenarios defined by the Steering Group (of RSWG)." It was to concentrate on options which could be adequately assessed within eighteen months and also consider those that may require a longer time frame. The mandate also emphasized the necessity of considering the social, economic and environmental implications of technology and policy options on national, regional and international levels.

The EIS was originally given the following five tasks:

Task 1 Review past and current work on technology and policies related to GHGs.

Task 2 Select appropriate analytical tools for assessing social, economic, and emission level implications of policy options.

Task 3 Categorize options by the timing of their potential application.

Task 4 Analyze nearer-term options for their economic, social, and emission reduction implications and prepare a report for the RSWG.

Task 5 Prepare a plan to define and further develop and analyze longer-term options based on the information derived from Task 1.

Subsequently, the IPCC at its Third Plenary asked the EIS to conduct analyses related to the remits from the Ministerial Conference in Noordwijk, Netherlands in November 1989, particularly with regard to the feasibility of alternative emission targets.

Although the issues fell within its original mandate, it was agreed that these new, specific tasks would require more time, data, and analysis in order to be dealt with properly. It was decided, therefore, that the results of the deliberations of the EIS on these remits could not be fully included in this report but only treated in an incomplete and preliminary way. A progress report would be presented to the fourth IPCC Plenary following an international workshop to be hosted by the United Kingdom in June 1990.

3.1.3 THE ACTIVITIES UNDERTAKEN

Since the first meeting, the EIS has met on four occasions to discuss the tasks stated above. A series

TABLE 3.1: Primary Energy Consumption and CO_2 Emissions
(Exajoules and Billion Tonnes Carbon)

		PRIMARY ENERGY					
					Average Annual Growth Rate		
	1985	2000	2010	2025	1985–2000	1985–2010	1985–2025
Global Totals	328.2	462.1	572.1	776.9	2.3%	2.2%	2.2%
Developed	234.7	308.1	357.2	434.6	1.8%	1.7%	1.6%
North America	85.4	108.2	120.7	142.1	1.6%	1.4%	1.3%
Western Europe	54.7	64.8	71.2	81.3	1.1%	1.1%	1.0%
OECD Pacific	19.2	29.6	34.6	42.2	2.9%	2.4%	2.0%
Centrally Planned Europe	75.5	105.4	130.7	169.0	2.2%	2.2%	2.0%
Developing	93.4	154.0	215.0	342.3	3.4%	3.4%	3.3%
Africa	13.5	21.0	31.3	52.9	3.0%	3.4%	3.5%
Centrally Planned Asia	31.2	47.0	61.6	91.9	2.8%	2.8%	2.7%
Latin America	19.1	27.5	35.8	55.0	2.5%	2.5%	2.7%
Middle East	8.0	19.2	27.7	43.2	6.1%	5.1%	4.3%
South and East Asia	21.6	39.3	58.6	99.2	4.1%	4.1%	3.9%
		CARBON DIOXIDE					
					Average Annual Growth Rate		
	1985	2000	2010	2025	1985–2000	1985–2010	1985–2025
Global	5.15	7.30	9.08	12.42	2.3%	2.3%	2.2%
Developed	3.83	4.95	5.70	6.94	1.7%	1.6%	1.5%
North America	1.34	1.71	1.92	2.37	1.6%	1.4%	1.4%
Western Europe	0.85	0.98	1.06	1.19	0.9%	0.9%	0.8%
OECD Pacific	0.31	0.48	0.55	0.62	3.0%	2.3%	1.8%
Centrally Planned Europe	1.33	1.78	2.17	2.77	2.0%	2.0%	1.9%
Developing	1.33	2.35	3.38	5.48	3.9%	3.8%	3.6%
Africa	0.17	0.28	0.45	0.80	3.5%	4.1%	4.0%
Centrally Planned Asia	0.54	0.88	1.19	1.80	3.3%	3.2%	3.1%
Latin America	0.22	0.31	0.42	0.65	2.4%	2.6%	2.7%
Middle East	0.13	0.31	0.44	0.67	5.8%	4.9%	4.1%
South and East Asia	0.27	0.56	0.89	1.55	5.1%	4.9%	4.5%

Note: Totals reflect rounding

of expert group meetings was held to discuss various issues concerning methodological tools and analytical approaches. There was a general consensus that no single approach would be adequate. A combination of modeling and non-modeling approaches was suggested. A hybrid approach was taken by which "bottom-up" national approaches would be combined with "top-down" global approaches to produce integrated assessments in order to exploit the advantages of both approaches. The issue of cost-effectiveness analysis was identified as one of the areas of future work.

It was decided to limit the scope of the EIS analysis to carbon dioxide, nitrous oxide, and methane emissions from the energy and industry sectors (including transportation and waste management). Emissions estimates for these GHGs were made, but additional future analysis of emission scenarios would be useful for evaluating response strategies in the energy and industrial sectors.

3.1.4 THE MATERIAL USED IN PRODUCING THE REPORT

This report is based on the country case studies presented to the IPCC by experts. These studies are preliminary and not necessarily official government positions.

Individual national case studies, to be based in part on IEA oil price scenarios, were solicited. These studies, often drawn from work conducted for domestic policy purposes, were submitted by national delegations. All but one were for industrialized market economies. Only Canada, the Federal Republic of Germany, the Netherlands, France, Japan, the United Kingdom and Switzerland examined response options. Only the Netherlands, Switzerland, and the United Kingdom included some assessment of costs. Furthermore, the national case studies often differed with respect to time frame, assumptions, and other factors.

The national case studies were supported by additional materials submitted to EIS. These included: a regional study by the Commission of the European Communities; a global/regional study from the IEA; a joint comparative study of their members from the IEA and OECD and a technical analysis of the potential of nuclear power by the International Atomic Energy Agency and ten independent studies of developing and Eastern European countries commissioned by the U.S. EPA. Response Options studies were included for China, Brazil, India, Indonesia, the Republic of Korea, Mexico, Venezuela, the USSR and Poland. Other valuable contributions were received from all EIS participants—for example, the RSWG Task A emission scenarios.

In total, the countries for which studies were received by EIS accounted for around 80 percent of global CO_2 emissions.

3.1.5 THE NATURE OF THE REPORT

This report represents only a first tentative step toward the goal of identifying the paths and strategies needed to ensure that energy and industry related greenhouse gas emissions are compatible with the concept of sustainable development. Such strategies should be economically efficient and compatible with other policy goals.

As discussed above, the report relies on various materials submitted to the EIS. However, their coverage in terms of all greenhouse gases and regions remains short of producing consistent global analysis. Adequate data were not available on the emissions of greenhouse gases at all stages of the fuel

cycle and no methodology was available to compare the effects of various gases on a standard basis, (e.g., CO_2 equivalence). The report therefore puts a heavy emphasis on CO_2, for which there were the most definitive data. Further, most of the country case studies submitted pertain to reference cases, rather than policy options, and thus little material was available on the socio-economic consequences of emission controls.

Two broad conclusions are supported by the national case studies from which the EIS reference scenario is drawn:

- First, the nature of the problem varies significantly depending on each country's economic structure, the situation of its energy sector, and its stage of development. The national case studies showed that economic growth and reductions from the reference case in the growth of greenhouse gases can co-exist and that policies and technologies can make a substantial contribution to limiting GHGs.
- Second, there is no single quick-fix technological option; improving efficiency on both the demand and supply side should be a priority; technological solutions must be cost effective; non-economic barriers to diffusion of attractive technologies were a fruitful area for further analysis; and technological research and development was a prime area for international cooperation. Overall, there was a consistent emphasis on energy efficiency and conservation.

3.2 ROLE OF THE ENERGY SECTOR

The energy sector plays a vitally important role in economic development for all nations. Energy policies need to ensure that sustained economic growth occurs in a manner that also preserves the global environment for future generations.

3.2.1 CURRENT CONTRIBUTION TO GREENHOUSE GAS EMISSIONS

The energy sector is the most important single source of greenhouse gases, accounting for approximately 57 percent of radiative forcing from anthro-

TABLE 3.2: Examples of Short-Term Options

I. IMPROVE EFFICIENCY IN THE PRODUCTION, CONVERSION, AND USE OF ENERGY

ELECTRICITY GENERATION	INDUSTRY SECTOR	TRANSPORT SECTOR	BUILDING SECTOR
· Improved efficiency in electricity generation: –repowering of existing facilities with high efficiency systems; –introduction of integrated gassification combined cycle systems; –introduction of atmospheric fluidized bed combustion; –introduction of pressurized fluidized bed combustion with combined cycle power systems; –improvement of boiler efficiency. · Improved system for cogeneration of electricity and steam. · Improved operation and maintenance. · Introduction of photovoltaics, especially for local electricity generation. · Introduction of fuel cells.	· Promotion of further efficiency improvements in production process. · Materials recycling (particularly energy-intensive materials). · Substitution with lower energy-intensive materials. · Improved electro-mechanical drives and motors. · Thermal process optimization, including energy cascading and cogeneration. · Improved operation and maintenance.	· Improved fuel efficiency of road vehicles: –electronic engine management and transmission control systems; –advanced vehicle design: reduced size and weight, with use of lightweight composite materials and structural ceramics; improved aerodynamics, combustion chamber components, better lubricants and tire design, etc.; –regular vehicle maintenance; –higher capacity trucks; –improved efficiency in transport facilities; –regenerating units. · Technology development in public transportation: –intra-city modal shift (e.g., car to bus or subway); –advanced train control system to increase traffic density on urban rail lines; –high-speed inter-city trains; –better intermodal integration. · Improved driver behavior, traffic management, and vehicle maintenance.	· Improved heating and cooling equipment and systems: –improvement of energy efficiency of air conditioning; –promotion of introduction of area heating and cooling, including use of heat pumps; –improved burner efficiency; –use of heat pumps in buildings; –use of advanced electronic energy; management control systems. · Improved space conditioning efficiency in house building: –improved heat efficiency through highly efficient insulating materials; –better building design (orientation, window, building, envelope, etc.); –improved air-to-air heat exchangers. · Improved lighting efficiency. · Improved appliance efficiency. · Improved operation and maintenance. · Improved efficiency of cook stoves (in developing countries).

TABLE 3.2 (*continued*): Examples of Short-Term Options

II. NON-FOSSIL AND LOW EMISSION ENERGY SOURCES

ELECTRICITY GENERATION	OTHER SECTORS
· Construction of small-scale and large-scale hydro projects. · Expansion of conventional nuclear power plants. · Construction of gas-fired power plants. · Standardized design of nuclear power plants to improve economics and safety. · Development of geothermal energy projects. · Introduction of wind turbines. · Expansion of sustainable biomass combustion. · Replacement of scrubbers and other energy-consuming control technology with more energy efficient emission control.	· Substitution of natural gas and biomass for heating oil and coal. · Solar heating. · Technologies for producing and utilizing alternative fuels: —improved storage and combustion systems for natural gas; —introduction of flexible-fuel and alcohol-fuel vehicles.

III. REMOVAL, RECIRCULATION, OR FIXATION

ENERGY/INDUSTRY	LANDFILLS
· Recovery and use of leaked or released CH_4 from fossil fuel storage, coal mining. · Improved maintenance of oil and natural gas and oil production and distribution systems to reduce CH_4 leakage. · Improved emission control of CO, SO_X, NO_X, and VOCs to protect sinks of greenhouse gases.	· Recycle and incineration of waste materials to reduce CH_4 emissions. · Use or flaring of CH_4 emissions. · Improved maintenance of landfill to decrease CH_4 emissions.

pogenic sources in the 1980s. The major greenhouse gases produced by the energy sector include CO_2 and CH_4 from combustion of fossil fuels and CH_4 from coal mines and oil and gas facilities. In 1985, approximately 5.3 billion tonnes of CO_2 as carbon (BTC) were released from fossil fuel combustion, and 50–95 million tonnes (MT) of CH_4 were released due to fossil fuel production and consumption. Another source of greenhouse gas emissions is solid waste landfill, which produces CH_4. Additionally, biomass burning for heating and cooking produces CH_4 and when not based on sustainable resources will produce net CO_2 emissions. Nitrous oxide, another important greenhouse gas, is also released from the energy sector, but the exact contribution is uncertain.

Other gases that contribute indirectly to greenhouse gas concentrations include CO, NO_x, and NMHC (non-methane hydrocarbons). Comparing relative contributions to radiative forcing across the various gases is complicated and dependent on a number of key assumptions, particularly time horizon. The relative importance of current emissions of different gases varies considerably depending on the time horizon over which the contribution to radiative forcing is considered. These differences are due to the significant differences in the average atmospheric lifetimes of the radiatively important gases. However, integration over longer time horizons tends to increase the importance of energy as a source category.

Table 3.7 on page 68 shows the current contribution to radiative forcing from all sources and from the energy sector.

During the period 1950–1985, global energy consumption in total and per capita from fossil fuel increased by nearly a factor of four, while CO_2 emissions increased from 1.5 to 5.3 BTC, or nearly 3.5 times. During this time the liquid fuel share increased from 31 to 45 percent, the coal share declined from 60 to 33 percent, and natural gas increased from 9 to 22 percent. See Figures 3.1 and 3.2.

FIGURE 3.1:   Global Fossil Fuel Carbon Emissions, 1950–1985

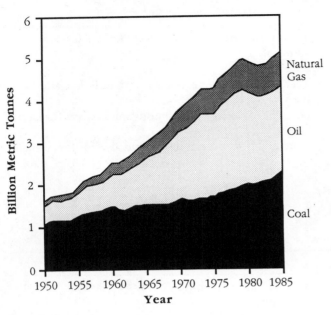

Source: Marland, 1988

FIGURE 3.2:   Global Fossil Energy Consumption, 1950–1985

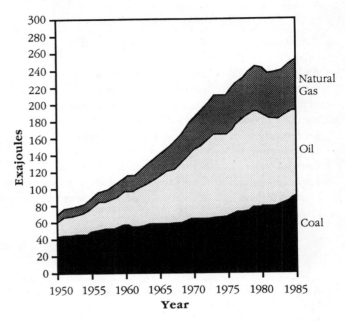

Source: Marland, 1988

FIGURE 3.3: CO_2 Emissions per Capita

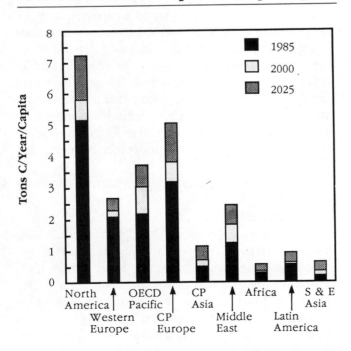

The historical growth in energy consumption and CO_2 emissions differs by region of the world (for per capita carbon emission, see Figure 3.3). For example:

- Fossil fuel energy use in developed market economies grew by 4 percent per year for the period 1950–1970. From 1970 to 1985 fossil energy consumption increased by an amount equal to only a 0.4 percent per year average growth rate. This was due to slow economic growth and energy price increases that accelerated structural changes and other shifts in energy use and production. Over the entire period 1950–1985, the average annual rate of growth in fossil energy use was 2.6 percent. Annual CO_2 emissions from these countries accounted for approximately 2.5 BTC in 1985 and on a per capita basis was 2.95 TC per person (down slightly from 3.1 TC per person in 1970).
- Fossil fuel energy use in centrally planned Eastern Europe and the USSR increased steadily

TABLE 3.3: Examples of Medium-/Long-Term Options

I. IMPROVE EFFICIENCY IN THE PRODUCTION, CONVERSION, AND THE USE OF ENERGY

ELECTRICITY GENERATION	INDUSTRY SECTOR	TRANSPORT SECTOR	BUILDING SECTOR
· Advanced technologies for storage of intermittent energy. · Advanced batteries. · Compressed air energy storage. · Superconducting energy storage.	· Increased use of less energy-intensive materials. · Advanced process technologies. · Use of biological phenomena in processes. · Localized process energy conversion. · Use of fuel cells for cogeneration.	· Improved fuel efficiency of road vehicles. · Improvements in aircraft and ship design. –advanced propulsion concepts; –ultra-high-bypass aircraft engines; –contra-rotating ship propulsion.	· Improved energy storage systems: –use of information technology to anticipate and satisfy energy needs; –use of hydrogen to store energy for use in buildings. · Improved building systems: –new building materials for better insulation at reduced cost; –windows that adjust opacity to maximize solar gain. · New food storage systems that eliminate refrigeration requirements.

II. NON-FOSSIL AND LOW EMISSION ENERGY SOURCES

ELECTRICITY GENERATION	OTHER SECTORS
· Nuclear power plants: –passive safety features to improve reliability and acceptability. · Solar power technologies: –solar thermal; –solar photovoltaic (especially for local electricity generation). · Advanced fuel cell technologies.	· Other technologies for producing and utilizing alternative fuels: –improved storage and combustion systems for hydrogen; –control of gases boiled off from cryogenic fuels; –improvements in performance of metal hydrides; –high-yield processes to convert ligno-cellulosic biomass into alcohol fuels; –introduction of electric and hybrid vehicles; –reduced re-charging time for advanced batteries.

III. REMOVAL, RECIRCULATION, OR FIXATION

· Improved combustion conditions to reduce N_2O emissions.

· Treatment of exhaust gas to reduce N_2O emissions.

· CO_2 separation and geological and marine disposal.

TABLE 3.4*: **CO$_2$ Emissions from the Energy Sector and Comparison of Emissions Reductions (from the Reference Scenario)**

	CO$_2$ EMISSIONS IN BILLION TONNES CARBON			
	1985	**2000**	**2010**	**2025**
Global Totals	**5.15**	**7.30**	**9.08**	**12.43**
Developed	3.83	4.95	5.70	6.95
North America	1.34	1.71	1.92	2.37
Western Europe	0.83	0.98	1.06	1.19
OECD Pacific	0.31	0.48	0.55	0.62
Centrally Planned Europe	1.33	1.78	2.17	2.77
Developing	1.33	2.35	3.38	5.48
Africa	0.17	0.28	0.45	0.80
Centrally Planned Asia	0.54	0.88	1.19	1.80
Latin America	0.22	0.31	0.42	0.65
Middle East	0.13	0.31	0.44	0.67
South and East Asia	0.27	0.56	0.89	1.55
Stabilize		−2.14 (29%)	−3.92 (43%)	−7.26 (59%)
20% Reduction		−3.17 (44%)	−4.95 (55%)	−8.29 (67%)

	CO$_2$ EMISSIONS IN TONNES CARBON PER CAPITA			
	1985	**2000**	**2010**	**2025**
Global	**1.06**	**1.22**	**1.36**	**1.56**
Developed	3.12	3.65	4.02	4.65
North America	5.08	5.73	6.11	7.12
Western Europe	2.11	2.29	2.44	2.69
OECD Pacific	2.14	3.01	3.29	3.68
Centrally Planned Europe	3.19	3.78	4.32	5.02
Developing	0.36	0.51	0.64	0.84
Africa	0.29	0.32	0.41	0.54
Centrally Planned Asia	0.47	0.68	0.85	1.15
Latin America	0.55	0.61	0.71	0.91
Middle East	1.20	1.79	2.11	2.41
South and East Asia	0.19	0.32	0.44	0.64
Stabilize		−0.16 (13%)	−0.30 (22%)	−0.50 (32%)
20% Reduction		−0.38 (31%)	−0.51 (38%)	−0.71 (46%)

* This table should be read in conjunction with Table 2 of the Policymakers Summary, which also provides data on carbon intensity by region, another important index of CO$_2$ emissions. Table totals reflect rounding.

between 1950 and 1985 at an average growth rate of 5.2 percent per year, from approximately 12 exajoules to approximately 70 exajoules. Annual CO$_2$ emissions increased from about 300 to 1400 MTC, while on a per capita basis emissions were nearly 3.3 TC per person.

• Fossil fuel energy growth in the centrally planned economies of Asia was not steady during the period 1950–1985, but it did increase dramatically from about 1 exajoule to over 23 exajoules, a growth rate of 9.8 percent per year. CO_2 emissions increased from around 20 MTC to over 500 MTC during this period, but on a per capita basis emissions were only 0.5 TC per person or roughly one sixth of developed country levels.

- Energy consumption in the developing market economies grew at an average rate of 5.7 percent per year, or by sevenfold, during the period 1950–1985 (from about 5 to 35 exajoules) even during the period of rapidly increasing oil prices. Total CO_2 emissions reached about 700 MT in 1985 but on a per capita basis were only 0.3 TC per person.

The historical growth in methane emissions also differs by region of the world, but is much more difficult to evaluate because of the lack of historical data.

3.3 SCENARIOS FOR FUTURE GREENHOUSE GAS EMISSIONS

The broad determinants of energy use that will affect greenhouse gas emissions:

- The population level of the country is an important determinant of overall energy requirements. Generally, countries with rapid population growth rates are likely to experience high growth rates in energy use.
- The level and structure of economic activity, often measured by annual income or product flows as gross domestic product (GDP). The energy intensity of the economy is measured by the amount of energy used per unit of aggregate income or product. The amount of energy used to create GDP depends on a number of factors, including the composition of economic activity between more or less energy intensive industries or sectors, climate, transportation distances, energy efficiency, etc. The energy efficiency of the economy is affected by the technologies used to produce goods and services. Management practices, behavioral attitudes and infrastructure also affect energy efficiency.

- The carbon intensity of energy facilities influences GHG emissions. Carbon intensity reflects the mix of fossil fuels used in the economy, the proportion of total energy requirements met through non-fossil energy such as nuclear and hydroelectric power, and the methods of resource production, distribution, transmission, and conversion.

Future levels of emissions are difficult to predict because of the inherent uncertainties in these and other factors, such as the introduction of new technologies. Therefore, scenarios of future emissions are necessary for assessing climate trends. The uncertainty surrounding such scenarios increases rapidly (and their usefulness decreases) as they are projected further into the future. Policy options need to be tested and costs assessed, therefore, against a range of possible future scenarios. This did not prove possible in the time available for production of this interim report.

3.3.1 THE EIS REFERENCE SCENARIO

The EIS Reference Scenario deals only with CO_2 and is presented in Table 3.1 on page 50. This was developed from the national case studies and other data submitted to EIS. It broadly reflects current trends but includes some limited measure of response to the climate change issue. Figure 3.5a on page 60 shows this scenario along with others produced by the IEA, EC, and WEC.

3.3.2 FUTURE CO$_2$ EMISSIONS IN THE EIS REFERENCE SCENARIO

The EIS Reference Scenario portrays a future where, in the absence of further policy measures, energy use and CO_2 emissions grow rapidly to over 9 BTC by 2010 and over 12 BTC by 2025. Global emissions rise faster than those in the high emission scenario provided by the RSWG Task A Group (see Figure 3.5a). Table 3.5 on page 59 shows that economic growth and the rate of improvement in energy intensity are modest at 3.0 percent and 0.8 percent a year respectively. Table 3.1 reveals that primary energy demand more than doubles be-

FIGURE 3.4 :   CO$_2$ Emissions by Region (Petagrams carbon/year)

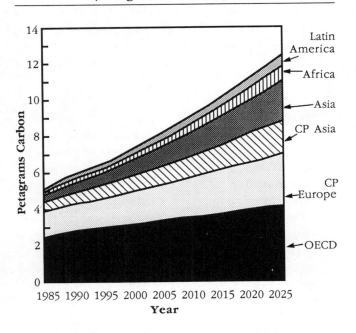

and global totals, in the EIS Reference scenarios and in selected RSWG Task A scenarios. The High Emissions Scenario envisages that equivalent CO$_2$ concentrations reach a value double that of pre-industrial atmospheric concentrations of CO$_2$ by 2030. The Accelerated Policies Scenario represents the largest emissions reduction projected by RSWG. Equivalent CO$_2$ concentrations in this scenario stabilize at a level less than double the pre-industrial atmospheric concentrations of CO$_2$. A line which represents a reduction to a level 20 percent below 1988 emission levels is also shown.

3.3.3   FUTURE METHANE EMISSIONS

Methane is emitted from coal mining, oil and natural gas systems, and waste management systems (i.e., landfills, wastewater treatment facilities). Growth in these emissions is highly dependent upon population and economic growth. Using the same assumptions as described above, methane emissions from these sources may increase by 85 percent by 2010 and 163 percent by 2025. Individually, emissions from coal mining may increase by 93 percent by 2010 and 186 percent by 2025; methane emissions from oil and natural gas systems by similar percent changes; and methane emissions from landfills by 50 percent by 2010 and 100 percent by 2025.

It should be noted that the future estimates of methane emissions from coal mining may be understated. It is likely that as developing countries intensify their coal mining activities to meet rapidly increasing demand for energy, they will mine more coal and coal that is deeper and more gaseous.

tween 1985 and 2025, reaching 777 EJ at the end of the period, an average annual growth rate of 2.2 percent.

The average annual rate of growth in CO$_2$ emissions over the period 1985–2025 varies from 0.8 percent in Western Europe to 3.6 percent in developing countries with 1.4 percent in North America and Pacific OECD countries. Overall the share of emissions from OECD countries declines from 48 percent in 1985 to 43 percent in 2000 and to 33 percent in 2025, and the share attributable to Eastern Europe declines from 26 percent to 24 percent in 2000 and to 22 percent in 2025. Meanwhile, emissions from developing countries rise from 26 percent in 1985 to 32 percent in 2000 and to 44 percent in 2025. Under this scenario, the per capita emissions in the developed countries increase from 3.2 TC per capita to 4.5 TC per capita in 2025. For the developing countries the per capita emissions rise from 0.4 TC per capita in 1985 to 0.8 TC per capita in 2025.

The estimated global growth in CO$_2$ emissions is higher than the high emission scenario provided by RSWG to WG1. Figure 3.5b on page 61 shows the 1990–2025 emissions from the OECD countries

3.4   THE SCALE OF THE EMISSION GAP

Table 3.4 summarizes the emission levels in the EIS Reference Scenario and illustrates possible alternative emission levels. It shows that stabilizing emissions at current levels poses a global policy challenge and that the capacity of regions to reduce emission levels varies greatly. Stabilizing emissions globally at 1985 levels would require reductions below the levels estimated in the Reference Scenario of 29 percent by 2000 and 43 percent by 2010. Table 3.5 summarizes trends in economic growth, energy intensity, and carbon intensity for the period 1985–2025.

A 20 percent reduction of emissions below estimated 1990 levels by 2025 is a much more significant policy challenge. Table 3.4 shows that, if implemented globally, it would require a 67 percent reduction (reduce 8.29 BTC from 12.4 BTC) from the reference emission levels in 2025.

3.5 RESPONSE OPTIONS AND MEASURES

There are a number of technical options available to reduce greenhouse gas emissions and policy measures to implement them. We define technical options to include not only the installation of new capital stock with lower emission characteristics (or the modification of existing stock) but also the managerial and behavioral changes which can reduce future emissions. Policy measures are the actions, procedures, and instruments that governments adopt to bring about additional or accelerated uptake of the technical measures beyond that in a reference scenario.

When analyzing the costs and benefits of alternative response strategies, a systematic approach is needed that identifies the overall effect not only of the technical options but also of the policy measures needed to bring them about. This task must take into account the international nature of energy markets as well as the individual characteristics of national economies. EIS has not, therefore, been able to complete such an analysis but some example resource costing was included in some of the national case studies submitted to EIS.

3.5.1 CRITERIA FOR RESPONSE STRATEGY EVALUATION

In deciding whether, how, and how much to accelerate the implementation of technologies, adoption

TABLE 3.5: Trends in Economic Growth, Energy Intensity, and Carbon Intensity, 1985–2025 (Average Annual Rate of Change, %)

	GDP	ENERGY INTENSITY	CARBON INTENSITY
Global Average	3.0	−0.8	0.0
Developed	2.6	−1.0	−0.1
North America	2.2	−0.9	0.1
Western Europe	2.3	−1.3	−0.2
OECD Pacific	3.1	−1.1	−0.2
Centrally Planned Europe	3.2	−1.1	−0.2
Developing	4.4	−1.1	0.3
Africa	4.0	−0.5	0.5
Centrally Planned Asia	5.3	−2.5	0.3
Latin America	3.3	−0.6	0.1
Middle East	4.9	−0.5	−0.2
South and East Asia	4.6	−0.7	0.6

of management techniques, and structural or behavioral change that could limit CO_2 emissions, a wide range of factors must be considered. These include:

- technical, economic and market potential of technologies;
- development status and time scale for implementation of technologies;
- implications for other GHGs;
- interaction between measures;
- resource costs and private costs;
- macroeconomic and microeconomic effects;
- implications for other policy goals, and social consequences;
- policy robustness;
- political and public acceptability, effectiveness, limitations, and effect of policy instruments.

Understanding the distinction between technical, economic, and market potential is important for developing realistic response strategies. The technical potential of an energy technology is its capacity to reduce potential greenhouse gas emissions, irrespective of the costs involved. This capacity is largely a function of technical feasibility and resource availability.

However, the economic potential may be significantly less. This occurs where there are positive resource costs when evaluated at social discount rates—allowing for second round effects. The inclusion, where possible, of the economic benefits of emission reduction might also influence the economic potential of measures. Finally, the market potential might be even less, due to market imperfections and the use of higher discount rates by private sector decision makers. Attitudes to risk and the presence of non-monetary costs will also be major influences.

The challenge for policymakers is to enhance the market uptake of appropriate options taking full account of all the interactions, second round effects, costs, and benefits.

3.5.2 TECHNICAL OPTIONS

The most relevant categories of technologies to reduce greenhouse gas emissions from energy systems are:

- efficiency improvements and conservation in energy supply, conversion, and end-use;
- fuel substitution by energy sources that have lower or no greenhouse gas emissions;
- reduction of greenhouse gas emissions by removal, recirculation, or fixation; and
- management and behavioral changes (e.g., increased work in homes through information technology) and structural changes (e.g., modal shift in transport).

To fully understand their present and future potential and the types of actions that might be taken to enhance their potential, technologies and consumer or producer actions must be viewed in terms of the time frame in which they can be effective. Changes in management and behavior that lead to energy conservation and emissions reduction can begin now. Many technologies are available now, whereas others need further development to lower costs or to improve their environmental characteris-

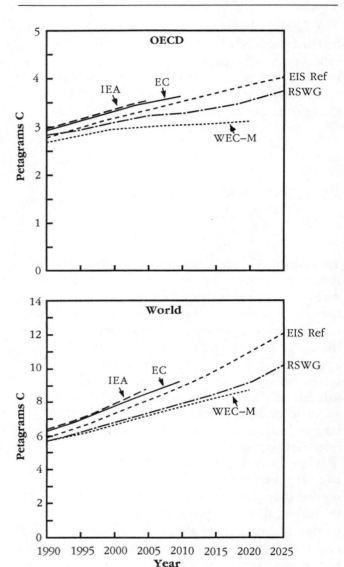

FIGURE 3.5a: CO$_2$ Emission Comparison EIS Reference Scenario vs. Other Emissions Scenarios

Emissions Scenarios:*

EIS Ref=EIS Reference Scenario

RSWG=RSWG Task A Scenario

WEC–M=World Energy Conference (Moderate Scenario)

IEA=International Energy Agency

EC=Commission of the European Communities

* CO$_2$ emission coefficients calculated from those found in Marland and Rotty, 1984

FIGURE 3.5b: CO$_2$ Emission Comparison
EIS Reference Scenario vs.
RSWG Task A Scenario

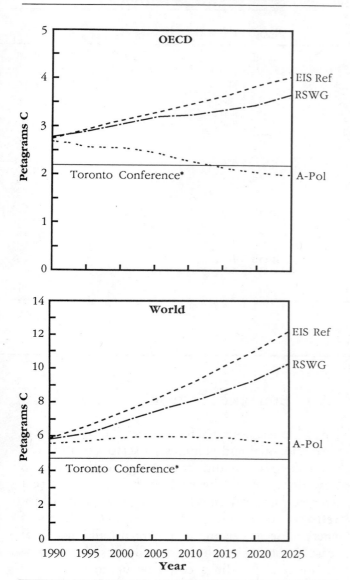

EIS Ref=EIS Reference Scenario

RSWG=RSWG Task A Scenario

A-Pol=Accelerated Policies

* Recommended 20% reduction from 1988 levels by 2005

tics. Tables 3.2 and 3.3 provide examples of technologies within each of the broad categories defined above, and their possible application in the short- and medium- or long-term time frame.

This distinction on time frame is suggested in order to comprehend the remaining technological needs of each category and to formulate a technological strategy. First-wave or near-term technologies are those that are or will be ready for introduction and/or demonstration by 2005. Second-wave technologies are available, but not yet clearly economic and thus would mainly be implemented in the medium-term time frame of 2005 to 2030. They could be introduced sooner if they were close to economic or particularly beneficial to the environment. Third-wave technologies are not yet available but may emerge in the long term or post 2030 as a result of research and development.

The technical, economic, and market potential of cross-cutting technological options will vary, depending upon the sector in which they are to be applied. Cross-cutting technologies include those for energy efficiency and conservation, natural gas fuel use, renewable energy, other non-fossil fuels, and energy storage. For this reason, the specific technological options within the three broad categories listed above are analyzed by sector. There is in general extensive information and data available on the technical potential of the many technological options. However, the economic and market potential of the options depends on specific circumstances (national, local, and even sectoral) in which the option is to be applied. Therefore, no figures for these potentials are provided. Rather, it is left to the country case studies to analyze economic and market potential of options in the context of national circumstances.

For management and behavioral changes regarding technologies in the first-wave stage as described above, the advisability of applying policy measures to accelerate their implementation should be determined. A phased approach to technology development and introduction into the market is offered as an important strategy to be considered for concerted national and international collaborative action. Near-term technological potential of particular relevance to the developed world is discussed by sector below. Technological potential for developing countries has not been developed in the same detail due to a lack of information.

TABLE 3.6a: Applicability, Effectiveness, Limitations, and Economic Effects of Information

MEASURE APPLICABILITY	EFFECTIVENESS	RELATIVE IMPACTS		LIMITATIONS I/P/C[1]
		DURATION	SIZE	
Exhortation	all responses, especially those making economic sense	short	small	I/P/C
Negotiation	development of charges, regulations or policies for all responses	long	large	p
Training	services/operations	long	small	P/C
Testing	industrial equipment, vehicles and other consumer products	medium	small	P/C

[1] I/P/C: Effectiveness to Influence Investment, Producer, or Consumer
[2] na: not assessed
Source: IEA Secretariat.

3.5.3 TRANSPORTATION SECTOR

Substantial technical potential exists for fuel substitution through the use of fuels derived from natural gas and of ethanol or other fuels derived from biomass. Substantial technical potential also exists for electric or hydrogen-fueled vehicles, which could also reduce emissions with appropriate primary energy sources. Presently, the economic and market potentials of most of these options are very low because petroleum fuels are relatively cheap, alternative non-CO_2-producing fuels are relatively costly, and some alternative vehicular technologies have performance drawbacks. The technical potential for vehicle efficiency improvements is very high despite the substantial improvements already made. Economic and market potentials are constrained by replacement rates of vehicles, consumer demand and preference for larger, more powerful and better-equipped cars, and higher incremental costs. Improved driver behavior, vehicle maintenance, traffic management, and promotion of public transportation could also reduce CO_2 emissions.

3.5.4 BUILDINGS SECTOR

The technical potential for energy efficiency gains in the residential and commercial sectors is also substantial. Space conditioning energy requirements in new homes could be roughly half of the current average for new homes. The technical potential for retrofits could average 25 percent. Reductions of energy use in existing commercial buildings by at least 50 percent may be technically feasible, and new commercial buildings could be up to 75 percent more efficient than existing commercial buildings. However, market potential is lower and depends on the replacement rate. The realization of significant gains in this sector requires the involvement of government, the many concerned institutions, and ultimately the individual residential or commercial consumers. This sector therefore requires special efforts in order to achieve desired levels of market penetration of energy efficient technologies. For these technologies to achieve their market potential, institutional barriers must be removed and careful attention given to the design of R&D programmes

| IMPLEMENTATION | | MICRO-ECONOMIC EFFECTS | MACRO-ECONOMIC EFFECTS | ECONOMIC EFFECTS |
COST TO GOVT.	DURATION			
low	quick	—easily erodes —usually not enough to achieve substantial effects	may alter consumption	n.a.[2]
low	quick	depends on willingness of industries to be regulated	may alter consumption	n.a.
low	quick	getting participation of "needless" groups	may alter consumption	n.a.
moderate	medium	requires facilities, constant update	may alter consumption	n.a.

for this sector. Improved operation and maintenance of buildings could also help. Fuel substitution usually occurs in this sector only when there are equipment replacements and a source exists for the alternative, more competitive fuel. Therefore, the possibilities in the sector include decreasing coal and oil use for heating and substituting natural gas use or district heating applications. However, consumers and firms in existing structures may find the infrastructure unavailable or highly expensive even if the fuel itself is cost-competitive.

3.5.5 INDUSTRY SECTOR

The technical potential for efficiency improvements in the industry sector ranges from 13 percent in some sub-sectors to over 40 percent in others. The most dramatic efficiency improvements over the last 15 years have been in the energy-intensive industries. Technical options exist for accelerating this trend and for achieving similar savings in other industries. Such options stem mainly from recent improvements in process technologies, as well as better design and materials. Considerable opportunities for energy savings also exist in the industrial sector by the recycling of energy-intensive waste. There may also be significant potential for reducing greenhouse emissions through industrial fuel switching, especially as many industrial boilers are already dual-fuel capable, with natural gas being the fuel typically substituted for fuel oil under present circumstances. Combined heat and power, cogeneration, combustion of biomass wastes, methane from landfill, and other renewable energy sources also have potential.

The technical potential for methane recovery and reduction exists in solid waste landfills (through gas recovery, flaring, and incineration), oil and natural gas production, gas transmission and distribution systems, and underground coal mines.

3.5.6 ELECTRICITY SECTOR

Under current price conditions, both efficiency and fuel substitution are largely dependent on the rate of

TABLE 3.6b: Applicability, Effectiveness, Limitations, and Economic Effects of Regulatory Measures

MEASURE APPLICABILITY	EFFECTIVENESS	RELATIVE IMPACTS DURATION	RELATIVE IMPACTS SIZE	LIMITATIONS I/P/C[1]
Mandatory Standards	equipment and appliances, buildings, "add-on" technologies	medium to long	varies with stringency	I/P/C
Voluntary Standards	equipment and appliances, buildings, "add-on" technologies	medium to long	varies	I/P/C
License/Permit	—siting new facilities —tradeable permits	medium to long	varies	I/P/C

[1] I/P/C: Effectiveness to influence Investment, Producer, or Consumer
[2] Self: Can be made self-funding

retirement of existing units, the growth in demand for electricity, and the cost of the replacement units. The technical potential for greater efficiency in generation is in the range of 15 to 20 percent. Fuel substitution could achieve CO_2 reduction in CO_2 emissions from electrical generation in the range of 30 percent (from oil to natural gas) to 100 percent (from fossil fuel to non-fossil fuel). The electricity sector has a potential to substantially increase its efficiency where cogeneration of electricity and heat or combined cycle power generation can be applied. The economic potential for greater fuel efficiency is considerably lower and for substitution from coal or oil to natural gas or non-fossil fuel is critically dependent on the relative prices and availability of the fuels in question. In evaluating switching between fossil fuels it is necessary to account for any potential increases in methane emissions from production and transmission of the fuels in calculating the net benefit of such strategies. Overall, efficiency of the electricity system can be improved through the use of least cost utility planning. The goal of least cost utility planning is to meet energy-service requirements through the least cost combination of supply additions and demand management.

3.5.7 POLICY MEASURES

The pool of policy measures is broadly similar for all nations. The measures fall generally into three groups:

- *Information measures* include all efforts to better inform the public on greenhouse gas emissions and the means available to the public for their reduction. This includes research, development, and demonstration programs for emerging technologies and education and training of professional experts in all sectors. These programs should be targeted at particular sections of the populace and should emphasize the present and potential future costs and benefits of such actions. The transfer of information and expertise between countries requires particular attention in order to ensure the relevance and the applicability of the information to local conditions. This is especially true in the case of technology transfer to developing countries.
- *Economic measures* include the broad areas of taxes, charges, subsidies, and pricing policies that include incorporating environmental costs into energy prices (both imposition and re-

IMPLEMENTATION		MICRO-ECONOMIC EFFECTS	MACRO-ECONOMIC EFFECTS	ECONOMIC EFFECTS
COST TO GOVT.	DURATION			
moderate	medium to long	implementation and enforcement require technical expertise, authority	−internalizes externalities, −raises producer costs and prices	can affect trade patterns
none	medium	not enforceable by government	−internalizes externalities, −raises producer costs and prices	can affect trade patterns
moderate (self)[2]	medium	implementation and enforcement require technical expertise, authority	can create new markets and establish prices for environmental goods	

moval). Such measures may also be used to complement regulations, making them more effective in meeting environmental goals. Economic measures may also be used to support the research, development, demonstration, or application of technologies for enhanced energy efficiency, fuel substitution, or pollution control. The use of tradeable emission permits or emission compensation on a global scale for greenhouse gas abatement might take advantage of the ability of certain regions to control emissions more cheaply than others.

• *Regulatory measures* include a broad array of control mechanisms, and standards regulations used for environmental protection have ranged from emission standards to requirements for environmental impact assessments. Regulations used to enhance energy security have ranged from end-use efficiency standards to requirements for the use of certain fuels in certain sectors. Such regulations may of course help to protect the environment as well.

Tables 3.6a–c illustrate policy measures in these three groups with some of the criteria listed in Sec-

tion 3.5.1. More careful exploration and examination will be required in the future.

3.6   COSTS

It is essential that the costs of emission abatement strategies are fully assessed. Anecdotal evidence suggests that the cost of some strategies could be high and that, from a given starting point, the more ambitious the strategy the higher the associated costs. The EIS is not yet able, however, to provide informed advice on the costs associated with the measures and response strategies discussed in this report. Further work in this area is essential.

Some preliminary indications are available from the individual Country Studies submitted to EIS, which suggest:

a) Significant emission abatement potential is available at low or negative resource cost when tested at social discount rates. By 2020 this might amount to around 20 percent of global

TABLE 3.6c: Applicability, Effectiveness, Limitations, and Economic Effects of Economic Measures

| MEASURE APPLICABILITY | EFFECTIVENESS | RELATIVE IMPACTS | | LIMITATIONS I/P/C[1] |
		DURATION	SIZE	
Taxes	−fuel quality −fuel choice −technology development	as long as in effect and some lags thereafter	depends on magnitude and elasticities	I/P/C
Charges (Sub-set of Taxes)	−reimburse common services (solid waste water treatment) −emissions reduction	—	depends on magnitude and elasticities	I/P/C
Subsidies	−technology development or introduction −infrastructure investments	—	depends on magnitude and elasticities	I/P/C
Market Prices	−commodities, quality or type of technology choice	n.a.[3]	n.a.	I/P/C

[1] I/P/C: Effectiveness to Influence Investment, Producer, or Consumer
[2] Self: Can be made self-funding
—: same as above
[3] n.a.: not assessed

emissions in the EIS Reference Scenario for that year and is primarily attributable to the accelerated implementation of energy efficiency and conservation measures. Intervention by governments would, however, be required to realize this potential.

b) A significant further tranche of emission abatement is potentially available at relatively moderate resource costs. This is attributable primarily to additional fuel substitution and energy conservation measures.

c) As the scale of abatement rises, marginal abatement costs will escalate. Marginal resource costs would be high if emissions were to be stabilized at levels significantly below current values by 2020.

d) The costs of achieving any particular level of abatement will vary among nations, as will the preferred options for achieving such a goal.

Achieving a 20 percent reduction from current emission levels would require major changes in

| IMPLEMENTATION | | | | |
COST TO GOVT.	DURATION	MICRO-ECONOMIC EFFECTS	MACRO-ECONOMIC EFFECTS	ECONOMIC EFFECTS
low to moderate	medium	political unacceptability of taxes high enough to be effective	−raises consumer prices so lowers consumption of taxed goods −raises producer costs and so internalizes externalities -tax forgiveness does converse	−diversions of investment and consumption and output −redistribution of tax burden creates cross-subsidies
low to moderate (self)[2]	—	—	raises producer costs and so internalizes externalities	may improve efficiency of investment, consumption, and output
high	—	−difficult to eliminate once relied upon −unanticipated spin-off	−inappropriate signal to polluter/users −excess output and demand −inefficient output	−diversions of investment and consumption and output −redistribution of tax burden creates cross-subsidies
n.a.	n.a.	externalities may not be captured initially	efficient pricing tends to result in efficient allocation of resources, i.e., efficient use and production	efficient investment, consumption and output

global energy markets, plans, and infrastructure, and intervention by governments. Maintaining this emission reduction goal would require continued technological improvements, structural changes in the global economy, and changes in the proportion of carbon-intensive fuels utilized over the remainder of the next century.

The full evaluation of costs and benefits is vital to the proper development of policy. Such estimates must include not only the resource costs of technical options but also the cost of government policy implementation, macroeconomic and second round effects, social and environmental costs and benefits, and private and non-monetary costs.

The EIS is aware of the existence of a number of reports and assessments undertaken. These have not been presented to the subgroup or examined by it. It may be appropriate for the group to review these reports in the next phase of its work.

TABLE 3.7: **Contribution by Greenhouse Gas to Radiative Forcing During the 1980s**

	TOTAL %	ENERGY ACTIVITIES %
CO_2	49	76
CFCs	17	0
CH_4	19	7
N_2O	5	3
Other[1]	10	14
Total	100	100

[1] Primarily Tropospheric Ozone due to CO, NO_X and VOCs.

3.7 THE GLOBAL POLICY CHALLENGE

3.7.1 RESPONSE CASE STUDIES

The EIS received preliminary policy studies from a number of countries. The European Commission submitted a response scenario for the European Community. The IEA secretariat submitted a preliminary study. The IAEA presented a paper on nuclear power.

The criteria for the examination of options selected differ among studies, so that the results in terms of emissions reductions achieved are not comparable and cannot be fully integrated. Therefore, only preliminary analysis is possible. Further work needs to be done. More studies from additional countries are needed to cover a larger fraction of current CO_2 emissions. Moreover, the comparability of the results should be enhanced.

According to the material and scenarios submitted to EIS it appears that the capacity of regions and countries to limit emissions varies greatly. However, some broad generalizations are possible.

- Some developing countries may be able to reduce the annual growth in CO_2 emissions from over 3 percent to around 2 percent while maintaining economic growth. The largest opportunities in developing countries appear to be increased efficiency in both energy supply and demand.

- East European countries and the USSR may be able to slow the growth or to stabilize CO_2 emissions over the next two decades, if policies to restructure their economies, increase efficiency, and promote economic development and substitution are implemented.

- West European countries including the EC may be able to stabilize or reduce CO_2 emissions by early in the next decade through a variety of measures including taxes, energy efficiency programs, nuclear power, natural gas, and renewables, without macroeconomic drawbacks. A few of these countries (Norway, the Netherlands, and Sweden) have formally adopted policies to limit emissions.

- North American and Pacific OECD countries may be able to slow the growth in CO_2 emissions by increased efficiency in energy supply and demand, fuel switching to nuclear, natural gas, and renewables, and other measures. Further policy actions on the part of countries in this group are undergoing further analysis.

The material available to the EIS demonstrates the important role industrialized countries' emissions play in total global emissions in the near term. The material also indicates that the technical potential for reductions is large. Therefore, in the near term, without actions in the industrialized countries, no significant progress in limiting global emissions will occur. However, the costs, and the extent to which this potential can be achieved, are uncertain.

3.8 RESPONSE STRATEGIES

Climate change offers an unprecedented challenge to energy policy development. Many uncertainties remain about both the impacts of climate change itself and our response to it.

It is very important that countries begin the task of developing flexible and phased response strategies. The underlying theme of any strategy must be economic efficiency—achieving the maximum benefit at minimum cost. Strategies that focus only on one group of emission sources, one type of abatement option, or one particular greenhouse gas will not achieve this. Energy policy responses should therefore be balanced against alternative abatement options in the forestry and agricultural sectors, and adaptation options and other policy goals where applicable at both national and international levels. Ways must be sought to account for consequences for other countries, and intergenerational issues, when making policy decisions.

Responses must also balance increasing understanding of the science and impacts of climate change with increasing efforts to avoid as much as possible its negative consequences. In parallel, we must develop a clearer understanding of the full social and economic implications of various response options available.

Encouragement for accelerated implementation of energy efficiency measures (on both the demand and supply side) should be a major common focus of initial policy responses. This will need to be supported by enhanced R&D if momentum is to be maintained. Encouragement for additional use of natural gas and low-cost renewable or less greenhouse gas producing energy technologies is also likely to be a common feature.

The appropriate mix of policy instruments will require detailed evaluation in the light of individual national circumstances. Initially, the highest priority must be to review existing policies and remove inappropriate conflicts with the goals of climate change policy. New initiatives will, however, be required. The international implications of some policy instruments (e.g., trade and competitiveness issues associated with carbon taxes, energy efficiency standards, and emission targets) will need to be resolved quickly if effective responses are not to be hampered.

The recommendations presented below suggest increasing levels of response with increasing knowledge and post-hoc evaluation of previous actions. It is predicated on a determined drive to take actions now to start with measures that make sense for other policy reasons; to promote energy efficiency and lower greenhouse gas emission technologies; and to accelerate R&D aimed at evaluating future options, developing new alternatives and reducing the cost of those options already available.

This is, of course, a simplified summary of a complex process. It does, however, contain the key

points of a phased, planned response with regular review of both previous actions and outcomes and of future options. It is inevitable that some countries will progress faster than others, particularly perhaps in the early stages. But it is vital that all nations begin the journey now.

3.9 RECOMMENDATIONS

IPCC countries have made a commitment to negotiate a framework convention as soon as possible. That convention will provide the international community with a legal mechanism for considering and developing subsequent agreements and protocols. The energy sector is a major source of greenhouse gas emissions. The consideration of energy sector emissions, reduction opportunities, policies, and costs should be an important part of a convention process.

The Subgroup is of the view that because of the critical role of the energy sector in the economic development process and because of the strong linkages with other sectors, there is a need to develop both general policy recommendations and specific short- and long-term recommendations.

3.9.1 GENERAL RECOMMENDATIONS

Despite the fact that many uncertainties remain, we recommend that *all* individual nations should:

1) Take steps now to attempt to limit, stabilize, or reduce the emission of energy-related greenhouse gases and prevent the destruction and improve the effectiveness of sinks. One option that governments may wish to consider is the setting of targets for CO_2 and other greenhouse gases.

2) Adopt a flexible progressive approach, based on the best available scientific, economic, and technological knowledge, to action needed to respond to climate change.

3) Draw up specific policy objectives and implement wide-ranging comprehensive pro-

grammes that cover all energy-related greenhouse gases.

4) Start with implementing strategies that have multiple social, economic, and environmental benefits, are cost effective, are compatible with sustainable development, and make use of market forces in the best way possible.

5) Intensify international, multilateral, and bilateral cooperation in developing new energy strategies to cope with climate change. In this context, industrialized countries are encouraged to promote the development and the transfer of energy-efficient and clean technologies to other countries.

6) Increase public awareness of the need for external environmental costs to be reflected in energy prices, markets, and policy decisions to the extent that they can be determined.

7) Increase public awareness of energy efficiency technologies and products and alternatives, through public education and information (e.g., labeling).

8) Strengthen research and development and international collaboration in energy technologies, economic and energy policy analysis, which are relevant for climate change.

9) Encourage the participation of industry, the general public, and NGOs in the development and implementation of strategies to limit greenhouse gas emissions.

While the specific recommendations for action that follow apply in general to all countries, industrialized countries in particular should seek to implement such measures as soon as possible, given their greater economic and technological capacity to act in the shorter term.

3.9.2 SHORT-TERM STRATEGY

As short-term strategies all individual nations should:

1) Establish interim policy objectives to limit energy-related greenhouse gases, draw up programmes to meet the objectives and mon-

itor the effectiveness and cost of the programmes against the objectives.

2) Start on measures that are technically and commercially proven, and beneficial in their own right.

3) Focus on economic instruments that could have an important role in limiting greenhouse gas emissions. We note with approval the work on these issues under way within the IPCC process and urge all countries to contribute to progress on that work.

4) Identify and take immediate steps to remove inappropriate regulatory barriers. Countries should review energy-related price and tariff systems, with the aim of removing disincentives to the efficient use of energy.

5) Promote the market penetration of:
   • improved efficiency in the production, conversion, and use of energy;
   • non-fossil and low greenhouse emission energy sources; and
   • technologies to remove, recirculate or fix methane emissions from landfills, coal mines and other sources.

   Actions should be taken on a number of the options identified in Table 3.2, e.g., in the transport sector. The choice of options should be based on cost-effectiveness analysis on a national and international level.

6) Integrate consideration of environmental costs into policy decisions at all levels of energy planning, both public and private.

7) Improve efficiency standards for mass produced goods—e.g., cars, trucks, buses, electrical appliances, buildings, air conditioners, ventilators, industrial motors, pumps for heating systems.

8) Start to develop and make widely available tools to assist in the evaluation and development of options and strategies to reduce energy-related greenhouse gas emissions (e.g., analyzing and quantifying the full fuel cycle effects, least-cost energy planning, developing a measure to facilitate comparison such as CO_2 equivalence, constructing a

framework for multisectoral policy decisions).

9) Collaborate on development, validation, and monitoring of national and global energy-related emission data.

10) Encourage the effective transfer of appropriate technology and information on the effectiveness of policies that are successful in promoting energy efficiency.

11) Contribute to an international common understanding of how to limit or reduce energy-related greenhouse gas emissions.

3.9.3 LONG-TERM STRATEGY

As our understanding of climate change develops, policies and strategies should be kept under review. It is not possible to forecast how they will develop in any detail, but it is clear that implementing the concept of sustainable development should be a central theme.

In this context all individual nations should:

1) Accelerate work on the longer-term options identified in Table 3.3 including:
   - improved efficiency in the production, conversion, and use of energy;
   - increased use of non-fossil and low greenhouse gas emission energy sources; and
   - reduction of greenhouse gas emissions by removal, recirculation, or fixation.

2) Formulate and implement strategies achieving sustainable emission levels that take account of the factors listed in Tables 3.6a–c regarding the impacts on energy prices, long-term economic growth and risk/security aspects of energy supply.

3) Evaluate the relative cost effectiveness of limitation and adaptation climate change strategies, and seek ways to account for international and intergenerational consequences.

4) Encourage the development of new technologies to limit, reduce, or fix greenhouse gas emissions associated with economic and energy activities.

5) Encourage infrastructural improvements—e.g., in transport, electrical grids, and natural gas distribution systems.

3.9.4 FURTHER WORK TO BE DONE

There is much work remaining to be done. A brief list of items related to the mission of the EIS is presented below.

1) An area of high priority is to assess the feasibility of different targets and strategies for limiting climate change and their costs, benefits and effectiveness. Such assessments should take a full fuel cycle approach and consider trade-offs among all greenhouse gases. Secondary effects should also be considered.

2) There is an urgent need to improve the data available from the developing and East European countries and for additional studies in the future, through the participation of a larger number of countries in the EIS.

3) No single analytical tool or model is sufficient to analyze the many issues discussed in this report. A broad set of tools ranging from macroeconomic models, technology assessment tools, and policy models of specific sectors—e.g., the transportation, utilities, and residential/commercial sectors—need to be developed in the future.

4) There is an important need to collaborate on development, monitoring, and validation of national and global energy-related emission data. Common methods of measuring, monitoring and evaluating energy-related greenhouse gases that can accommodate such issues as accounting for greenhouse gas emissions from bunkers, non-energy fuel use and energy trade, CO_2 equivalence, and CH_4 and N_2O, should be developed.

5) Further comparisons of target options and response strategies in the energy and industry sectors using "top down" and "bottom up" approaches are needed.

6) There is a need to move beyond simply assessing broad emission strategies to assessing the specific technologies and options open to individual countries. This will require the development and exchange of more detailed information both at the country and international levels than has been possible to date.

7) All nations should analyze the feasibility of arrangements in a worldwide context to discourage the movement of high emitting production facilities from countries with high environmental control standards to countries with lower standards.

4

Agriculture, Forestry, and Other Human Activities

CO-CHAIRS
D. Kupfer (Germany, Fed. Rep.)
R. Karimanzira (Zimbabwe)

CONTENTS

AGRICULTURE, FORESTRY, AND OTHER HUMAN ACTIVITIES

EXECUTIVE SUMMARY

INTRODUCTION

Existing forests serve a multitude of functions vital for mankind in addition to providing wood as a renewable resource. Thus, there is a paramount need to conserve forest resources and to implement measures to increase forest biomass at the same time.

The total area of forests (excluding other wooded lands) at present amounts to about 4 billion ha, roughly half of it tropical forests, and of the remainder, temperate and boreal forests account for one third and two thirds respectively. During the course of human history, roughly 2 billion ha have been lost due to various human activities, mostly in the temperate zones.

The amount of carbon presently stored in forests is equivalent to about the amount in the atmosphere, namely, approximately 700 billion tonnes of carbon. This means that 1 ha of forest contains on a global average between 100 and 200 tonnes of carbon, while afforested areas may fix on average 5–10 tonnes carbon per ha per year. Land uses involving conversion of forests through burning of biomass or felling contribute about 9 percent of total carbon equivalent greenhouse gas emissions, and about 15–30 percent of anthropogenic CO_2 emissions. Agricultural production systems provide both sources and potential sinks for atmospheric greenhouse gases. It is estimated that agricultural activities currently contribute about 14 percent of total carbon equivalent of greenhouse gas emissions, including emissions of carbon dioxide, methane, and nitrous oxide, and emissions of gases that contribute indirectly to global warming, such as nitrogen oxides and carbon monoxide.

World population is projected to grow at an average of 1.3 percent per year reaching about 8.2 billion by 2025. To meet the increased food requirements, agricultural production will also need to increase. Agricultural crop area in developing countries is expected to grow by 1.2 percent per year in combination with increased yields from existing crop acreage obtained largely from increased use of nitrogen fertilizers. Production of meat and dairy products is expected to increase by over 45 percent in this period. Achievement of food production requirements is and will remain the dominating goal in many areas around the world, and actions in response to climate change must recognize economic and social impacts in addition to environmental considerations.

Organic matter in waste and wastewater is converted into methane by various types of methane bacteria under anaerobic conditions. Anaerobic conditions exist in most landfill sites and in most lagoons used for treating organic-loaded wastewater. Total global methane emissions from waste disposals and from wastewater lagoons are estimated to be 20–70 million tonnes per year, or on average 8 percent of total anthropogenic methane emissions.

CONTRIBUTION OF AGRICULTURE, FORESTRY, AND OTHER HUMAN ACTIVITIES TO GREENHOUSE GASES

The agriculture and forestry sector is an important source of greenhouse gases accounting for approximately 23 percent of total carbon equivalent greenhouse gas emissions from anthropogenic sources in the 1980s. These sources include rice production, ruminant animals, fertilizers, loss of soil organic matter, land conversion, biomass burning, and other non-energy activities.

Deforestation contributes between 0.4 and 2.8 billion tonnes of carbon (BTC), and biomass burning (forests, savanna, and shrub-fallow) between 20 and 80 million tonnes (MT) methane per year. The scientists who addressed the IPCC Tropical Forest Workshop, São Paulo, Brazil, January 1990 were reasonably certain that in 1980 emissions were between 1.0 and 2.0 billion tonnes of carbon (BTC), and in 1989 emissions were between 2.0 and 2.8 BTC.

Ruminant animals produce methane as part of their natural digestive process. Total methane emissions from domestic ruminant animals have been estimated to be between 65 and 100 million tonnes per year. In addition, animal wastes from anaerobic waste management systems are likely to yield on the order of 15 million tonnes globally.

Flooded rice fields produce methane due to microbial decay of organic matter. While uncertainty exists, they appear to account for between 25 and 170 million tonnes, or on average 20 percent of global methane emissions. Rice production is expected to increase from the current level of 458 million tonnes to over 750 million tonnes by the year 2020.

Loss of soil organic matter from agricultural soils is uncertain but could amount to up to 2 billion tonnes of carbon (BTC) per year.

Use of nitrogen fertilizers results in emissions of nitrous oxide equivalent to 0.01–2.2 million tonnes of nitrogen per year.

Biomass burning for land use conversion and the burning of agricultural wastes is estimated to account for over half of all biomass burned annually. These agriculture-related activities therefore contribute over 5–10 percent of total annual methane emissions, 3 to 8 percent of nitrous oxide emissions, 10–20 percent of carbon monoxide emissions, and 5–20 percent of NO_x emissions.

Landfill sites and wastewater treatment plants emit about 20–80 million tonnes of methane per year.

Future Greenhouse Gas Emissions

Future greenhouse gas emissions are difficult to predict because of uncertainties in estimating economic and population growth rates, and changes in for-estry agriculture practices, and climate sensitivity. Scenarios of emissions, which must be used with caution, suggest that emissions are likely to grow well into the future without mitigating policy measures (see Executive Summary Table 4.1). These estimates suggest that CO_2 emissions from deforestation could range between 1.1 and 3.9 billion tonnes of carbon in 2020, that methane emissions from flooded rice will increase to about 150 million tonnes in 2025, and that methane emissions from managed livestock (including their wastes) will increase to about 185 million tonnes. Nitrous oxide emissions from use of nitrogen fertilizers will probably increase by up to about 3.5 million tonnes. Emissions from biomass burning are highly uncertain and have been assumed to remain constant at 55 million tonnes as a minimum.

Emissions of methane from landfill sites and wastewater treatment plants will probably increase to about 50–90 million tonnes per year by the year 2020.

Policy Options, Technologies, and Practices to Reduce Emissions

Currently available policies, technologies, and practices in forestry, agriculture, and waste disposal are likely to be only partially effective in reducing the predicted growth in emissions, unless they are coupled with emission reductions in the energy and industry sector. However, many practices and technologies are available today that, if utilized, could modify the rate of growth in emissions and that appear to make sense for economic and environmental reasons. Other options have been identified that require additional research and demonstration. Policies should address not only technical options but also instrumental (economic, regulatory, information, etc.) and institutional options in order to become effective. Although uncertainties about the rate and extent of climate change remain, it is recommended that all countries take steps to:

- adopt clear objectives for the conservation, reforestation, and/or sustainable development of forests in national development plans;
- amend national policies to minimize forest loss associated with public and private development

EXECUTIVE SUMMARY TABLE 4.1: Estimates and Projections of Annual Anthropogenic Emissions of Greenhouse Gases from Agriculture, Forestry, and Waste Management Activities

	1985			2020–2025		
	CO_2 (BTC)	CH_4[a] (MT-CH_4)	N_2O (MT-N)	CO_2 (BTC)	CH_4 (MT-CH_4)	N_2O (MT-N)
Land Use Changes[b] (Including Deforestation[c])	1.0–2.0	50–100	—	1.1–3.9	50–100	—
Biomass Burning[b]	3.9	20–80	0.2	—	—	—
Animal Systems	—	65–100	—	—	170–205	—
Rice Cultivation	—	25–170	—	—	100–210	—
Nitrogen Fertilizer	—	—	0.01–2.2	—	—	—
Loss of Soil Organic Matter	0–2	—	2.9–5.2	—	—	—
Waste Management	—	20–70	—	—	50–90	—
Total Annual Anthropogenic Emissions from All Sources (Including Energy Use)	6	540	12	12	760	16

[a] CH_4 can be expressed as tonnes carbon by multiplying the CH_4 estimate by 0.75.

[b] Land use changes and biomass burning estimates overlap and are not additive.

[c] A recent preliminary report on tropical deforestation (Myers, 1989) estimates emissions from deforestation to be 2.0–2.8 BTC per year for 1989, with a mean working figure of 2.4 BTC.

Sources: IPCC Working Group I Final Report, Summer 1990; IPCC AFOS Tropical Forestry Workshop, São Paulo, 1990; Andreae, 1990; IPCC-AFOS Agriculture Workshop, Washington, D.C., 1990.

projects (e.g., roads, dams, resettlement projects, mining, logging);

• adopt forest management solutions integrated across sectors to policies on environment, agriculture, transportation, energy, economics, poverty, and landlessness;

• implement policies to promote increasing productivity in sustainable agriculture and to protect natural resources; and

• improve public awareness of all these points through education programmes.

Policy options need not be implemented in sequence, i.e., policies should be implemented as it becomes apparent that they meet a number of national needs. In addition, policies associated with production, processing, storage, transportation, and marketing need to be examined to derive the optimum effectiveness from research, technological developments, and land use practices. To promote sustainable agriculture and forestry, analyses are needed of economic incentives, taxes, pricing and trade barriers, cultural practices, technology transfer measures, education and information programmes, and international cooperation and financial assistance measures.

CRITERIA FOR THE SELECTION OF POLICY OPTIONS

It is important that these options be pursued without undue economic disruption. Policies should also:

• be of widespread applicability;

• be compatible with the social and economic life of communities dependent on agriculture and forestry;

• be equitable in the distribution of the burdens of action between developed and developing countries, taking into account the special situation of the latter;

• result in the spread of knowledge, management skills, and technologies;

• result in net environmental gain; and

• take account of the fact that emissions in this sector largely comprise many small sources or diffuse sources from large areas.

NEAR-TERM OPTIONS

The opportunities for reducing greenhouse gas emissions and enhancing carbon sinks in the near term include the following:

Forestry

- Develop and implement policies that will reduce current deforestation and forest degradation to reduce greenhouse gas emissions and enhance forest areas as carbon sinks.
- Improve efficiency of use of fuelwood.
- Introduce sustainable harvesting and natural forest management methods to reduce tree damage.
- Partially replace fossil energy sources by sustainably managed sources of biomass which would reduce net emissions of additional CO_2.
- Increase efforts to replace high energy input materials with wood, and encourage further recycling of forest products in order to provide for long-term storage of carbon.
- Strengthen the use of remote sensing in forest management and in the determination of forest removal and emission patterns, by developing and coordinating data collection and analyses among relevant institutions.
- Determine the feasibility of implementing the Noordwijk Declaration aim of achieving net global forest growth of 12 million ha per year, through conservation of existing forests and aggressive afforestation, and develop appropriate national strategies.
- Perform analyses for and begin to implement large-scale national forestation and forest protection plans.
- Identify and eliminate inappropriate economic incentives and subsidies in forestry and non-forest sectors that contribute to forest loss.
- Develop enhanced regeneration methods for boreal forests in order to cope with change in the climate.
- Ensure the health of existing forests, in particular by reducing air pollution, e.g., acid deposition, tropospheric ozone, SO_2, NO_x, NH_3, VOCs, etc., by adopting appropriate policies to reduce site-specific and regional pollution problems.
- Use available surplus agricultural land for forestry as appropriate.
- Fully integrate the requirements of forest conservation and sustainable management in all relevant sections of national development planning and policy.
- Evaluate the implications of population growth and distribution for the achievement of national forestry measures and take appropriate action.

Agriculture

- Reduce biomass burning through fire control, education and management programmes, as well as the introduction of the use of appropriate alternative agricultural practices.
- Modify agricultural systems dependent on the removal of biomass by burning to provide opportunities for increasing soil organic matter and reduction of greenhouse gas emissions.
- Reduce methane emissions through management of livestock wastes, expansion of supplemental feeding practices for livestock and increased use of production and growth-enhancing agents with appropriate safeguards for human health and taking into account legitimate consumer concerns.
- Reduce nitrous oxide emissions by using improved fertilizer formulations and application technologies, and through judicious use of animal manure.
- Introduce where appropriate minimum- or no-till systems that are recommended for those countries currently using tillage as part of the annual cropping sequence, to yield additional benefits such as direct energy savings, improved soil tilth, and increased soil organic matter.
- Use areas marginally suitable for annual cropping systems for perennial cover crops for fodder or pastoral land uses, or forests if soils are suitable. This would increase carbon uptake, both in the vegetation and soil, and would yield other benefits such as reduced soil erosion, improved water infiltration and quality, and delayed streamflow.

Waste Management

- Consider using landfill gas collection systems and flaring to reduce methane emissions in developed countries where practical.

- Use biogas systems to treat wastewater in developing countries in order to reduce methane emissions and provide inexpensive sources of energy.
- Promote maximum recycling of wastes.

LONGER-TERM OPTIONS

Several opportunities for reducing greenhouse gas emissions and enhancing carbon sinks have been identified for the longer term. In general, these opportunities must be developed, demonstrated, and assessed in terms of greenhouse gas reductions and the full range of potential costs and benefits. These alternatives must maintain or enhance the productivity of agriculture and forestry systems. This will require substantial research efforts focused on better understanding of the processes by which these gases are emitted, further investigation of promising options, and better field measurement devices.

Forestry

- Develop and implement standardized methods of forest inventory and bio-monitoring to facilitate global forest management and to make production ecology studies and cost/benefit analyses comparable across forest areas.
- Increase wood production and forest productivity by silvicultural measures and genetically improved trees, thus helping to increase the forest carbon sink, to meet increasing demand for wood as well as to support replacement of fossil fuels and other materials by wood and to avoid inappropriate land use conversion.
- Incorporate forest protection strategies for fire, insects, and diseases into future management and planning under various climate change scenarios.
- Develop and implement silvicultural adjustment and stress management strategies to maintain existing forests.
- Develop and implement environmentally sound management practices for peatlands that restrict the release of carbon, especially as methane.
- Make use of the specific opportunities of intensively managed temperate forests to mitigate the greenhouse effect by expanding their biomass.

Agriculture

- A comprehensive approach including management of water regimes, development of new cultivars, efficient use of fertilizers, and other management practices could lead to a 10–30 percent reduction in methane emissions from flooded rice cultivation, although substantial research is necessary to develop and demonstrate these practices. A 10 percent reduction in emissions from rice systems may contribute about 15–20 percent of the total reduction required to stabilize atmospheric concentrations of methane.
- Through a number of technologies it appears that methane emissions from livestock systems may be reduced by up to 25–75 percent per unit of product in dairy and meat production. The net effect of such improvements depends upon the extent to which such methods can be applied to domestic ruminant populations, which will vary greatly from country to country. However, each 5 percent reduction from this source could contribute 6–8 percent of the reduction necessary to stabilize methane in the atmosphere.
- Fertilizer-derived emissions of nitrous oxide potentially can be reduced (although to what extent is uncertain) through changes in practices such as using controlled-release nitrogen, improving fertilizer-use efficiency, and adopting appropriate alternative agricultural systems.
- Trace gas emissions from biomass burning, land conversion, and cropping systems may be reduced through widespread adoption of sustainable agricultural practices, optimizing use of fertilizer and organic amendments, and improved pasture management and grazing systems. These gains may be offset by pressures of increasing population and increased demand for food and fiber production.

GENERAL ISSUES AFFECTING THE AGRICULTURE, FORESTRY, AND WASTE MANAGEMENT SECTORS

Emissions from these sectors are intimately related to the ability of countries to provide for national and global food security and raw materials for export. Policies for emission control and sink enhancement therefore may affect the economic basis of some countries, especially developing countries. Efforts to prevent deforestation and to promote afforestation will have multiple impacts that in many cases may enhance the abilities of indigenous agricultural communities to feed themselves and to earn a living from local surplus. Flexibility should be provided to governments to develop least-cost implementation strategies, and countries should pursue those options that increase productivity, make economic sense, are environmentally sound, and are beneficial for other reasons. Agriculture, forestry, and waste treatment sources of greenhouse gas emissions consist of very large numbers of small sources and/or of diffuse sources from large land areas, and as such represent a major challenge.

Controls on existing sources of pollution that may affect agricultural and forestry lands are an important component in reducing emissions and protecting sinks. Measures to prevent land degradation, hydrological problems, and loss of biodiversity and to improve productivity will generally complement efforts in this sector. Furthermore, analysis is needed about the costs and benefits of individual policy measures.

It is desirable to consider a range of measures in the context of a comprehensive response strategy. Currently available options in the agriculture and forestry sectors are likely to be only partially effective unless coupled with action to reduce emissions from the energy and industry sector. There is an urgent need to improve all relevant data, especially on deforestation rates and on the socio-economic factors that lead to the use of deleterious practices in agriculture and forestry. In order to achieve the full potential of identified measures, research is needed into the routes by which new technologies can be introduced while preserving and enhancing social and economic development.

In the past, the agriculture and forestry sector has proved efficient at introducing new methods of production, not always with beneficial ecological consequences. The pace of technological development (even in the absence of global warming) is likely to remain high in this sector for the foreseeable future. This may afford new opportunities for emission reduction and sink enhancement, provided efficient transfer of knowledge, advice, and technology occurs.

INTERNATIONAL OPTIONS

A broad range of institutional issues must be addressed in order to ensure that the objectives of increased global food production and reduced greenhouse gas emissions can be met in the future. Among the options that governments, international organizations and intergovernmental bodies such as IPCC should consider are the following:

- A series of agricultural workshops and symposia to assess new information on agricultural production and emission forecasts, exchange information on the effectiveness of new technologies and management practices, and evaluate the potential impact of agricultural policies. These meetings should address the institutional and economic issues particular to the major agricultural sources of emissions, such as biomass burning, temperate agricultural practices, livestock emissions, and emissions from rice cultivation.

- Guidelines for future research by FAO, CGIAR (Consultative Group on International Agricultural Research), IUFRO (International Union of Forestry Research Organisations), and others to address the need to investigate impacts of climate change and options for reducing emissions.

- Strengthen, support and extend the Tropical Forestry Action Plan process to all countries with tropical forests, and expand support for immediate implementation of completed plans.

- Strengthen the role of ITTO to develop international guidelines to encourage:
  a) sustainable forest management techniques, including national legislation requiring management of forest for sustained wood production;
  b) an assessment of incentives for sustainable forest management including feasibility of labeling.

- Strengthen the role of development banks, IMF, FAO, UNEP, and other multilateral or bilateral international organizations in helping developing countries achieve conservation and sustainable development of forests and agriculture by:

a) requiring analyses of climate change implications, potential greenhouse gas emissions, and response programmes in their review of project proposals;

b) expanding greatly aid and investment flows to both sectors;

c) expanding debt relief via renegotiation of debt, and debt for conservation exchanges; and

d) linking structural adjustment measures to alleviation of climatic impacts and reduction of gas emissions.

DEVELOPMENT OF A WORLD FOREST CONSERVATION PROTOCOL

The following declaration was agreed upon in the workshop on tropical forests in São Paulo in January 1990:

Consideration of forestry issues, and of tropical forestry issues in particular, must not distract attention from the central issue of global climate change and the emission of greenhouse gases attributable to the burning of fossil fuels by developed countries. No agreement on forests and global climate change will be reached without commitments by developed countries on greenhouse gas emissions. The groups recognized that the conservation of tropical forests is of crucial importance for global climatic stability (particularly having regard to the important contribution of tropical forest destruction to global warming, carbon dioxide, methane, and other trace gases), but of more crucial importance for national economic and social development, for the conservation of biodiversity, and for local and regional climatic and environmental reasons. The Workshop recommended that the IPCC support the development of a forestry protocol in the context of a climate convention process that also addressed energy supply and use. The Workshop concluded that the specific elements of such a protocol are a matter for international negotiations. These elements may include: fundamental research, tropical forest planning, measures to use, protect, and reforest, international trade, fi-

nancial assistance, and the advantages and disadvantages of national and international targets. The objective should be to present more concrete proposals on the occasion of the UN Conference on Environment and Development, to be held in 1992.

In light of the above, it is recommended that a World Forest Conservation Protocol, covering temperate, boreal, and tropical forests, be developed in the context of a climate convention process that also addresses energy supply and use, as noted by the January 1990, IPCC/RSWG Tropical Forest Workshop in São Paulo, Brazil, and that in accordance with UNGA Resolution 44/207, operative paragraph 10, a meeting of interested countries from both the developed and developing world and of appropriate international agencies be held to identify possible key elements of such protocols and practical means of implementing them. Such a meeting should also develop a framework and methodology for analyzing the feasibility of the Noordwijk remit, including alternative targets as well as the full range of costs and benefits.

All countries should make a contribution to the solution of the global warming problem. The São Paulo Workshop stated:

Although forests can assist in mitigating the effects of atmospheric carbon build-up, the problem is essentially a fossil fuel one and must be addressed as such. In this way, and as a general principle, the final report of the present IPCC Workshop on Tropical Forests, while putting tropical forests in the overall context of global warming, should make it clear that the burden of response options is not to be placed on developing countries and thus should state clearly that all countries should make a contribution to the solution of the global warming problem. The temperate forest dieback (caused by acid rain), as analogous to tropical deforestation (caused by tropical people's attempts to satisfy basic human needs), could be specifically mentioned in such a context.

It should, however, be noted that recent findings indicate that forests may be adversely affected throughout the world by a variety of causes, only one of which is acid deposition.

The São Paulo Workshop further noted:

> Forests cannot be considered in isolation, and solutions must be based on an integrated approach which links forestry to other policies, such as those concerned with poverty and landlessness. The forest crisis is rooted in the agricultural sector and in people's needs for employment and income. Deforestation will be stopped only when the natural forest is economically more valuable than alternative uses for the same land.

4.1 INTRODUCTION

This report examines the roles of agriculture, forestry, and human activities other than energy production and industry as sources of greenhouse gases, and possible measures for reducing their emission into the atmosphere.

Working Group I of IPCC suggests in its report that the global average temperature will rise within the next 80 years by about 2.4–5.1 degrees C above pre-industrial, with a best estimate of 3.5 degrees C. Predictions of concomitant changes in precipitation and wind patterns are much less certain. Little information is currently available on the rate of change, transient conditions, and changes in the occurrence of extreme events such as storms and forest fires.

Agriculture- and forestry-related activities contribute to the emission of greenhouse gases, notably of carbon dioxide, methane, and nitrous oxide. These sectors account for approximately 23 percent of greenhouse gases from anthropogenic sources in the 1980s. These sources include rice production, ruminant animals, fertilizers, land conversion, biomass burning, and other non-energy activities.

Deforestation contributes between 0.4 to 2.8 billion tonnes of carbon (BTC), and biomass burning (forests, savanna, and shrub-fallow) between 40 and 75 million tonnes (MT) methane per year. The scientists who addressed the IPCC Tropical Forest Workshop in São Paulo were reasonably certain that in 1980 emissions were between 1.0 and 2.0 BTC and that in 1989 emissions were between 2.0 and 2.8 BTC.

Ruminant animals produce methane as part of their natural digestive processes. Total methane emissions from domestic ruminant animals have been estimated to be between 60 and 100 million tonnes. In addition, animal wastes from anaerobic waste management systems are likely to yield on the order of 15 million tonnes globally.

Flooded rice fields produce methane due to microbial decay of organic matter. While uncertainty exists, they appear to account for approximately 110 million tonnes, or 20 percent of global anthropogenic methane emissions. Rice production is expected to increase from the current level of 458 million tonnes to over 750 million tonnes by the year 2020. Use of nitrogen fertilizers results in emissions of nitrous oxide on the order of 2 million tonnes.

Loss of soil organic matter from agricultural soils is uncertain but could amount to up to 2 billion tonnes of carbon (BTC) per year.

Biomass burning for land use conversion and the burning of agricultural wastes is estimated to account for over half of all biomass burned annually. These agriculture-related activities therefore contribute over 5–10 percent of total annual methane emissions, 3–8 percent of nitrous oxide emissions, 10–20 percent of carbon monoxide emissions, and 5–20 percent of NO_x emissions.

Landfill sites and wastewater treatment plants emit about 30–70 million tonnes of methane per year.

Future greenhouse gas emissions are difficult to predict because of uncertainties in estimating economic and population growth rates and changes in forestry and agriculture practices. Scenarios of emissions, which must be used with caution, suggest that emissions are likely to grow well into the future without policy measures (see Table 4.1). These estimates suggest that CO_2 emissions from deforestation could range between 1.1 and 3.9 billion tonnes of carbon in 2020, that methane emissions from flooded rice will increase to about 150 million tonnes in 2025, and that methane emissions from managed livestock (including their wastes) will increase to about 185 million tonnes. Nitrous oxide emissions from use of nitrogen fertilizers will

TABLE 4.1: **Estimates and Projections of Annual Anthropogenic Emissions of Greenhouse Gases from Agriculture, Forestry, and Waste Management Activities**

	1985			2020–2025		
	CO_2 (BTC)	CH_4[a] (MT-CH_4)	N_2O (MT-N)	CO_2 (BTC)	CH_4 (MT-CH_4)	N_2O (MT-N)
Land Use Changes[b] (Including Deforestation[c])	1.0–2.0	50–100	—	1.1–3.9	50–100	
Biomass Burning[b]	3.9	20–80	0.2	—	—	—
Animal Systems	—	65–100	—	—	170–205	—
Rice Cultivation	—	25–170	—	—	100–210	—
Nitrogen Fertilizer	—	—	0.01–2.2	—	—	—
Loss of Soil Organic Matter	0–2	—	2.9–5.2	—	—	—
Waste Management	—	20–70	—	—	50–90	—
Total Annual Anthropogenic Emissions from All Sources (Including Energy Use)	6	540	12	12	760	16

[a] CH_4 can be expressed as tonnes carbon by multiplying the CH_4 estimate by 0.75.

[b] Land use changes and biomass burning estimates overlap and are not additive.

[c] A recent preliminary report on tropical deforestation (Myers, 1989) estimates emissions from deforestation to be 2.0–2.8 BTC per year for 1989, with a mean working figure of 2.4 BTC.

Sources: IPCC Working Group I Final Report, Summer 1990; IPCC AFOS Tropical Forestry Workshop, São Paulo, 1990; Andreae, 1990; IPCC-AFOS Agriculture Workshop, Washington, D.C., 1990.

probably increase to about 3.5 million tonnes. Emissions from biomass burning are highly uncertain and have been assumed to remain constant at 55 million tonnes, as a minimum.

Emissions of methane from landfill sites and wastewater treatment plants will probably increase to about 50–90 million tonnes per year by the year 2020.

Agriculture and forestry may also contribute to gases such as NO_x, ozone, and carbon monoxide, which contribute indirectly to the greenhouse effect. While significant, these sectors are of course not the only sources and also provide sinks for carbon dioxide. As a consequence, any consideration of measures and policy options to control emissions from these sectors must take account of the overall effects of greenhouse gases.

The interactions between climate change and agriculture and forestry management complicate precise evaluation.

• Climate change in itself will affect forests and agriculture. In an altered climate, change will tend to increase stress on the biosphere in a variety of ways. For instance, seasonal rainfall patterns might change. Similarly, extreme weather events could occur more frequently. Further, the risk of forest fires and diseases might increase.

• Altered forest and agriculture patterns will in turn affect the climate. The areas where the cover changes from forest to agriculture will reflect light from the sun differently, and this altered albedo, by changing the radiation balance, might also change climate. Similarly, climate-related changes in needs for irrigation will affect hydrological cycles, including cloud formation, which could significantly affect climate.

• Anthropogenic emissions that affect climate might also directly affect forests and agriculture. In the temperate zone, where most of the tropospheric ozone precursors are emitted, ozone, which contributes to the greenhouse effect, can cause direct damage to vegetation.

While these complex mechanisms need to be clarified, the need for further research is no excuse for delaying action.

Counter-measures may pose challenges for food and fiber security. Options to reduce greenhouse gas emissions should be balanced with the growing demand for food and fiber and other material goals. The approach for developing options should be to maintain and increase agricultural production in a sustainable manner while reducing greenhouse gas emissions.

This report also addresses the role and control of organic matter decomposition in wastes and waste-waters, which can, under anaerobic conditions, contribute significantly to methane emissions (approximately 15 percent of global anthropogenic release).

AFOS organized a series of workshops on:

- Boreal Forests (October 9–11, 1989 in Finland)
- Temperate Forests (October 30 to November 1, 1989 in the Federal Republic of Germany)
- Agriculture (December 12–14, 1989, in the United States)
- Tropical Forests (January 9–11, 1990, in Brazil)

to assess the significance of emissions from these sectors and possible options for their control. These workshops were attended by representatives from all concerned regions. The following sections on forestry and agriculture are based on the reports of these workshops. The section on methane from landfill sites and wastewater treatment plants is based on a report prepared by one of the participating countries.

4.2 FOREST RESPONSE STRATEGIES

The total area of forests (excluding other wooded lands) at present amounts to approximately 4 billion ha, roughly half of it tropical forests, the other half temperate or boreal forests, one third and two thirds respectively. During the course of human history, roughly 2 billion ha have been lost due to various human activities.

The amount of carbon presently stored in forests is equivalent to the amount in the atmosphere—namely, approximately 700 billion tonnes carbon. This means that 1 ha of forest contains on a global average between 100 and 200 tonnes of carbon, while afforested areas fix on average 5–10 tonnes carbon per ha per year.

Land uses involving conversion of forest through burning of biomass or felling contribute about 9 percent of greenhouse gas emissions, and about 15–30 percent of anthropogenic CO_2 emissions.

The Noordwijk Declaration on Atmospheric Pollution and Climate Change states in point 21 that,

The Conference . . .

- agrees to pursue a global balance between deforestation on the one hand and sound forest management and afforestation on the other. A world net forest growth of 12 million hectares a year in the beginning of next century should be considered as provisional aim.
- requests the IPCC to consider the feasibility of achieving this aim. . . .

The outcomes of the forestry workshops clearly demonstrate the necessity both to conserve forest resources and to implement measures to increase forest biomass at the same time. Existing forests serve a multitude of functions vital for mankind in addition to providing wood as a renewable resource. However, it is important to recognize that reforestation of areas where forests have previously been destroyed does not immediately fulfill all the functions of an intact forest. In addition to the dangers of loss of biodiversity and soil erosion that arise with deforestation, deforestation releases around 100 tonnes of carbon per hectare both immediately, through biomass burning, and in the longer term, through decomposition of biomass and humus. Newly afforested areas fix on average around 5–10 tonnes of carbon per ha per year. Thus, it takes decades to fix again the carbon released by deforestation. Conservation of forests is therefore of paramount importance.

At present, around 1 billion tonnes of carbon are released into the atmosphere annually by forest destruction, primarily through anthropogenic activities as a consequence of growing population and rising demand for food and fiber. The total release of carbon by anthropogenic activities is of the order of 7–8 billion tonnes, of which about 6 billion tonnes are released due to burning fossil fuels.

Reforestation of deforested areas is a valid counter-measure. Currently, only one million hectares per year are being afforested. In addition to the area needed to balance deforestation (according to estimates for 1980–85 about 11–12 million ha per year), it is estimated that in 20–50 years, very roughly, up to 200 million hectares might be planted, which would require an average rate of 4–10 million ha per year. Such an area could store up to about 20 percent of present carbon dioxide emissions for a limited time. The area that could be afforested, however, depends on several other factors. Providing food or housing for a growing world population will also require areas that cannot then be afforested; soil degradation, possibly amplified by climate change, will reduce the area that could usefully be afforested; and the time needed to implement these measures might in practice be longer than 50 years. These points make it difficult to attain a yearly net forest growth goal of 12 million hectares per year as set by the Noordwijk Declaration.

Afforestation costs are highly dependent on the specific area, the tree species and local wages. A rough global estimate ranges from U.S. $200 to $2,000 per hectare. Detailed studies are needed to obtain more reliable figures; costs and carbon dioxide reduction figures obtainable by afforestation should also be compared with other measures possible in industrialized countries—for example, energy sector counter-measures. However, this in no way limits the usefulness of conservation measures. The amount of afforestation required to balance total anthropogenic carbon dioxide emissions—7–8 billion tonnes—would be about 1 billion ha, or an area the size of Europe from the Atlantic to the Urals. Afforestation on such scale is impracticable, so that forestry measures can play only a minor but significant contributory role in reducing the buildup of greenhouse gases. It might be realistic to achieve a 10–15 percent reduction of such emissions. This role is small in comparison to the potential emissions of CO_2 implied by a continuing (and even increasing) global reliance on fossil fuels.

Forests may be able to adapt to a rate of temperature change of around 1 degree C in 100 years, but the more rapid changes predicted are likely to have deleterious effects, especially in combination with additional stress factors such as air pollution.

The response of forest ecosystems to temperature change depends, however, on a great number of other parameters (e.g., precipitation, extreme climates) on which no information is available at the moment. Shifts may occur in the ranges and species composition of forest communities, together with crisis-type disturbances in development and changes in biodiversity and genetic diversity of individual species.

Policies and Measures

It must be accepted from the outset, however, that the ultimate contribution of forestry can only be to facilitate a transition from our reliance on fossil fuels over the next 50–75 years. Since global warming could affect the distribution, productivity, and health of forests, it is necessary that measures to control greenhouse emissions should be taken both within and outside this sector. Almost half of the man-made increase in atmospheric greenhouse gas levels is caused by the burning of fossil energy sources. Conservation of fossil fuels and the use of alternative energy sources including the use of renewable energy sources is therefore essential, as well as measures to reduce emissions from fossil fuel sources. Changes in energy policies must take account of the environmental consequences of particular energy options.

Forestry policies can be identified which could contribute to a global warming response strategy. In this context, reducing current deforestation and forest degradation should be a first priority. It is also vital to make a contribution to the reduction of the man-made increase in the greenhouse effect by making greater use of wood. The partial replacement of fossil energy sources by wood will reduce the emission of additional CO_2. Wood in competition with building material and other material whose production requires many times more energy contributes to energy savings and to maintaining carbon sinks. Wood production from genetically improved trees and intensively managed forest can replace fossil fuel carbon with the aid of silvicultural and breeding measures. As these measures can be carried out on substantial areas of existing forested land, a significant potential will be available in the long run.

Monitoring and Research

There is a need to strengthen bio-monitoring of forests to complement the atmospheric monitoring already under way. Standardized methods of forestry inventory are required to provide the data necessary to manage forests globally. This would also mean that studies of production ecology and cost/benefit analyses would be comparable across forest areas. Major gaps in knowledge include soil properties, primary nutrient cycling, and decomposition in the forest floor. Provided that the impacts of other environmental factors on tree growth are not too great and do not counteract the direct effects of CO_2, it is theoretically possible that increased atmospheric CO_2 concentrations might increase yield. Appropriate species and provenances would be required and suitable silvicultural measures would have to be adopted to take maximum advantage of such a situation.

Forest Protection

Fire protection strategies based on fire hazard information should be developed and incorporated into future management and planning under climatic change scenarios.

Better forest pest protection strategies should also be developed. The activity, abundance, and distribution of most forest insect and disease species, especially in temperate and boreal areas, are expected to increase with increasing temperatures.

Forest Management

Forest and woodland biomass can be managed by maintaining yields of existing forests and by increasing the productivity of native, exotic, and genetically improved species.

Forest managers and researchers have believed that forest renewal, productivity, health, and diversity would be assured by prudent silvicultural practices. Much of our knowledge and procedures for strategic planning and management have been based on empirical data such as growth and yield models, and the validity of these may be suspect under a change of climate. Therefore, a review of current silvicultural practices and data is required.

Land availability and the difficulty of securing forests against unsustainable exploitation are important constraints. However, possibilities exist for enhancing land productivity through reforestation and improved forest management. These constraints can be overcome, provided that local communities share adequately in the benefits from forest conservation and reforestation. These include provision of employment, access to produce from the forest, and benefits associated with effective land-use management of adjacent agricultural areas that include appropriate agroforestry development.

Forestry and Agriculture

Research is needed into the interactions of forestry and agriculture in terms of biological, agronomic, and economic aspects. In particular, the possibilities of encouraging farm woodland and agroforestry activities need further consideration. Increasing experience with silvi-pastoral and silvi-arable systems is becoming available and may have useful environmental and economic benefits. Particularly in temperate areas, many countries are now looking to farm forestry as a means of controlling agricultural surpluses. This trend should also be encouraged because of its potential effects on greenhouse gas emissions.

End Uses and Biomass Conversion

The need and demand for raw materials from forests has increased globally, especially in the pulp and paper industry. If timber could be utilized more effectively and a larger percentage of timber products recycled, less wood would need to be cut from forests without resulting in a decrease in industrial production. This would leave more standing biomass in the forest over a longer time to sequester carbon dioxide. Extension of rotation periods may be possible in managed forests. An alternative would be to grow trees for shorter rotations than at present, as trees would then be exposed for a shorter time to stressful and changing conditions, and the carbon could be stored in the end product. Materials such as concrete, steel, and aluminum could often be replaced by wood, which would have the added advantage of saving on the energy required in the production of these materials and on the consequent carbon emissions.

Climate stress and damage caused by insects and

diseases may produce wood with different properties than the materials now used. How to harvest and use such wood needs to be investigated. Carbon dioxide emissions from electrical energy production are lowest with wood biomass, due to cycling and refixation of carbon. Using a sustainably managed source of biomass, such as intensive, short-rotation fuelwood plantations, biomass energy could produce net reductions in carbon dioxide emissions. However, the use of forest biomass for energy should not be promoted to the detriment of forest ecosystems.

Use of logging and milling residues (bark, timber residues) should also be considered—for instance, in energy production. In many countries, wood is still utilized in large quantities for energy purposes, and indeed, shortage of wood for firewood is a problem in many tropical countries. One aim could be to maintain or even increase the usage of wood for energy through the development of technology under a sustainable management policy.

Sections 4.2.1–4.2.3 address the issues particular to boreal, temperate, and tropical forests.

4.2.1 SPECIAL ISSUES ON BOREAL FORESTS

4.2.1.1 *Introduction*

The boreal forest is one of the world's major vegetation regions, occupying 1,200 million square ha. Coniferous species dominate the zone that encircles the globe through northern Eurasia and North America from about 45N to 75N. The boreal forest in North America and eastern Asia contains many species not found in the Euro-Siberian region. The majority of its soils are covered by ericaceous plants, lichens, and moss species in an understory that is usually cold, and permafrost is common. The dependence of national economies and indigenous people on forestry in the boreal region is strong. Even with new technology and urbanization, forests and forestry still play major roles in peoples' lives, by providing subsistence or recreation. The cultures and economies of forest peoples, such as the Athabaska Indians and Eskimo of Alaska or Nordic Laplanders, are intricately adapted to the natural environment and depend upon it for self-perpetuation.

4.2.1.2 *Carbon Sinks of the Boreal Region*

An estimated 25 percent of the global soil carbon is contained in boreal forest soils, and significant amounts of this carbon will be lost if the lands overlying these soils are converted to other uses. In addition, drainage of boreal muskegs and peat bogs would release more nitrous oxides and carbon dioxide as water tables drop, while maintaining high water tables in these systems would release more methane.

The boreal forests may be divided into (1) managed, (2) intermediate, and (3) non-managed forests. Nordic countries within the boreal zone (Finland, Norway, and Sweden) fit into the first category. The total boreal forest area of these countries is about 50 million ha, and annual removals of industrial roundwood and fuelwood are approximately 100 million cubic meters.

In Canada approximately 55 percent of its boreal forest (210 million ha) has been classified as productive forest. However, only a small part of the productive forest is under active forest management; thus it is classed as an intermediate forest. Total removals of industrial roundwood and fuelwood, during 1986, were approximately 115 million cubic meters. Within the boreal forest region, the highest proportion of non-managed forests are in Alaska (USA) and Yakutia (USSR). According to figures, Alaska, for instance, has 9 million ha of commercial forest land, which is equivalent to less than 10 percent of the total forest area. In Yakutia, the total forest area is approximately 150 million ha. Recent information from Soviet sources suggests that the approximate total boreal forest area in the Soviet Union is 600 million ha.

4.2.1.3 *Consequences of Climate Change on Emissions*

The prevailing climate determines the structure and function of the boreal forest ecosystem, including components of the carbon balance, productivity, physiology, insects, diseases, and fire. Current predictions suggest that in the boreal zone the change in temperature will be relatively low in Norway, because of its maritime climate, but high in northwestern continental Canada. However, the direct and combined effects of CO_2 enrichment and temperature change on boreal tree species are not fully understood. Nutrient cycling, for example, will be accelerated by an increase in temperature and could increase productivity if accompanied by appropriate moisture. Results obtained in a study carried out by Environment Canada (1988) using growth indices, indicated that there would be a northward shift of major forest ecosystems by up to several hundred kilometers. Similar effects may be expected in other countries with boreal forests.

The incidence of forest fire is strongly connected to climatic events. Fire has an important role in the carbon cycle because it returns carbon to the atmosphere faster than any other process. In Canada the occurrence of wildfires has increased dramatically over the last ten years. During 1989, over 6 million hectares of forest land were consumed by fire. On the basis of the data from previous years the ten-year average has been about 2 million hectares, twice that being burned two decades ago. In addition to the above-ground biomass, a massive expanse of peatlands and permafrost in northern latitudes contains large reserves of carbon. According to the latest estimates, the boreal and arctic peatlands contain about 200 billion tonnes of carbon (Sjors 1981, Ahlholm and Silvold 1990). The warming of these areas will lead to unpredictable emissions of carbon in the form of methane and carbon dioxide.

4.2.1.4 *Possibilities to Refix Carbon Dioxide: A Case Study*

A case study of Finland prepared for the IPCC shows the effects of forest management strategies on the carbon storage of existing forests. It shows that in intensively managed commercial forests it is possible to affect considerably the carbon storage in forests through forest management strategies.

In the case study the time horizon is fifty years, from 1989 to 2039. It is assumed that sequestration of fossil fuel emissions during this period could help offset fossil fuels used during a transition to alternative sources of fuel. It is assumed in all the calculated alternatives that forests will continue to be utilized economically. The production of stemwood during the simulation period was not less than that presently recommended by forest management policies. The case study used the forest simulation model

YSI, developed by Dr. Timo Pukkala, University of Joensuu, Finland. A forest management planning model based on the YSI simulation model was developed and it required extensive physical and biological data from the forest. The basic data for the calculations were taken from Finland's national forest inventory (1977–84).

In Finland's forests, the carbon content of the stemwood is about 320 million tonnes with a carbon content of the total above-ground tree biomass of about 590 million tonnes. Under a no-change scenario, about 270 million tonnes more carbon could be stored in above-ground biomass by selecting the appropriate forest management method supported by the model.

This increment equals the carbon emissions from the use of fuel oils in Finland during the next thirty-nine years, if the annual C-emissions remain at the 1989 level. If the climate warms up by 2 degrees C according to the example and assuming that the trees are capable of adapting to the new equilibrium, 470 million tonnes more of carbon can be stored in forests by changing management methods. This would equal the total use of fuel oils in Finland during the next 67 years. There are alternative strategies for forest management, even in already intensively managed forests, which can considerably increase the mean volume and the carbon storage of existing forests with little change in economic incomes or cutting amount. The net incomes in the calculation example will decrease due to delayed harvestings by less than 20 percent when discounted at 4 percent interest (costs and incomes are calculated with present prices). Beneficial strategies for carbon management could also be found using shorter rotations and increased growth. This case study demonstrates that it is possible to increase carbon storage by changing forest management strategies. The study suggests that forest management strategies are important response strategies to cope with the predicted climate change. It should be pointed out that this study offers the best-case scenario.

4.2.1.5 *Measures and Policy Options*

Forests may be considered as major indicators of the environmental and economic health of boreal forest countries.

4.2.1.5.1 Forest Protection

Protection activities related to fire, insects, and diseases in the boreal forests will have to be intensified to compensate for the changes in species composition due to northward migration of insects, and diseases need to be investigated to ensure timely action in maintaining forest health.

4.2.1.5.2 Forest Management

Results obtained from genetic trials in different countries have indicated that northern seed lots transferred southward have survived fairly well; but poorer than local trees in warmer climate conditions. This could be directly related to adaptation to longer days in more northern regions. There is considerable debate on whether forest land should be regenerated artificially or naturally. The approaches to this issue depend on whether one deals with managed or unmanaged forests and is related to economic questions. Recently, in Nordic countries, where artificial regeneration has been widely used, there have been extensive planting stock losses. Given these losses, artificial regeneration must be improved to ensure that native species can survive in a rapidly changing climate. While our present silvicultural practices have favored monocultures, growing conditions could change and mixed stands may be favored over the monocultures. It has been shown by Mielikainen (1985) that mixed stands could be economically valuable in Finland.

These areas will also serve as carbon reservoirs along with other large-scale reforestation projects. It should be noted that the market value of the forests varies from country to country and changes according to the end use of forest products. Within the boreal region lie extensive areas of peatlands. In Finland, 45 percent of the total forest land is wetland, of which about 50 percent has been drained. In addition, most boreal regions have discontinuous or continuous areas of permafrost and constitute a large source of carbon. In these areas, management practices should be implemented that restrict the release of carbon, especially methane, which is 16 times more effective as a greenhouse gas than carbon dioxide. It should be noted that forestry practices change the relation of anaerobic to aerobic conditions in soils that regulate the proportions of methane and carbon dioxide emissions.

4.2.1.5.3 End Uses and Biomass Conversion

During the ten-year period 1976–86, the demand for raw forest materials has increased globally, especially in North America. To meet the continuous demand for these products, the needs of the pulp and paper industry will have to be reflected in future forest resource management scenarios. Some of that demand will be filled through recycling of forest products. In the boreal forest the use of a sustainable managed source of biomass, such as an intensive short rotation fuelwood plantation, biomass energy can produce net reductions in CO_2. In addition, using logging residues (bark, timber residues) and black liquor from forest industrial activities should be considered in energy production.

4.2.2 SPECIAL ISSUES ON TEMPERATE FORESTS

This section deals with temperate forests located in Europe (including the Mediterranean area and excluding Northern Scandinavia), the Middle and Near East, Central and East Asia, Australasia, most of North America, Chile, and Argentina. The total area of temperate zone closed forests is estimated to be about 600 million ha or 17 percent of the Earth's total closed forest area (3,600 million ha). This section focuses on how and to what extent temperate forest destruction contributes to the man-made increase in the greenhouse effect and how climate change may influence such forest ecosystems, and considers measures to protect the forests and to increase their capacity to absorb CO_2 from the atmosphere.

4.2.2.1 Greenhouse Gas Emissions from Temperate Forests

A natural or sustainably managed temperate forest is neither a source nor a sink of CO_2. At present, over large regions the carbon balance is achieved by forest growth in one area being offset by loss of biomass in another (e.g., by harvesting, wildfires, severe storms, etc.). Thus, as long as their total area is relatively stable, their contribution to emissions is negligible.

4.2.2.2 *Global Warming: Impacts and Effects on Temperate Forests*

Climate changes in the post-glacial period have significantly altered the composition of forest communities in temperate zones. Tree species and forest types were displaced or changed composition. Forest communities have been able to adapt within certain limits and rates of climate changes, with some communities adapting intact and others undergoing changes in species composition. However, the rate of climate change now predicted may exceed these limits by far.

In temperate zones, human activities have had multiple impact on the natural range and composition of forests. During the course of human history, cumulative forest losses amount to about 2 billion ha mostly in temperate zones (equivalent to the present total tropical forest area). Some regions have become poor in forest resources. The remaining forests are thus all the more important, as they continue to fulfill multiple functions (e.g., watershed protection, habitats, timber production, etc.) and maintain biogeochemical cycles. The expected climate change is likely to have adverse effects on temperate forests and these will be augmented by additional risk factors such as air pollution and biotic and abiotic stresses.

The new type of forest damage occurring chiefly in the industrialized countries of the temperate zones also contributes to the man-made increase in the greenhouse effect to an extent not yet quantified, through premature dying of forest stands and a reduction of humus and biomass of surviving forest stands. Irrespective of the man-made increase in the greenhouse effect, measures must be undertaken or strengthened to reduce air pollution and its effects on forests. The aim of adjustment strategies is to secure sustained and comprehensive functioning of the forest with simultaneous maximum CO_2 fixation capacity of forest ecosystems. Forestry measures, such as selection of site-adequate tree species, use of suitable provenances, achievement of stable, diverse mixed stands, and preservation of genetic diversity, can counteract crisis-type disturbances in development to a limited extent only and over long periods. But nevertheless, they have to be introduced and/or strengthened now to be effective in time.

Because temperate forest stands tend to be intensively managed, and surplus agricultural land may be available, there may be promising opportunities to mitigate the greenhouse effect by expanding forest biomass in temperate zones.

4.2.2.3 *Costs of Forestry Countermeasures*

Some forestry adjustment strategies to changed climatic conditions can be carried out without large additional costs by adjusting silvicultural measures that have to be carried out in any case. The situation is different, however, if climate changes result in severely reducing the stability of stands and if selected measures for the stabilization of the stand have to be taken. Such costs are estimated at approximately U.S.\$5 per hectare per annum. Furthermore, there might be a drastic increase in these costs if climate changes require a conversion to other tree species. As the degree of the damage strongly differs from region to region and as this is also true for the type of required forestry adjustment strategies, average costs for such adjustment measures cannot be reasonably assessed.

Determination of costs of reforestation and forest management in temperate zones is quite difficult for many reasons, including the diversity of situations found in countries. For example, besides the objective of carbon fixation, in many countries afforestation has to fulfill a productive, protective, and recreational function, all at the same time. The specific objectives and conditions for the respective measure strongly influence costs. Afforestation costs vary considerably, ranging from approximately U.S.\$200 up to \$2,000 per hectare, excluding possible opportunity costs of eliminating the present use.

Stand volume in forests with an open stocking can be increased by underplanting measures; the costs for such measures amount to approximately 50 to 75 percent of regular plantation costs. In forests with an already relatively closed stocking, the stand volume can be increased by an extended rotation period. Increased expenditure for the required stand tending can at a later stage be offset by a considerably higher revenue on account of an improvement in the structure of the assortment and the value. Short rotation wood crops may also provide high yields and give more flexibility in times of climate change.

4.2.2.4   Constraints on Forestry Measures

In the temperate zones there are only a few physical constraints on afforestation. They may result from continued pollution and, on some sites, also from the expected climate changes. The essential constraints—above all, in densely populated industrialized nations with intensive land use—are of a social nature, and opportunity costs for land use are high. Competing demands on land use by nature conservation, recreation, settlement, and traffic can make a transition to forestry use or afforestations considerably more difficult and expensive. As noted earlier, surpluses of agricultural products have also led to proposals to transfer significant areas of agricultural land to other uses, including forestry.

Overall, however, many questions remain to be answered before the potential contribution of forestry response strategies in temperate zones can be better quantified.

4.2.3   SPECIAL ISSUES ON TROPICAL FORESTS

4.2.3.1   Introduction to Tropical Deforestation and Climatic Concerns

The development and ecological stability of countries with tropical forests is linked to the health of those forests. They contain half of the world's species of fauna and flora and provide raw materials, sustain rainfall on downwind agricultural land, maintain healthy fisheries and water supplies, and prevent soil erosion, dam siltation, and flash flooding. Global, regional, and local climate are linked to the health of tropical forests. Deforestation may be responsible for one-quarter to one-fifth of global anthropogenic carbon dioxide emissions, one to four-tenths of all (natural and anthropogenic) methane releases, as well as contribute to concentrations of nitrous oxides, ozone, carbon monoxide, and other gases implicated in global warming. Regionally, deforestation may interrupt moisture and latent heat transferred from the tropics to higher latitudes, influencing the climate of the temperate zone. Locally, partial or total removal of forest cover can cause drying of the microclimate, leading to an increase in natural fires that prevent natural forest regeneration of forests.

4.2.3.2   Forest Carbon Pools and Forest Cover Statistics

Tropical open forests (relatively dry scrublands and woodlands) and closed forests (90 percent of which are moist) contain about half of the terrestrial carbon pool. The closed forests represent the greatest store of carbon. In 1980 closed tropical moist and dry forests occupied an area of about 1.2 billion hectares. The distribution of this was 679 million, 217 million, and 306 million ha in Latin America, Asia, and Africa, respectively. A small number of countries, including Brazil, Burma, Colombia, Gabon, Guyana, Indonesia, Papua New Guinea, Venezuela, and Zaire, contain 80 percent of closed moist forest. Open forests occupied about 1.4 billion hectares. The distribution of this was 929 million, 363 million, and 67 million ha in Africa, Latin America, and Asia, respectively. Secondary forests (forest regrown after clearing), fallow forest (woody vegetation derived from forest clearing for shifting agriculture), and plantations do not store as much carbon as primary forest. In 1980 regenerating forest fallow occupied about 408 million ha, with 169 million, 166 million, and 73 million ha of this distributed in Latin America, Africa, and Asia, respectively. Plantations occupied 11.5 million ha, the distribution of which was about 5.1 million, 4.6 million, and 1.8 million ha in Asia, Latin America, and Africa, respectively (FAO, 1982).

4.2.3.3   Estimates of Current Rates of Forest Loss

Estimates for the period 1980–85 indicate that about 11.3 million ha of tropical forests were annually cleared and converted to other land uses (Table 4.2). Approximately 7.5 million ha of the annual clearing was in closed forest, with 57 percent, 25 percent, and 18 percent of this clearing occurring in Latin America, Asia, and Africa, respectively. About 3.8 million ha of the clearing was in open-canopy formations, with 62 percent, 33 percent, and 5 percent of this clearing occurring in Africa, Latin America, and Asia, respectively (FAO, 1988). Logging (mostly selectively in closed-moist forests) disturbed an additional 5 million ha per year in the same period (FAO, 1986). In 1980 the forestation (both reforestation and afforestation) rate was about one tenth of the deforestation rate (about 1.0

million ha/yr). While forestation has nearly doubled, the ratio to deforestation has remained at about one tenth the current rate of deforestation.

The current rates of forest loss, natural regeneration, reforestation and afforestation and forestation in 1990 are subject to debate. One recent, but unconfirmed study by Myers (1989) indicates that in 1989 the rate of clearing of closed tropical moist forests had doubled to 14 million ha/year. The FAO is developing a new survey of forest cover to be released in 1991 and by 1992 satellite data assessing the status of the world's tropical forests should be available from a joint space agency programme led by Brazil.

Table 4.3 lists the closed tropical moist forest deforestation rates for the top ten countries in the late 1970s based on FAO data and in 1989 based on Myers. One recent study (Myers, 1989) undergoing further review suggests that there is a trend of increasing deforestation rates, although differences in methodologies between the FAO and Myers studies may also account for the increase. The change in countries on the list and their relative positions may indicate the effect of economic, social, and political factors on rates of forest loss.

4.2.3.4 *Patterns and Causes of Deforestation*

The three main determinants of deforestation are: land conversion for agriculture and pasture; wood removals for fuelwood and inappropriate timber utilization; and public and private development projects (e.g., logging, mining, roads and dams— see Table 4.4). In the semi-arid tropics forest loss and land use conflicts are associated with natural and man-induced desertification. The extent of desertification in Africa, Asia (excluding USSR) and Latin America was about 1,536 million ha in the early 1980s (WRI et al., 1988).

Deforestation in the humid tropics is mainly due to clearance for agriculture, while in mountain and dry regions a substantial portion of the deforestation may be caused by the need to fell trees for fuelwood and fodder.

Land Conversion for Agriculture and Pasture

The clearing of tropical forests for cropland and pastures is the greatest cause of deforestation. An estimated 6.8 million ha of fallow and 3.9 million ha

TABLE 4.2: Estimates of Rates of Tropical Deforestation for 1980–85 and 1989 (Square Kilometers*/Year)

	FAO 1980–85 (%)	MYERS 1989 (%)
Tropical closed forest		
America	4.339 (57)	7.68 (55)
Africa	1.331 (18)	1.57 (11)
Asia	1.826 (25)	4.43 (32)
Total	7.496	13.68
Tropical open forest		
America	1.272 (33)	
Africa	2.345 (62)	
Asia	0.190 (5)	
Total	3.807	

* Square Kilometer = 10⁶ ha

TABLE 4.3: Top Ten Countries in Terms of Closed Moist Forest Deforestation Rates (Square Kilometers/Year)

LATE 1970s (FAO, 1981)	(KM²/YR)	1989 (MYERS, 1989)	(KM²/YR)
Brazil	14,800	Brazil	50,000
Indonesia	6,000	Indonesia	12,000
Mexico	5,950	Burma	8,000
Colombia	5,100	Mexico	7,000
Philippines	2,900	Colombia	6,500
Côte d'Ivoire	2,900	Thailand	6,000
Nigeria	2,850	Malaysia	4,800
Peru	2,700	Zaire	4,000
Thailand	2,450	Nigeria	4,000
Malaysia	2,550	India	4,000

TABLE 4.4: Percent of Conversions of Tropical Forests Due to Different Land Uses (1980–1985)

	AMERICA	AFRICA	ASIA
Cropland	20	16	41
Pastures	17	−19	−1
Shifting Cultivation	25	46	49
Degraded Lands	38	57	11

Source: FAO (1987), FAO (1982).

Note: Mixed forest-pasture in Africa and Asia is being converted into more intensive land uses, usually agriculture, thus resulting in negative numbers in the table.

of closed forest may be cleared annually for agriculture; 1.3 million ha of fallow and 0.7 million ha of closed forest may be cleared for pasture annually (Houghton et al., 1985). Figures for open forests are not available. The increase in croplands and pastures may have accounted for 40 percent of the primary tropical forest conversion between 1980 and 1985 (FAO, 1987). According to 1980 figures of the U.S. Interagency Task Force on Tropical Forests, 100 million ha of land, much in the forest, could be brought into cultivation by the year 2000. The rate of expansion of croplands and pastures does not explain the rapid rate of loss of forests. Over-cropping and over-grazing may be permanently degrading productive lands, so that they must be permanently abandoned and replaced with new land cleared from primary forest to maintain a given level of agricultural activity (FAO, 1987) (Table 4.4).

Traditional, sustainable long-rotation shifting cultivation clears only fallow, so it is in balance with respect to carbon. It has largely been replaced by short-rotation shifting cultivation that degrades soil and other forms of unsustainable agriculture, which are expanding into carbon-rich primary forest, producing a net source of carbon, in response to growing populations.

Wood Removals for Fuelwood and Commercial Logging

Wood removed from tropical forests is used for two main purposes: 83 percent is mainly for fuelwood and charcoal, and 17 percent is for industrial roundwood (13 percent for local use and 4 percent for export); however, commercial logging practices damage a significant percentage of the trees left standing, resulting in total forest damage far greater than the removal rates entailed in selective logging (EC, 1989).

FUELWOOD

Studies have led to two contradictory conclusions about whether fuelwood collection causes deforestation, contributing to the greenhouse effect. One recent study suggests that fuelwood harvesting accounts for little permanent loss of tree stock, with most fuelwood coming either from sustainable fuelwood production or from by-products of natural death or forest clearing. Considering the net balance

of greenhouse gas production from wood burning, using more energy-efficient wood-burning cooking stoves to reduce fuelwood demand may actually increase emissions of greenhouse gases overall, since these stoves typically raise heat transfer efficiency at the expense of reducing combustion efficiency. As a result, emissions of carbon monoxide and more potent greenhouse gases such as methane and nitrous oxide may increase relative to the production of carbon dioxide, which is a less potent greenhouse gas (Leach, 1990).

By other estimates, demand for wood outstrips supply, especially in drier and mountain regions. About 0.5 million ha of closed forest and 2.0 million ha of fallow forest may be degraded or cleared for fuelwood annually (Houghton et al., 1985).

Whether or not fuelwood collection is presently contributing to a net decline in forest cover, it could potentially become a more significant source of emissions in the future if village wood lot establishment does not keep pace with the anticipated demand. In 1980 planting for fuelwood averaged 550,000 ha per year, a fifth of the 2.7 million ha that was needed to satisfy demand. To meet the projected fuelwood demand of 55 million ha in 2000, 2.7 million ha would need to be planted per year, given 1980 as a base year (Brown et al., 1988, citing John Spears of the World Bank).

LOGGING

Tropical forests and plantations supply about 10 percent of the world's demand for timber and pulp. Logging for timber occurs mainly in closed forests of the humid tropics. By the early 1980s about 13.25 percent of humid tropical forests had been logged. Commercially valuable species account for less than 10 percent by volume of tropical forests, so logging in the tropics is mostly selective over extensive areas. Only 2–10 commercially valuable trees are removed per hectare, but as many as 30–70 percent of the remaining trees are left damaged. If forests are allowed to regenerate, the net flux of carbon from selective logging may be near zero. Destructive logging practices, however, may damage forest productivity and slow regeneration. If logged areas are subsequently colonized and used as croplands, as happens in many African and Asian countries, the net effect is a loss of carbon from the land. Logging is expected to increase with growing timber demand in the future.

Public and Private Development Projects

Although it is recognized that building roads is necessary to develop remote regions, the associated timber harvesting, in many cases, may damage the forest by opening it to penetration by farmers who burn the forest for agriculture. Large-scale development projects in the forest, such as flooding for hydropower, mining, and use of forest for charcoal production to fuel industrial processes such as steel smelting, may also cause large amounts of forest loss. No estimates of forest lost from large-scale projects are available.

4.2.3.5 *Estimates of Current Emissions from Forest Land Clearing*

Estimates of the quantity of carbon dioxide, methane, nitrous oxide, carbon monoxide, ozone, and other gases released when forests are cleared are imprecise due to uncertainty about emission factors, the rate of deforestation, and the amount of carbon stored in vegetation and soils, and the effect of soil disturbance on the flux of methane and oxides of nitrogen. The data are, however, adequate to estimate the range of emissions.

CARBON

Generally used estimates of the flux of carbon from tropical deforestation range from 0.4 to 2.6 billion tonnes of carbon (BTC) per year in 1980 (Houghton et al., 1987). In 1989 emissions from tropical deforestation may have been 2.0 BTC to 2.8 BTC according to certain preliminary estimates by Myers (1989). The scientists who addressed the São Paulo IPCC/RSWG Workshop in January 1990 were reasonably certain that in 1980 emissions were between 1.0 and 2.0 billion tonnes of carbon and that in 1989 emissions were between 2.0 and 2.8 billion tonnes of carbon. This range of estimates of 0.4 to 2.8 BTC is 6–33 percent of total anthropogenic emissions and represents 3–16 percent of the contribution to total greenhouse gas emissions in the 1980s (Houghton, 1989). Soil carbon losses may cause up to one third of carbon emissions associated with deforestation (Marland, 1988). Over half of 1980 emissions from deforestation were produced by six countries: Brazil, Indonesia, Colombia, Côte d'Ivoire, Thailand, and Laos (Figure 4.1). Those

emissions totaled about 1.2 BTC in 1980. The actual emissions listed are subject to debate due to differences in assumptions about forest biomass and deforestation rates.

METHANE

Methane emissions from total tropical biomass burning (i.e., forests, savanna, and shrub-fallow) may range from 40 to 75 million tonnes of carbon per year (Houghton, 1989, based on Crutzen). Tropical biomass burning may contribute 8–11 percent of the annual global methane flux. This may represent about 1–2 percent of the contribution to total greenhouse gas emissions in the 1980s.

FIGURE 4.1: **Estimated Net Release of Carbon from Tropical Deforestation in 1980 (Teragrams Carbon)**

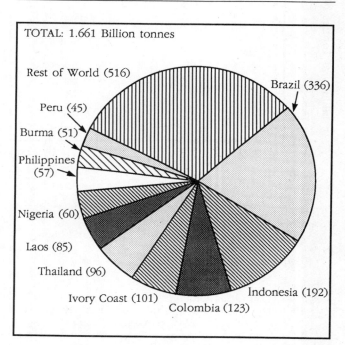

Note: The deforestation rates were derived from three estimates: FAO/UNEP (1981), Myers (1980, 1984), FAO (1983). The forest biomass was derived from two estimates: Brown and Lugo (1984), which is based on FAO/UNEP (1981), and Brown and Lugo (1982). The midpoint of total emissions was taken and the point estimate of country contribution to that total was determined based on FAO/UNEP (1981) rates and wood volumes.

Source: USEPA, 1989, based on Houghton et al., 1987.

FIGURE 4.2: **Deforestation of Tropical Forests, 1980–2050: Two Base Case Projections of Increasing Carbon Emissions and Two Policy Option Cases for Decreasing Carbon Emissions**

Long-term projections indicate that if current trends continue, as many as 1.5 to 2 billion ha of accessible tropical forests could be deforested in 50 to 100 years, releasing 120 to 335 BTC, depending on whether high or low biomass estimates are taken (Houghton, 1989).

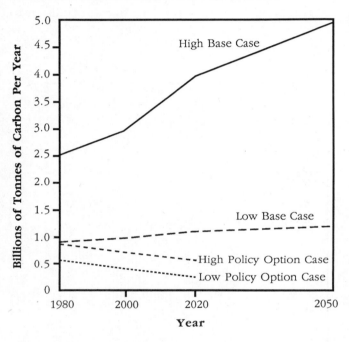

Base Case Projections (Houghton, 1990b)
  High: Population based deforestation rate and high forest biomass estimates
  Low: Linear deforestation rate and low forest biomass estimates

Policy Option Cases (Grainger, 1989a)
  High: Low increase in agriculture productivity relative to change in per capita consumption
  Low: High increase in agriculture productivity relative to changes in per capita consumption

Note: See Appendix 4.1 for explanation.

NITROUS OXIDE

Nitrous oxide is naturally released from undisturbed soils although emission rates are uncertain. Higher emission rates are associated with fertilized, cleared, and pasture soils. Nitrous oxide is also produced during uncontrolled biomass burning.

CARBON MONOXIDE AND OZONE

About 5–10 percent of the carbon emitted in biomass burning is in the form of carbon monoxide. Although not a greenhouse gas, it indirectly increases the concentration of methane. The oxidation of carbon monoxide also produces ozone, a greenhouse gas in the lower atmosphere (Houghton, 1989). Burning to clear forest in the Amazon during the dry season almost triples the levels of carbon monoxide and ozone in the atmosphere (Kirchoff et al., 1990).

4.2.3.6 *Estimates of Future Forest Loss and Emissions*

Future greenhouse gas emissions are difficult to predict because of uncertainties about forest biomass, as well as population and economic growth rates. Two base case emissions projections by Houghton (1990b) suggest that emissions could range between 1.1 and 3.9 BTC in 2020 and 1.2 to 4.9 BTC in 2050. The low-bound estimates are from the low base case, which assumes low biomass and linear deforestation rates. The high-bound estimates are from the high base case, which assumes high biomass and population based deforestation rates (Figure 4.2 and Appendix 4.1). The country-by-country emissions for these two base cases for 1980 through 2050 are given in Appendix 4.2. In the slowly changing world scenario (i.e., continuing poverty, unsustainable agriculture, and population growth) the deforestation rate increases with population to 34 Mha/yr by 2050, releasing 2 BTC per year. As a result, Asia's unprotected forests are exhausted. After this date, emissions decline, and then fall sharply around 2075 with the loss of remaining African and then Latin American forests.

In a rapidly changing world (i.e., increased agricultural productivity, etc.) deforestation rates might stabilize at 15 million ha per year, releasing 1 BTC per year from 2000 to 2100, when emissions

FIGURE 4.3: CO₂ Emissions from Deforestation

Figure 4.3 illustrates three long-term emission scenarios developed by USEPA (1989) based on low forest biomass estimates. One scenario is based on a slowly changing world, one is based on a rapidly changing world and one involves a massive forestry response strategy to combat the buildup of carbon dioxide in the atmosphere.

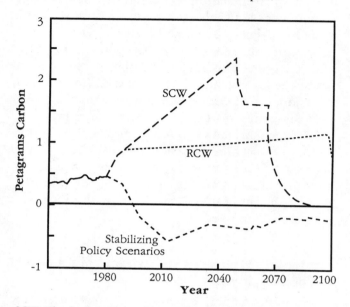

SCW: Slowly changing world scenario, i.e., continuing poverty, unsustainable agriculture, and population growth.
RCW: Rapidly changing world, i.e., increased agricultural productivity, etc.
1 Petagram = 1 billion tonne

Sources: Lashof and Tirpak, 1989.

would fall as all remaining forests are exhausted.

Under an ambitious stabilizing policy stopping deforestation by 2025, and planting 1000 Mha of trees by 2100 in tropical and temperate zones, the biosphere would become a carbon sink by 2000, with absorption of carbon from the atmosphere peaking at 0.7 BTC per year in 2025.

Under policy option cases modeled by Grainger, decreases in the rate of deforestation would be possible. The high policy option case and low policy option case assume declining deforestation based on differing levels of improvements in agricultural productivity in relation to changes in food consumption per capita. Under the low and high policy case

options, carbon emissions might be reduced by at least 40–70 percent, from between 0.5 (low case) to 0.8 (high case) BTC in 1980 to between 0.2 (low case) to 0.5 (high case) BTC per year in 2020 (Grainger, 1989a) (Figure 4.2; See Appendix 4.1 for assumptions).

4.2.3.7 *Strategies to Reduce Emissions: Types of Response Options*

Climate change offers an unusual challenge to forest planning and policy development. It is very important that countries begin the task of understanding the full social and economic consequences of continued deforestation, and the need to promptly examine options to slow forest loss. Quantitative and qualitative analyses of the costs and benefits of current forest policies, and their consequences for greenhouse gas emissions, need to be undertaken in the next few years. Policy options to control greenhouse gas emissions through forestry practices fall into three major categories:

1) Options to reduce forest sector sources of greenhouse gases:
   - reduce forest clearing for shifting agriculture by substituting sustainable, intensified, sedentary techniques, including agroforestry;
   - reduce frequency and amount of forest and savanna consumed in biomass burning to create and maintain pasture/grasslands;
   - reduce forest loss due to public development projects, through environmental planning and management;
   - improve efficiency of cookstove and industrial biomass use;
   - reduce damage to standing trees and soils during timber harvest; and
   - reduce soil carbon loss, via soil conservation farming practices and other management techniques.

2) Maintain existing sinks of carbon in forest systems:
   - conserve standing forest as stocks of carbon through establishment of protected areas and sustainable management;
   - introduce sustainable harvesting methods to reduce tree damage; and
   - establish sustainable extractive reserves and natural forests.

3) Expand carbon sinks through sustainable forest management:
   • improve productivity of existing forest lands;
   • establish plantations on available pasture/savanna and cropland, marginal land, degraded land;
   • expedite natural regeneration of deforested land; and
   • increase soil carbon storage through soils management.

Analyses of the benefits of potential response options under these three general strategies are summarized below.

Strategy 1: Options to Reduce Forest Sector Sources of Greenhouse Gases

Reduce forest clearing for shifting and sedentary agriculture by substituting sustainable cropping systems. One important policy option to slow deforestation is to reduce forest clearing for agriculture. One scenario of halting deforestation envisages replacing 80 percent of slash-and-burn with sustainable agriculture (Lashof and Tirpak, 1989). Expansion of the agricultural frontier into tropical forest could be curbed by:

• introducing crop mixes, planting and management systems, and improved genetic strains to increase productivity per unit area faster than increasing food consumption per capita. Inputs of fertilizers, water, and capital at varying levels will be necessary to achieve adequate agricultural intensification and sustainability; however, emphasis should be placed on R&D in "higher knowledge, lower external input" agricultural systems, since high-input systems are expensive and not feasible in many areas. Development of cash crops as well as subsistence food crops should be emphasized, so that farmers may acquire cash needed to invest in inputs;
• focusing agricultural development efforts on sites with adequate soils that are alternatives to tropical forests, such as savanna, pasture, and underutilized croplands;

• intensifying management on existing pasture to increase site productivity, by introducing appropriate technologies such as optimized foraging strategies, fertilization, mechanization, and improved livestock management.

Long-term research in the Peruvian Amazon indicates that for every ha in crop production converted to sustainable practices, 5–10 ha of forest might be saved (Table 4.5). Research on sustainable systems in Peru, Nigeria, Brazil and elsewhere is promising, but needs to assess the potential of large-scale introductions on a variety of soil types, and to assess the net greenhouse gas emission (tradeoffs in emissions among gases) benefits of introducing various competing systems. Wet rice cultivation, while very sustainable in the humid tropics, is also a major source of methane emissions. Cross sectoral analyses are needed to evaluate the new management and biotechnology techniques suggested in the AFOS Agriculture Workshop, before specific practices are recommended.

Strategy 2: Maintain Existing Sinks of Carbon in Forest Systems

Slowing or halting deforestation may offer the greatest net social and ecological benefits, most likely at comparatively low costs per ton of carbon emissions avoided, of the response options reviewed by AFOS (see Figure 4.4). Halting the conversion of tropical forests would immediately reduce CO_2 emissions by perhaps 1–3 BTC per year, depending on actual current emission rates.

TABLE 4.5: Forest Saved by Sustainable Agricultural Techniques in the Peruvian Amazon

SUSTAINABLE AGRICULTURAL PRACTICES	NUMBER OF HA OF FOREST SAVED PER HA IN SUSTAINABLE USE
Flooded Rice	11.0
Legume-Based Pasture	10.5
High Input	8.8
Low Input	4.6
Agroforestry	not determined

Source: Sanchez, 1988.

Response options to protect standing forest include:

1. **CONSERVE STANDING FOREST AS STOCKS OF CARBON, BY EXPANDING PROTECTED FORESTS AND EXTRACTIVE RESERVES.**

The expansion of protected tropical forest has accelerated since 1970 from very low levels. Substantial new areas of existing forest could be designated for protection from encroachment. Expansion of protected areas could also assist developing countries to conserve the tropical biodiversity for long-term socio-economic benefits.

Non-destructive resource use in extractive reserves, where rubber, nuts, and other products are harvested for market without cutting forest, could be greatly expanded, if markets are developed and land is legally protected, following Brazil's progressive experiments in Amazonia. Identification of new protected area should take into account the effects of and changes in forest conservation needs that would accompany climate change, such as the need for species migration corridors.

2. **INTRODUCE SUSTAINABLE HARVESTING AND NATURAL FOREST MANAGEMENT METHODS TO REDUCE TREE DAMAGE.**

Widespread introduction of sustainable timber and forest product extraction techniques from the primary forest that damage fewer trees left standing than current methods would help assure long-term carbon storage. Improved forest management could increase forest productivity. Approximately 137 million ha of logged forest could benefit from enrichment planting and regeneration because current selective logging techniques reduce long-term productivity by damaging 30–70 percent of the species left behind (Grainger, 1989b).

Strategy 3: Expand Carbon Sinks Through Sustainable Forest Management and Intensified Forest Management

Options under review by analysts include the following:

1. **ESTABLISH PLANTATIONS ON AVAILABLE PASTURE/SAVANNA AND CROPLAND, MARGINAL LAND, DEGRADED LAND.**

The increasing demand for wood products and biofuels provides an opportunity to promote the estab-

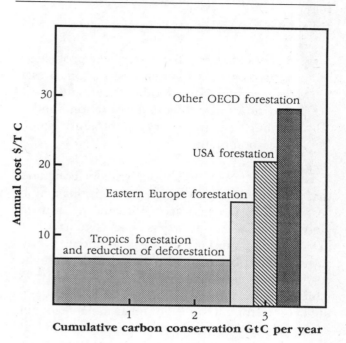

FIGURE 4.4: **Regional Average Costs of Forest Management Measures***

* The figure is taken from: McKinsey & Company: *Protecting the Global Environment: Funding Mechanisms (Appendices)*, which was prepared for The Noordwijk Conference, The Netherlands, November 1989.

Source: FAO, WRI, McKinsey analysis.

lishment of plantations on degraded lands to play a role in absorbing carbon dioxide from the atmosphere.

There are two estimates of land available for forestation. According to Grainger (1989b), a total of 621 million ha of land may be available for forestation in the tropics. Of this total, 418 million ha are in the dry and montane regions and 203 million ha are forest fallow in the humid areas. According to Houghton (1990a), up to 865 million ha of land in the tropics are available for forestation. Of this total, there may be about 500 million ha of abandoned lands that previously supported forest in Latin America (100 million ha), Asia (100 million ha), and Africa (300 million ha). The additional 365 million ha could be available only if increases in agricultural productivity allowed this land to be removed from production. If this upper estimate of 865 million ha of potentially available land were reforested, a total of 150 million BTC could be removed from the atmosphere after forest maturation (Houghton, 1989).

Two types of forest could be created:

- protection forests (for watershed protection/ erosion control; ecosystem restoration; preservation; carbon stock; biodiversity), and
- production forests (biofuel plantations; industrial timber; agroforestry; community woodlots; carbon pump).

The São Paulo workshop members concluded that local fuelwood and shelterbelt options may be cost effective, but may lack potential for a major global impact. Agroforestry is believed to have both low costs and substantial global potential. Biofuel and industrial options are believed to have higher costs, but large global potential. Unfortunately, establishing plantations in the next few decades is unlikely to stop the spread of unsustainable logging in natural forest, since the growth of mature species of wood quality comparable to natural forests may take 30 to 70 years.

2. AFFORESTATION SCENARIOS
A wide variety of afforestation options could be pursued on a global basis. For example, three scenarios of replanting evaluated by Grainger (1989b) include planting 6 million ha/year over 10 years, 8 million ha/year over 20 years, or 10 million ha/year over 30 years to offset 5 percent, 13 percent, or 26 percent of the 5.5 BTC currently released from burning fossil fuels annually (Table 4.6). The estimated costs total $2.4, $3.0 or $4.0 billion per year, respectively (Table 4.7). Options trading-off time and money to reach specific planting targets are presented in Table 4.7. The shorter the time-frame to reach a given level of planting and carbon uptake, the higher the annual cost of planting. The 13 percent reduction scenario achieved over 20 years would lead to an additional annual harvest of 2.6 billion cubic meters, which would allow global forest production to meet the anticipated demand for wood in 2030 (Grainger, 1989b).

National and global scale forestation scenario building to offset climate change is currently limited by incomplete and not widely representative tree growth rate, forest standing biomass, and establish-

TABLE 4.6: Afforestation Strategies to Achieve a Range of Reductions in Annual Fossil Fuel CO_2 Emissions of 5.5 Billion Tonnes of Carbon

% CO_2 REDUCTION	TOTAL AREA PLANTED (HA 10^6)	PLANTING RATES/YR (HA 10^6)			
		10	20	30	50
26	300	30	15	10	6
13	150	15	8	5	3
5	60	6	3	2	1

Note: Assumes average growth rate of 15 cubic meters/yr/ha or 3–4 tonnes carbon
Source: Grainger, 1989b.

ment cost data. Present estimates may be high or low by a factor of 3–10, especially if economies of scale are realized. For example, recent cost estimates in the United States for a 20-million-ha federal planting programme have declined by about 40 percent during programme analysis and design, as better data became available. In the United States, estimated tree establishment costs for a 5 percent CO_2 offset programme involving about 12 million ha have now come down to around U.S.$320/ha. Such a programme could capture carbon at about $9 per ton carbon (most likely a very low cost per ton, compared with other options available), according to the U.S. Forest Service.

TABLE 4.7: Costs of Planting Scenarios to Achieve a Range of Reductions in Annual Carbon Dioxide Emissions from Fossil Fuels

% CO_2 REDUCTION	TOTAL AREA PLANTED (HA 10^6)	TOTAL COST	COSTS (US$ BILLION/YEAR) OVER PLANTING PERIOD			
			10	20	30	50
26	300	120	12.0	6.0	4.0	2.4
13	150	60	6.0	3.0	2.0	1.2
5	60	24	2.4	1.2	0.8	0.5

Note: Costs of establishing a forest plantation vary from U.S.$230 to $1,000 per ha, with an average of $400 per ha. Harvest cost per ha may be U.S.$6,750 (Sedjo and Solomon, 1988), but probably would be offset by revenues. Maintenance costs would be extra, and land rental or purchase costs relatively low.
Source: Grainger, 1989b.

3. IMPROVE PRODUCTIVITY OF EXISTING FOREST LANDS

If more species and size classes of trees were used for timber, then efficient harvest and management would be more feasible, and less extensive areas would be logged. Development of technologies and marketing methods to promote greater use of lesser-known species, using harvests based on detailed forest inventories and management plans, would improve yields per unit area.

4.2.3.8 *Policy Options*

National Level Policy Options Available to Improve Forest Management in Response to Climate Change

Because of the important role of the forestry sector in ecological cycles and economic development, there is a need to develop recommendations at both national and international levels. Although uncertainties about the rate and extent of climate change remain, the following are a number of steps that nations could take:

1) In the context of the review of the Tropical Forestry Action Plan currently under way, adopt clear objectives for the conservation, and/or sustainable development of tropical forests in national development plans, including forests associated with agricultural lands, commercial-use forests, and extractive reserves and other protected areas; and reconcile conflicts between current forest management and programmes proposed and initiated under TFAP.

2) Strengthen tropical forest management use of near real-time remote sensing analysis of biomass burning and of forest loss patterns. Coordinate national and multilateral space and development agency imagery programme data collection and analysis with national forestry regulatory and enforcement programmes.

3) Estimate current and future emissions of greenhouse gases from the tropical forest sector, and potential reductions of emissions from various offset planting and management scenarios.

4) Produce and implement national forestation plans for degraded tropical forest lands, based

on cost, feasibility, and true net social benefits analyses.

International Response Options Available to Slow Forest Loss and Stimulate Forestation

A number of response options are available that were specifically identified at the Brazil workshop. These include:

1. DEVELOPMENT OF A WORLD FOREST CONSERVATION PROTOCOL IN THE CONTEXT OF A CLIMATE CONVENTION PROCESS THAT ALSO ADDRESSES ENERGY SUPPLY AND USE.

The AFOS São Paulo Workshop final statement recommended, after discussion and agreement in plenary, that the IPCC support the development of a global forestry protocol to apply to tropical forests as well as temperate and boreal forests in all countries, in the context of a climate convention process that also addressed energy supply and use. The statement agreed upon at the São Paulo workshop for a workshop recommendation calling for development of a World Forest Conservation Protocol was:

> Consideration of forestry issues, and of tropical forestry issues in particular, must not distract attention from the central issue of global climate change and the emission of greenhouse gases attributable to the burning of fossil fuels by developed countries. No agreement on forests and global climate change will be reached without commitments by developed countries on greenhouse gas emissions. The groups recognized that the conservation of tropical forests is of crucial importance for global climatic stability (particularly having regard to the important contribution of tropical forest destruction to global warming, through emissions of carbon dioxide, methane, and other trace gases), but of more crucial importance for national economic and social development, for the conservation of biodiversity and for local and regional climatic and environmental reasons. The Workshop recommended that the IPCC support the development of a forestry protocol in the context of a climate convention process that also addressed energy supply and use. The Workshop concluded that the specific elements of such a protocol are a matter for international negotiations. These elements may include: fundamental research, tropical forest planning, measures to use, protect, and reforest,

international trade, financial assistance and the advantages and disadvantages of national and international targets.

The objective should be to present more concrete proposals on the occasion of the UN Conference on Environment and Development, to be held in 1992.

In light of the above, it is recommended that a World Forest Conservation Protocol, covering temperate, boreal, and tropical forests, be developed in the context of a climate convention process that also addresses energy supply and use, as noted by the January 1990, IPCC/RSWG Tropical Forest Workshop in São Paulo, Brazil and that in accordance with UNGA Resolution 44/207, operative paragraph 10, a meeting of interested countries from both the developed and developing world and of appropriate international agencies be held to identify possible key elements of such protocols and practical means of implementing them. Such a meeting should also develop a framework and methodology for analyzing the feasibility of the Noordwijk remit, including alternative targets, as well as the full range of costs and benefits.

All countries should make a contribution to the solution of the global warming problem. The São Paulo Workshop stated:

> Although forests can assist in mitigating the effects of atmospheric carbon build-up, the problem is essentially a fossil fuel one and must be addressed as such. In this way, and as a general principle, the final report of the present IPCC Workshop on Tropical Forests, while putting tropical forests in the overall context of global warming, should make it clear that the burden of response options is not to be placed on developing countries and thus should state clearly that all countries should make a contribution to the solution of the global warming problem. The temperate forest dieback (caused by acid rain), as analogous to tropical deforestation (caused by tropical people's attempts to satisfy basic human needs), could be specifically mentioned in such a context.

It should however be noted that recent findings indicate that forests may be adversely affected throughout the world by a variety of causes, only one of which is acid deposition.

The São Paulo Workshop further noted:

> Forests cannot be considered in isolation, and solutions must be based on an integrated approach which links forestry to other policies, such as those concerned with poverty and landlessness. The forest crisis is rooted in the agricultural sector and in people's needs for employment and income. Deforestation will be stopped only when the natural forest is economically more valuable than alternative uses for the same land.

2. SUMMARY OF OTHER OPTIONS IDENTIFIED BY SÃO PAULO WORKSHOP SESSIONS
   a) Increase Support for Tropical Forestry Action Plan (TFAP).
      TFAP offers the framework within which national plans for heightened forest management can be developed. Priorities within the TFAP could be reviewed by member countries and FAO to consider how they can be strengthened to address climate change concerns. Meeting resource needs for TFAP should be a priority.

   b) International Tropical Timber Organization (ITTO).
      Consideration should be given to strengthening the role of ITTO to develop international guidelines to encourage:
      • sustainable forest management techniques, including national legislation requiring management of forest for sustained wood production;
      • an assessment of incentives for sustainable forest management including the feasibility of labeling.

   c) Development Assistance Organizations and Development Banks.
      The development banks—IMF, FAO, UNEP—and other multilateral or bilateral international organizations could help tropical forest countries achieve conservation and sustainable development of forests by:
      • requiring analyses of climate change implications, potential greenhouse gas emissions, and forestry response programmes in their review of project proposals;
      • expanding greatly aid and investment flows to forestry;

- expanding debt relief via renegotiation of debt, and debt for conservation exchanges; and
- linking structural adjustment measures to alleviation of climatic impacts and gas emissions reductions.

d) Role of International Forestry Organizations. International forestry organizations have, as a rule, not yet developed programmes or guidelines to member states on forestry and climate change. Potential institutional initiatives could include:

- adding forestry and climate research and development programmes to the CGIAR system (Consultative Group on International Agricultural Research), at least one on each major tropical continent, to conduct and disseminate research results and beneficial field practice guidelines to foresters remanaging forests for adaptation to and mitigation of climate change;
- urging International Union of Forestry Research Organizations (IUFRO) to develop a coordinated set of research programmes on technical silvicultural, ecological, and management issues pertaining to climate change.

4.3 AGRICULTURE RESPONSE STRATEGIES

Agricultural activities currently produce approximately 14 percent of the greenhouse gases emitted globally. Agriculture is a significant contributor to the increasing atmospheric concentrations of carbon dioxide, methane, and nitrous oxide, and to emissions of nitric oxide and carbon monoxide. In the development of options to reduce greenhouse gas emissions from agricultural activities, it is important to recognize that the economies of many developing countries are strongly dependent on agriculture.

Agricultural systems need to be assessed as whole systems. Their potential as sinks for carbon and nitrogen must be evaluated, as well as their role as sources. Trade-offs of the various gases need to be

evaluated, and the systems' capacities to alter carbon and nitrogen cycles must be examined. In addition, the systems must be evaluated in terms of energy inputs and losses from sectors providing goods and services, production, storage, transportation, processing, and marketing.

4.3.1 SUMMARY OF AGRICULTURAL EMISSIONS OF GREENHOUSE GASES

Agriculture contributes to the emissions of greenhouse gases through the following practices: flooded rice cultivation, nitrogen fertilizer use, ruminant animals, improper soil management, land conversion, and biomass burning. The appendix to this section illustrates the level of emissions and agricultural activities on a country by country basis. Globally, rice cultivation, ruminant animals, and biomass burning are estimated to contribute approximately 15, 9, and 8 percent respectively, of total methane production; methane, in turn, accounts for about 20 percent of current greenhouse gas emissions. The use of nitrogen fertilizers is estimated to account for between 0.2 and 20 percent of the current global source of nitrous oxide, nitrous oxide representing 5 percent of current greenhouse gas emissions. Land clearing and biomass burning contribute between 10 and 30 percent of current greenhouse gases, contributing to increases in emissions of carbon dioxide, methane, nitrogen oxides, and carbon monoxide. Emissions from these sources are expected to increase substantially over the next several decades due to population and economic growth and the associated expansion and intensification of agricultural activities.

- Population is estimated to grow at an average of 1.3 percent per year, with global population reaching 8.2 billion by 2025.
- The land area in cropland or pasture is increasing while global forest area is declining.
- Area under cultivation in developing countries is projected to increase at a rate of 1.2 percent per year through 2025, which would result in a 50 percent increase in current levels by 2025.
- Area devoted to rice cultivation is projected to increase from 148 million hectares in 1984 to 200 million hectares by 2025 globally. Methane

from rice cultivation may increase nearly 35 percent by the year 2025.

- Production of meat and dairy products is projected to increase by over 45 percent between 1990 and 2025 and to result in a similar increase in methane emissions.
- In general, land under cultivation is being more intensively cultivated, and the intensity of nitrogen fertilizer use is increasing. Nitrogen fertilizer use is estimated to increase by a factor of five over 1985 levels by 2025 in developing countries (see Table 4.8). Globally, fertilizer use is projected to nearly double over the same time frame. Nitrous oxide emissions could increase by 70–110 percent, although technological advances in fertilizer formulations may significantly reduce these potential increases.
- Emissions of methane are estimated to increase from 511 million metric tonnes in 1985 to about 730 million metric tonnes in 2025.
- Agricultural sources will be contributors to these increases. In addition, emissions of nitrous oxide may increase from 12.4 million tonnes in 1985 to some 16 million tonnes in 2025. Fertilizer use may play a large role in this increase (Figure 4.5).

TABLE 4.8: Projected Levels of Agricultural Activities in 93 Developing Countries*

	PROJECTED AREA FOR CROPS (MILLION HA)		PROJECTED NITROGEN FERTILIZER USE (MILLION TONNES)	
	1985	2025	1985	2025
Rice	106	152	5.2	22
Wheat	61	106	2.3	15
Maize	59	96	1.1	7
Other Cereals	92	142	0.4	2
Legumes	82	161	0.5	3
All Others	163	288	2.1	9
Total	563	945	11.6	58

* Does not include China
Source: FAO projections given at workshop.

4.3.2 MEASURES AND POLICY OPTIONS

Future agricultural practices and policies can affect the levels of greenhouse gas emissions from agricultural sources and contribute to stabilization of their atmospheric concentrations. However, reducing emissions from one or two sources will not be sufficient. For example, in order to stabilize methane concentrations, reductions in methane emissions from flooded rice fields, livestock systems, and biomass burning as well as from other anthropogenic sources, will be necessary. Similarly, modifications in nitrogen fertilizer use, land use conversion, and crop systems will be necessary parts of multisectoral strategies to achieve the reductions to stabilize atmospheric concentrations of carbon dioxide and nitrous oxide.

The opportunities for reducing gas emissions in the near-term which appear to be economically viable in their own right consist of the following:

- *Biomass burning:* Biomass burning might be reduced through fire control, education and management programmes, as well as the introduction of the use of appropriate alternative agricultural practices. Agricultural systems dependent on the removal of biomass (by burning high-yield grain crops) may be modified to provide opportunities for increasing soil organic matter and reduction of greenhouse gas emissions or removal for use as an alternative fuel source.
- *Livestock systems:* Methane emissions can be reduced through management of livestock wastes; expansion of supplemental feeding practices for livestock; and increased use of production- and growth-enhancing agents, with appropriate safeguards for human health and taking into account legitimate consumer concerns.
- *Fertilizer use:* Nitrous oxide emissions may be reduced by using improved fertilizer formulations and application technologies, and through judicious use of animal manure.
- *Sustainable agricultural practices:* Where appropriate, minimum- or no-till systems are recommended for those countries currently using tillage as part of the annual cropping sequence. These tillage systems may yield additional ben-

FIGURE 4.5 : **Projected Emissions of Greenhouse Gases**

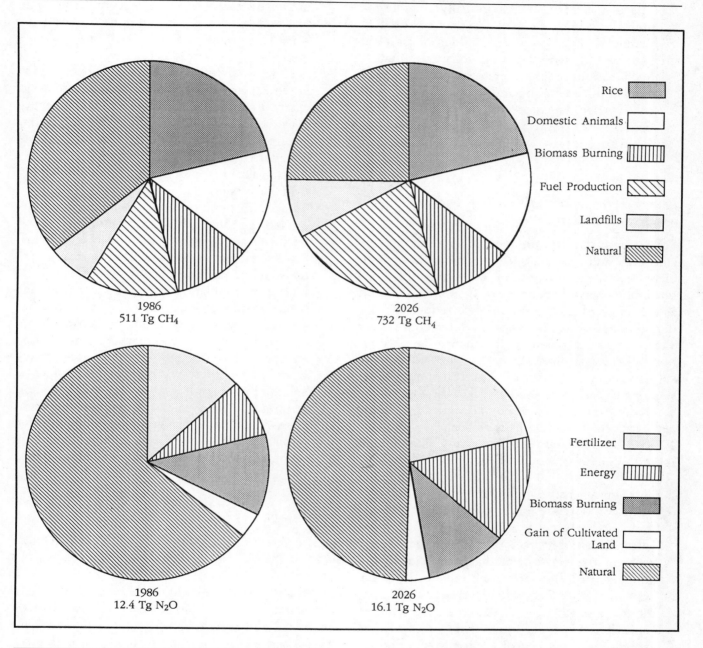

Rice

Domestic Animals

Biomass Burning

Fuel Production

Landfills

Natural

1986
511 Tg CH$_4$

2026
732 Tg CH$_4$

Fertilizer

Energy

Biomass Burning

Gain of Cultivated Land

Natural

1986
12.4 Tg N$_2$O

2026
16.1 Tg N$_2$O

efits such as direct energy savings, improved soil tilth and increase of soil organic matter.

- *Marginal lands:* Areas marginally suitable for annual cropping systems should be shifted to perennial cover crops for fodder or pastoral land uses of woodlands, if soils are suitable. Such actions could increase carbon uptake, both in the vegetation and soil, and could yield other benefits such as reduced soil erosion, improved water infiltration and quality, and delayed stream flow.

Longer-Term Options Requiring Research and Demonstration

Several opportunities for reducing greenhouse gas emissions and enhancing carbon sinks have been identified for the longer term. In general, these opportunities must be developed, demonstrated, and assessed in terms of greenhouse gas reductions and the full range of potential costs and benefits. These alternatives must maintain or enhance the productivity of the agricultural systems. This will require substantial research efforts focused on better understanding of the processes by which these gases are emitted, further investigation of promising options, and better field measurement devices.

General opportunities for reducing emissions of these gases have been identified:

- A comprehensive approach including management of water regimes, development of new cultivars, efficient use of fertilizers, and other management practices could lead to a 10–30 percent reduction in methane emissions from flooded rice cultivation, although substantial research is necessary to develop and demonstrate these practices. A 10 percent reduction in emissions from rice systems might contribute about 15–20 percent of the total reduction required to stabilize atmospheric concentrations of methane.
- Through a number of technologies it appears that methane emissions from livestock systems may be reduced by up to 25–75 percent per unit of product in dairy and meat production. The net effect of such improvements depends upon the extent to which such methods can be ap-

plied to domestic ruminant populations, which will vary greatly from country to country. However, each 5 percent reduction from animal systems could contribute 6–8 percent toward the reduction necessary to stabilize methane in the atmosphere.

- Fertilizer-derived emissions of nitrous oxide potentially can be reduced (although to what extent is uncertain) through changes in practices such as using controlled-release fertilizers, improving fertilizer-use efficiency, and adopting alternative agriculture systems.
- Trace gas emissions from biomass burning, land conversion, and cropping systems may be reduced through widespread adoption of improved agricultural practices, optimizing use of fertilizer and organic amendments, and improved pasture management and grazing systems. These gains may be offset by pressures of increasing population and increased demand for food and fiber production. In addition, policies associated with production, processing, storage, transportation, and marketing need to be examined to derive the optimum effectiveness from research, technological developments, and land use practices. Analyses are needed on economic incentives, taxes, pricing and trade barriers, cultural practices, technology transfer measures, education and information programmes, and international financial assistance measures.

4.3.3 International and Institutional Needs

A broad range of institutional issues must be addressed in order to ensure that the objectives of increased global food security and reduced greenhouse gas emissions can be met in the future. Among the options that governments, international organizations and intergovernmental bodies such as IPCC should consider are the following:

- A series of agricultural workshops to assess new information on agricultural production and emission forecasts, exchange information on the effectiveness of new technologies and man-

FIGURE 4.6: **Emissions of Methane by Domestic Animals, 1984 (Teragrams of Methane)**

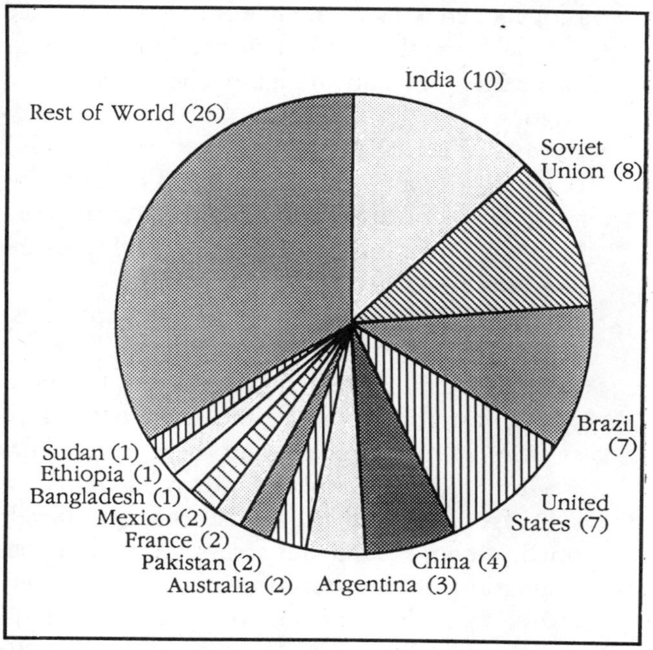

FIGURE 4.7: **Rice Area Harvested, 1984 (Million Hectares)**

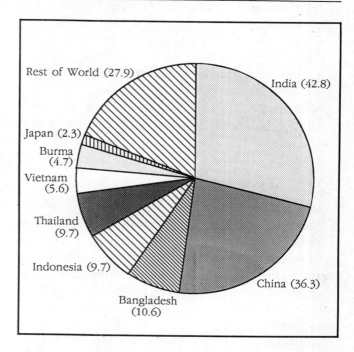

Distribution of the total methane emissions from domestic animals of 76 Tg CH_4. India, the Soviet Union, Brazil, the United States, China, and Argentina together account for just over 50 percent of the animal CH_4 emissions. Currently, approximately 20 percent of the total anthropologenic CH_4 emissions is due to domestic animal populations.

Source: Lerner et al., 1988.

Distribution of the total harvested rice paddy area of 148 million ha. Five Asian countries, India, China, Bangladesh, Indonesia, and Thailand accounted for 73 percent of the 1984 rice acreage.

Source: IRRI, 1986.

agement practices, and evaluate the potential impact of agricultural policies. These agricultural workshops should address the institutional and economic issues particular to the major agricultural sources of emissions, such as biomass burning, temperate agricultural practices, livestock emissions, and emissions from rice cultivation.

- An international symposium in 1991 or 1992 to assess current information on agricultural greenhouse gas emissions, practices, technology transfer, and extension service requirements.
- Guidelines for future research at national and international institutions such as FAO,

CGIAR, and others to address the need to investigate impacts of climate change and options for reducing emissions.

- An international symposium in 1991 or 1992 on biomass burning, its contribution to greenhouse gas emissions, and practices and policies to reduce emissions.

4.3.4 SUMMARIES FROM AGRICULTURAL WORKSHOP

More specific findings and recommendations are presented in the following sections. These are summaries of consensus documents produced during multiple-day discussions at the IPCC workshop on agricultural systems.

The sections are divided as follows:

- Greenhouse Gas Emissions from Flooded Rice Cultivation
- The Role of Managed Livestock in the Global Methane Budget
- Tropical Agriculture: Fertilizer Use, Land Use Conversion, and Biomass Burning
- Temperate/Boreal Agricultural Systems: Fertilizer Use, Land Use Conversion, and Soil Management

4.3.4.1 *Greenhouse Gas Emissions from Flooded Rice Cultivation*

Flooded rice fields are a major source of methane on a global scale, due to microbial anaerobic decay of organic matter. While uncertainty exists as to the exact contribution to the annual global emissions, they appear to be of the order of 25–170 million metric tonnes or 6–30 percent of annual global methane emissions.

In addition, methane emissions from flooded rice fields may increase by as much as 20 percent in the next decade, since rice production must increase to meet the rice requirements of growing human populations. Rice production is projected to increase from the current level of 458 million tonnes to over 550 million tonnes by the year 2000, and to some 760 million tonnes by the year 2020.

Measures and Policy Options

Reductions in methane emissions from flooded rice fields should be obtained while maintaining the productivity of the rice fields in all instances. In the long run, a comprehensive approach, including management of water systems, cultivar development, efficient fertilizer (both organic and mineral) use, and other management practices could achieve reductions of 10–30 percent. However, current understanding of the complex interaction between methane production and oxidation, and the flux between the atmosphere and the rice fields, is insufficient. This understanding is a prerequisite for determining potential options for reducing methane emissions from flooded rice fields.

Improved understanding of the process contributing to methane emissions from flooded rice fields can only be achieved by integrated, interdisciplinary projects that focus on process-related factors and that will allow for valid extrapolation. Research is needed on the following aspects:

- Biogeochemistry of methanogenesis in flooded rice fields including methane production, methane oxidation, and methanogenesis regulation factors.
- Factors affecting methane fluxes from flooded rice fields, such as climate, soil and water, cultivars, fertilizer application, and cultural practices and variations in fluxes spatially, seasonally, and diurnally.
- Effects of techniques to reduce methane emissions of nitrous oxide.
- Field level measurement techniques to assess spatial variability, and simulation models to synthesize the process and field level data. Technologies and practices for reducing emissions from flooded rice fields need to be developed, demonstrated and assessed, including an evaluation of the costs and benefits. Furthermore, to realize the full potential of the research, existing and possible agricultural policies regarding rice production need to be examined.

4.3.4.2 *The Role of Managed Livestock in the Global Methane Budget*

Ruminant animals produce methane as part of the digestive process. Total methane emission from the digestive processes of domestic ruminant animals have been estimated to be between 60 and 100 million metric tonnes per year, accounting for about 15 percent of the global methane emissions, second only to flooded rice field systems. These estimates suffer from several notable uncertainties that may cause overestimated emissions in some areas and underestimates in others, resulting in an overall underestimation of methane emissions, possibly by large amounts.

In addition, animal wastes from all sources are another potentially large source of methane emissions. Under anaerobic waste management systems (animal wastes under aerobic conditions do not produce methane), uncontrolled methane emissions from cattle wastes are likely to be of the same magnitude as methane emissions from the livestock di-

FIGURE 4.8: Nitrogen Fertilizer Consumption, 1984–1985 (Million Metric Tonnes Nitrogen)

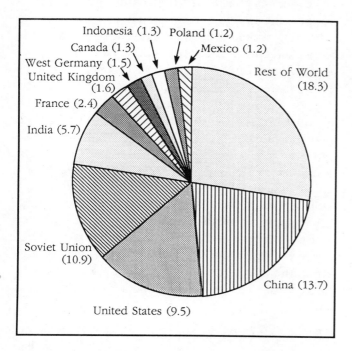

Distribution of the total agricultural fertilizer consumption of 70.5 million tonnes in China, the United States, and the Soviet Union together accounted for just over 50 percent of the 1984/1985 global fertilizer consumption. Currently 5–35 percent of the total anthropogenic N_2O emissions are attributed to agricultural fertilizer consumption.

Source: FAO, 1987.

gestive process. Preliminary analyses indicate that emissions from the source may currently be of the order of 65–100 million tonnes globally or about 15 percent of the methane from livestock digestive processes. Emission estimates from livestock systems should be refined, particularly in areas where interventions and methane control are most likely to be cost effective.

Measures and Policy Options

Reducing emissions from livestock is a particularly attractive methane reduction option because it is usually accompanied by improved animal productivity. While there are considerable differences in feed conversion efficiency between extensively and intensively managed animals, several techniques are available that might reduce emissions per unit of product (e.g., per kilo of meat or liter of milk) by up to 25–75 percent. However, the reductions actually achieved will of course depend on the extent to which and how effectively the appropriate technologies can be deployed. Furthermore, it is recognized that emission reductions achievable with the best technologies and their socio-economic consequences will vary within and among countries with variations in animal, management, and feeding characteristics.

Near-Term Options

Technologies available in the near term are:

- Animal wastes. Indications are that methane recovery through management of animal wastes can be economic in its own right under certain conditions.
- Supplemental feeding practices. Methane emissions can be substantially reduced (perhaps by 60 percent) from livestock on poor diets by strategic supplementation with locally produced feed additives. Experience in India indicates that the supplementation system can be self-sustaining and economic investment.
- Productive enhancing agents. Current economic evaluations indicate that the use of bovine somatotropin (BST) could improve feed conversion efficiency and is economically feasible at this time, taking into account human concerns on the use of BST.

While currently available, these technologies and practices must be further developed, demonstrated, and assessed in terms of the full range of cost reductions, other potential benefits and risks.

Longer-Term Options

Additional avenues exist for methane reduction in the longer term:

- Strategic supplementation of extensively managed cattle. Pastoral cattle management can result in relatively low animal productivity in some cases. Providing strategic supplementation of nutrients to these animals could reduce methane emissions by increasing efficiency and productivity per animal and achieving production targets with smaller herds.

- Diet modification for intensively managed animals. Current research indicates that methane emissions vary under different diets. Increasing the intake of animals and modifying the composition of their diet can reduce emissions per unit of product. Other feed inputs such as whole cotton seeds or polyunsaturated fats also appear to have promising impacts on methane emissions levels. Modifying feeding practices could potentially reduce methane emissions by large amounts in certain circumstances. However, size and location of animal populations for which this is a promising alternative must be identified.

- Reduction of protozoa. Recent studies indicate that reduced protozoa in ruminant digestive systems results in lower methane emissions and may enhance animal productivity. Further promising avenues consist of improvements in reproductive efficiency, which lead to smaller brood herd requirements and microbiological approaches to improved digestion processes. Again, these technologies and practices must be further developed, demonstrated, and assessed in terms of costs, reductions, and other potential benefits. Also, techniques for field level measurements of methane from livestock need to be developed and standardized.

4.3.4.3 Greenhouse Gas Emissions from Tropical Agriculture

Sources

Tropical agricultural systems include traditional fallow-cropping system, conventional high-production systems, agroforestry systems, and agropastoral systems.

An assortment of agriculture activities contribute to greenhouse gas emissions in the tropics. Tropical flooded rice fields and cattle are major sources of these gases.

The remainder of tropical emissions are due primarily to biomass burning, either to convert forest and savanna ecosystems into arable land or pastures, to return nutrients to the soil, to reduce shrubs on rotational fallow lands, or to remove crop residues. In all instances associated with biomass burning, emissions of greenhouse gases are not well estimated and no consistent measurement techniques are now in use. Furthermore, no estimates have separated temperate from tropical sources. Initial estimates put total nitrous oxide emissions at 5–15 percent of global emissions; contributions of CO_2 are estimated at 20–40 percent of global emissions, and contributions of NO_x at 10–35 percent of global emissions. Agriculturally related emissions of methane are perhaps best characterized and estimated. They are dominated by rice and livestock production, with a potentially large, but uncertain percentage from biomass burning. Methane contributions from burning are estimated at 20–80 million tonnes per year, or 5–13 percent of the total methane emissions. There is less certainty regarding the emission of carbon dioxide from agriculture, and the relative contribution of each of the major sources: land-use conversion, biomass burning, and soil degradation. The greatest uncertainties are in the area of nitrous oxide emissions, although initial estimates are of 5–15 percent of the global total.

Measures and Policy Options

Policy options must have value to the farmer beyond the greenhouse-gas-reducing benefits. Policies must not hamper national food security goals or distort competition on the world market. The pressure to convert land to crop and pasture use needs to be reduced, thus reducing emissions from burning, soil exposure, and erosion. Increasing the productivity of croplands on suitable soils using appropriate agricultural practices will have that effect. Reclaiming and restoring degraded agricultural lands should also be explored, in addition to enhancing the indigenous uses of native forests and establishing forest cropping systems that reduce the demand for further deforestation. Policies that increase efficiency in the use of water, fertilizer, and crop residues as well as the use of nitrogen-fixing crops should also be pursued. Incentives may be useful in encouraging the use of improved fertilizers and in abandoning the use of mechanized deforestation. Land-tenure policies need to be evaluated to provide an incentive to limit the conversion of moist forest ecosystems to agricultural ecosystems. Education programmes that teach improved organic-residue management and provide an understanding about the consequences of soil degradation need to be developed and proliferated. Some options for reducing greenhouse gas emissions include:

- Conversion of forest land to agricultural, as widely practiced in the tropics, may be reduced by adopting sustainable agricultural practices that optimize yields, or by adopting intensive practices on suitable agricultural soils.
- Reduction of emissions from burning crop residues and the routine burning of savannas through sustainable agriculture, including use of chemical and organic amendments, and improved forage species and management systems.
- Enhancement of soil-carbon storage by the use of different vegetation covers and cropping systems, including agroforestry systems. Agroforestry systems may also increase above-ground carbon storage, in addition to stabilizing soil and providing firewood.
- Limitation of fertilizer-derived emissions of nitrous oxide by use of fertilizers with slower conversion rates that are more in accord with crop requirements or that have been technically engineered to conform to rates of plant uptake. Improving fertilizer application for crops and adoption of alternative agricultural systems, such as agroforestry, also have some promise in this regard. Collaborative research among scientists in developing countries is needed to ensure consideration for regional and local physical and cultural factors, with special focus on carbon and nitrogen cycling, burning practices, and soils.

In addition, research is required in the following areas:

- Nitrous oxide emissions from fertilizer and leguminous crops across a broad range of cropping systems.
- Remote-sensing and monitoring methodology development is needed to evaluate the effects of policies to reduce these practices.
- Better estimates are needed on amounts of biomass burned annually, instantaneous emissions from the fire front, and longer-term biogenic emissions from a burn.
- Improved efficiency of technologies and devices for broadcast burning, charcoaling, and

use of fuelwood for heating and cooking. These devices and technologies need to be practical and affordable to indigenous populations.
- Appropriate tree species for agroforestry by sites and regions, and the effects of these trees on soils and cropping systems.
- Potential sinks for greenhouse gases in agricultural systems of the tropics and the interactions between sources and sinks. Long-term studies are needed to quantify the effects of different agricultural management systems on these sinks and especially on soil properties.

4.3.4.4 *Greenhouse Gas Emissions from Temperate/Boreal Agriculture*

Sources

An assortment of agricultural activities contribute to emissions of nitrous oxide, carbon dioxide, and methane in temperate and boreal regions.

Nitrous oxide emissions are estimated at 10–100 kg N/ha/yr globally. These emissions probably result from the significant amounts of nitrogen (of the order of 15–20 percent) that are lost from annual agricultural budgets and cannot be accounted for. In addition, agriculturally derived nitrogen converts to nitrous oxide off-site in water and eroded soil, and is carried in some cases by the movement of fertilizer and inorganic particles from newly converted lands. A significant factor in nitrous oxide emissions is the application of excess fertilizer when it cannot be fully utilized by growing plants, usually due to attempts to minimize uncertainty about crop yields. Nitrogen formulation, placement, and depth are important factors that influence the rate of nitrous oxide emissions, and their effects vary with fertilizer type, soil chemistry, and physical soil conditions. In addition, nitrous oxide emissions could potentially increase in far northern soils near the permafrost belt that now have low levels of denitrification if temperature and moisture regimes shift northward, and if cultivation increases.

Carbon dioxide fluxes are well-known processes, and the soil carbon relationships with crops and permanent vegetation have been modeled fairly well. No-till practice should show a marginal increase in soil carbon, though there is great uncertainty in the soil-carbon relationships involved in tillage practices and residue management. In temperate and boreal agricultural systems, soil carbon can usually be increased or decreased by about 20 percent from the existing continuous arable state when large amounts of organic matter are not added; thus these agricultural soils are not a significant carbon sink. Furthermore, cutbacks in nitrogen inputs from any source may reduce biomass accumulation and annual carbon sequestration. Use of crop rotation with organic matter amendments can increase or maintain soil equilibrium levels of organic carbon. However, in temperate and boreal agricultural systems the effect is very small relative to that which takes place in the conversion to grassland or forest ecosystems. An estimated 25 percent of the global soil carbon is contained in boreal forest soils, and significant amounts of this carbon will be lost if the lands overlying these soils are converted to agricultural use. In addition, drainage of peat bogs and boreal muskegs would release more nitrous oxides and carbon dioxide as water tables drop, while maintaining high water tables in these systems would release more methane.

The total energy demand of intensive systems contributes 3 percent of total anthropogenic carbon dioxide emissions. While significant changes in the energy demands of current agricultural systems are not expected, some marginal contributions to greenhouse gas stabilization may be realized with shifts from conventional tillage to conservation, and partial replacement of chemical fertilizer and pesticides with leguminous crops and biological control systems.

Methane fluxes in boreal and temperate systems do not appear to contribute significantly to current emissions, although these processes are not well known. Uncertainty exists as to whether agricultural soils serve as a sink for methane.

Measures and Policy Options

Agricultural policy formulation should include limitation and mitigation of greenhouse gas emissions. Two options are currently available for reducing emissions of nitrous oxide.

- Improved biological-utilization efficiency in fertilizer use. Farmers may be advised about concepts of nutrient balance with clear definitions about marginal and acceptable losses of nutrients, integrated with tillage and application requirements.
- Market and regulatory strategies. Allowable limits may be established for nitrous oxides and nitrate inputs and losses based on best available information, with the possible utilization of the concept of exchangeable loss units. Nutrient-sensitive areas may be designated such as the Chesapeake Bay area of the United States, where special sets of balance mechanisms must be developed. Also, the concept of stocking rates (in terms of intensity or capacity of the soils resource) may be utilized within intensive agricultural zones, using the capability-rating systems of worldwide soil surveys.

Research is needed on:

- Measurements of nitrogen and nitrous oxide emissions to determine: contributions of nitrates in ground water; the contribution of different legumes, the nitrification and denitrification processes, and the vadose zone; the contribution of arable lands to oxide production in surface waters and sediments; sources such as chemical nitrogen sources, livestock-manure sources, losses from storage, losses from sewage sludge, and losses in animal digestion; and the contribution from different regions of the world.
- Relationship of carbon-nitrogen ratios to gas emissions and factors affecting nitrous oxide-nitrogen ratios during denitrification.
- Mechanisms for reducing gas emissions while maintaining production levels, including optimal chemiforms of nitrogen to meet site-specific characteristics.
- The role of soil as a sink for nitrous oxides.
- Measurement technologies, including scale-dependent micrometeorological methods that are terrain transportable with fast-response sensors, landscape- and regional-scale extrapolation methods, improved nitrate nitrogen tests, and development of integrated data systems to integrate laboratory, field regional, and global scales.

The effectiveness of agricultural-management practice will have to be assessed in terms of emissions reductions, costs, and other benefits. In addition, policy research is needed to evaluate commodity-support programs and their impacts on gas emissions, including the impacts of set-aside policies on the intensive use of remaining lands.

4.4 METHANE FROM LANDFILL SITES AND WASTEWATER TREATMENT PLANTS

4.4.1 EMISSIONS

Organic matter in waste and wastewater is converted into methane by different types of methane bacteria under anaerobic conditions. Anaerobic conditions exist in most landfill sites and in most lagoons used for treating organic loaded wastewater. To estimate global methane production from landfill sites, the following assumptions have been used: Specific waste generation in developed countries is approximately 1 kg per capita per day, and in developing countries it is approximately 0.5 kg per capita and day. Methane generation in developed countries is estimated to be 86 kg methane per metric tonne of waste, and in developing countries 21.5 kg per tonne of waste. Disposal of domestic and commercial waste on landfill sites takes place mainly in urban centers. In rural areas, particularly in developing countries, waste disposal does not take place on large centralized landfills with anaerobic conditions. In order to estimate future emissions, UN statistics were used for urban populations. It is also assumed that for the period up to 2030, no changes take place in waste composition, generation per person and waste treatment. Figure 4.9 shows the estimated increase of methane emissions between the years 1985 and 2030.

Wastewater treatment plants in developed countries are not considered major sources of methane emissions because aerobic sewage treatment prevails and sludge digesters are equipped with gas-utilization facilities. In most developing countries, where land is comparatively cheap, domestic and industrial wastewater is normally treated in lagoons/oxidation ponds. As a general rule, the first ponds rely on anaerobic conditions, and up to 80 percent of the organic load is digested by methane

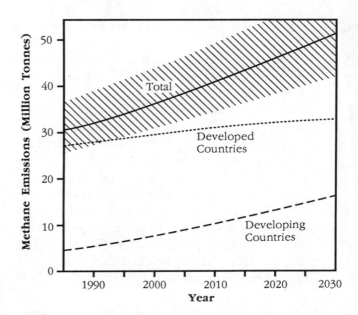

FIGURE 4.9: Methane Emissions from Landfill Sites in Developed and Developing Countries, 1985–2030

Note: Cross-hatched area indicates range of uncertainty.

bacteria in the following 5–15 ponds where oxidation processes take place.

An inventory of methane emissions from wastewater lagoons in the Kingdom of Thailand estimated that 0.5 million tonnes of methane are emitted per year. Little data is available for other countries with comparable conditions. On the basis of the Thai data, the global methane emissions from wastewater lagoons may be estimated to be roughly 20–25 million tonnes per year.

Total global methane emissions from waste disposals and from wastewater lagoons are estimated to be 55–60 million tonnes per year. This is 15 percent of the total methane emissions caused by human activities.

4.4.2 MEASURES AND POLICY OPTIONS

There are a number of technological options for reducing methane emissions from these sources.

Landfill Sites

- Collect and recycle waste paper separately in order to reduce organic matter in wastes.
- Collect and compost vegetable wastes separately in order to reduce organic matter in wastes.
- Utilize aerobic waste treatment (composting), especially in regions with a demand for soil conditioners.
- Utilize landfill gas collection system. It is estimated that 30–90 percent of landfill gas can be recovered by gas collection systems (built-in pipes and wells). The gas can either be burned in torches or be used as an energy source (heating, electricity production). Implementation of gas collection systems in landfill sites is recommended in developed countries. In developing countries, gas-collection systems are reasonable only in cases where methane contents in landfill gas are high enough to justify the high costs of the system.
- Replace landfills with solid waste incineration in developed countries. (Emissions—CO_2-equivalent—from landfills without gas collection are estimated to be 3–5 times higher than from waste incineration).

Wastewater Treatment

Most factories in the agricultural sector have both extremely loaded sewage and high energy demand (steam, electricity). Pretreatment of the sewage in biogas plants could therefore be recommended to reduce methane emission. By replacing bunker oil with methane, a capital return may be reached in about five years.

4.4.3 COSTS

The costs of landfill gas collection systems vary among countries and from site to site.

The figures in Table 4.9 give a rough estimation of cost.

TABLE 4.9

COUNTRY	US $ PER TONNE OF WASTE	US $ PER TONNE OF METHANE
UK	0.6	10–12
FRG	1–2	15–20
USA	1	10–25

The costs of wastewater pretreatment (biogasplants) mainly depend on type and organic load of the wastewater. Costs per cubic meter reactor volume: U.S.$250–400.

Note: Required reactor volume: 3–7 times the daily wastewater quantity.

4.4.4 RECOMMENDATIONS

- Landfill gas collection systems and flaring appear to be practical for many developed countries and should be considered a near-term option to reduce methane emissions. Additional demonstrations of alternative electrical generation systems could further reduce their costs.
- Biogas systems for wastewater treatment represent an inexpensive source of energy. These systems also reduce water pollution and agricultural waste disposal problems. It is recommended as a short-term option to introduce these systems in developed countries.
- Application of these systems in developing countries requires additional demonstration, training, and technology transfer and it would be a somewhat longer term policy option for these countries.

4.5 DISCUSSION AND CONCLUSIONS

4.5.1 GENERAL ISSUES AFFECTING THE AGRICULTURE, FORESTRY, AND WASTE MANAGEMENT SECTORS

- Emissions from these sectors are intimately related to the ability of countries to provide for national and global food security and raw materials for export. Policies for emission control therefore may affect the economic basis of some countries, especially developing countries. Efforts to prevent deforestation and to promote afforestation will have multiple impacts that in many cases may enhance the abilities of indigenous agricultural communities to feed themselves and to earn a living from local surplus. Flexibility should be provided to governments to develop least-cost implementation strategies, and countries should pursue those options that increase productivity, make economic sense, and are beneficial for other reasons.

- Agriculture and forestry sources of greenhouse gas emissions consist of very large numbers of small sources, and/or of diffuse sources from large land areas, and as such represent a major challenge.
- To meet this challenge it is desirable to consider a range of measures in the context of a comprehensive response strategy. Currently available options in the sectors are likely to be only partially effective unless coupled with action to reduce emissions from the energy and industry sector.
- Controls on existing sources of pollution which may affect agricultural and forestry lands are an important component in reducing emissions and protecting sinks. Measures to prevent land degradation, hydrological problems and loss of biodiversity, and to improve productivity will generally complement efforts in this sector. Furthermore, analysis is needed about the costs and benefits of individual policy measures.
- There is an urgent need to improve all relevant data, especially on deforestation rates and on the socio-economic factors that lead to the use of deleterious practices in agriculture and forestry. In order to achieve the full potential of identified measures, research is needed into the routes by which new technologies can be introduced while preserving and enhancing social and economic development.
- In the past, the agriculture and forestry sector has proved efficient at introducing new methods of production, not always with beneficial environmental consequences. The pace of technological development (even in the absence of global warming) is likely to remain high in this

sector for the foreseeable future. This may afford new opportunities for emission reduction, provided that efficient transfer of knowledge, advice, and technology occur.

- A number of technical options have been identified that could reduce emissions from and enhance sinks in forestry and agriculture, as well as reduce emissions from waste and wastewater treatment. In many cases, their application will require further research and development and large scale education and technology transfer programmes. In this regard, the papers produced by Working Group III on Legal and Institutional Measures, Public Information and Education, Technology Development and Transfer, Financial Measures and Economic Measures are particularly pertinent to developing recommendations for the implementation measures to meet these objectives.

4.5.2 INFORMATION ON EMISSION SOURCES

- Emissions of greenhouse gases, based on the best available estimates, are presented in Table 4.1. Large margins of uncertainty remain in these figures. Many are scaled up from regional estimates which may be unrepresentative at a global level.
- The figures indicate that agriculture and forestry contribute significant amounts of greenhouse gases. Where carbon dioxide is concerned, deforestation, especially in the tropics, is the major concern at present. Agriculture contributes significantly to global methane. Furthermore, the degradation of soil organic matter may release CO_2.
- These emissions are likely to increase over the next few decades, due to population and economic growth, the associated expansion and intensification of agriculture, increased demands on wood, especially for fuel, and pressure to use forestry land for agriculture.

4.5.3 RESEARCH

Clearly, there is a need for research to improve emission estimates, and, particularly, to develop standardized methods for measuring emissions at field, regional and global scales. Developing regional budgets that can be related to global scale events is a challenge. Research is also needed into biological processes contributing to emissions and sinks. Development of additional methods for sustainable agriculture and forestry is a priority.

4.5.4 POLICY OBJECTIVES

Nonetheless, policies must be developed not only for the long term, but also to initiate immediate actions to prevent further deterioration. It is necessary to be clear about the strategic objectives of such policies to reduce greenhouse gas emissions from forestry, agriculture, and wastes, and to maintain and provide sinks especially for CO_2:

Forestry

- To reduce the scale of destructive deforestation.
- To promote reforestation and afforestation.
- To promote sustainable forest management.
- To promote more complete utilization of forest products, including recycling.

Agriculture

- To reduce methane emissions from ruminants and rice cultivation.

- To reduce nitrogen loss from cultivation, especially in the form of nitrous oxide.

Wastes and Wastewaters

- To promote efficient waste treatment and collection and utilization of gaseous emissions.

4.5.5 CRITERIA FOR THE SELECTION OF POLICY OPTIONS

It is important that these objectives be met without major economic and social disruption, especially to developing countries. Policies must therefore:

- be of widespread applicability;
- be economic—that is, be compatible with the social and economic life of the communities dependent on agriculture and forestry;
- be equitable in the distribution of the burdens of action between developed and developing countries taking into account the special situation of the latter;
- result in the spread of knowledge, management skills, and technologies;
- result in net environmental gain; and
- take account of the fact that emissions in this sector largely comprise many small sources or diffuse sources from large areas.

REFERENCES

Andreae, M. 1990. Biomass burning in the tropics: Impact on environmental quality and global climate, presented at Chapman Conference on Global Biomass Burning, March 1990, Williamsburg, Virginia.

Becker M., G. Landmann, and G. Levy. 1989. Silver fir decline in the Vosges mountains (N. E. France): Role of climate and silviculture. *Water, Air, and Soil Pollution*, special issue. (In press).

Bolin, B. 1989. Preventive measures to slow down man-induced climate change. In Steering Committee of IPCC working group report, Geneva. May 1989.

Bonneau, M., and G. Landmann. 1988. De quoi la forêt est-elle malade? La Recherche 205 (19): 1942–1953. (English translation available 1st Trim. 1990.)

Brown, L., et al. 1988. *State of the World: 1988*. New York: W. W. Norton.

Brown, S., and A.E. Lugo. 1982. The storage and production of organic matter in tropical forests and their role in the global carbon cycle. *Biotropica* 14:161–187.

———.1984. Biomass of tropical forests: A new estimate based on volumes *Science* 223:1290–1293.

Bundesforschungsanstalt für Forst und Holzwirtschaft, ed., 1975. *Weltforstatlas*. Hamburg and Berlin: Paul Parey Verlag.

Bundesministerium für Ernährung, Landwirtschaft und Forsten. 1988. Schriftenreihe A: *Waldzustandsbericht-Ergebnisseder Wald—schadenserhebung 1988*; Angewandte Wissenschaft 364. Münster-Hiltrup: Landwirtschaftsverlag GmbH.

Bund-Laender-Arbeitsgruppe (BLAG). 1987. Erhaltung forstlicher Genres-sourcen—Konzept zur Erhaltung forstlicher Genressourcen in der Bunsderepublik Deutschland. *Forst und Holz* 44 (1989):379–404.

Burschel, P. 1988. Die Bäume und das Kohlendioxid. In *Forstarchiv* 59:31.

Burschel, P., and M. Weber. 1988. Der Treibhauseffekt—Bedrohung und Aufgabe für die Forstwirtschaft. In *Allgemeine Forst Zeit.*:37.

Cassens-Sasse, E. 1988. *Versaürung der Waldboeden.* Alfeld/Leine: Dobler-Druck.

Da Silva, C.P. 1982. Role of Canadian forests in the CO_2 issue. Academic paper, University of Waterloo, Canada.

Dickinson, R.E. 1986. How will climate change? In Bolin, B., B.R. Doos, J. Jaeger, and R.A. Warrick, *The greenhouse effect, climate change and ecosystems,* Scope 29, 207–270. Chichester, John Wiley & Sons.

Ellenberg, H., Sr. 1978. *Vegetation Mitteleuropas mit den Alpen.* 2nd ed. Stuttgart: Ulmer. 981 pp.

———.1979. *Zeigerwerte der Gefaesspflanzen Mitteleuropas.* 2nd ed. Göttingen: E. Goltze.

Ellenberg, H., Jr. 1988. Eutrophierung Veränderungen der Waldvegetation: Folgen für den Reh—Wildverbiss und dessen Rückwirkungen auf die Vegetation. *Schweizerische Zeit-schrift für das Forstwesen* 139(4):261–282.

———.1989. Auswirkungen einer Klimaveraenderung auf den Waldals ein aus Pflanzenarten zusammenge-setztes Oekosystem (internal study, BFH).

Enquete Commission of the Deutscher Bundestag "Vorsorge zum Schutz der Erdatmosphaere." 1989. Schutz der Erdatmosphaere: Eine internationale Herausforderung. 2nd ed. Bonn: Deutscher Bundestag.

Environment Canada Report. 1988a. Exploring the implications of climatic change for the boreal forest and forestry economics of western Canada. In *Climate Change Digest,* CCD 88–08.

Environment Canada Report. 1988b. The implication of climate change for natural resources in Quebec. In *Climate Change Digest,* CCD 88–08.

European Community (Commission of the European Communities). 1989. *The conservation of tropical forests: The role of the community.* The Official Journal of the European Community OJ 89/C264/01.

Falkengren-Grerup, U., 1986. Soil acidification and vegetation changes in deciduous forests in Southern Sweden. *Oecologie* 70:339–347.

Fanta, J. 1982. Dynamik des Waldes auf Sandboeden in den Niederlanden. In *Urwald Symposium* Wien 1982, ed. H. Mayer, 66–80. Wien: Waldbau-Institut.

———.1986. Forest site as a framework for forest succession. In *Forest dynamics research in Western und Central Europe,* Ed. J. Fanta, 58–65. Proceedings of the IUFRO workshop, held Sept. 17–20, 1985 in Wageningen, NL.

FAO (United Nations Food and Agriculture Organization). (1981). *Tropical forest resources assessment project: Forest resources of tropical Asia.* Rome: FAO.

———. 1982. *Tropical forest resources:* Forestry Paper no. 30. Rome: FAO.

———.1983. *Production Yearbook.* Rome: FAO.

———.1986. *FAO production yearbook 1985,* V-39 Rome: FAO.

———.1987. *Production yearbook.* Rome: FAO.

———.1988. *An interim report on the state of forest resources in the developing countries.* Forest Resources Division, Forestry Department. Rome: FAO.

Fosberg, M., J.G. Goldammer, C. Price, and D. Rind. 1990. Global change: effects on forest ecosystems and wildfire severity. In: *Fire in the tropical biota.* J.G. Goldammer., ed. Springer-Verlag. In press.

Gamlin, L. 1988. Sweden's factory forests. In *New Scientist* 28 (Jan): 41–47.

Glawion, R. 1986. Tropische Regenwaelder in Australien. In *Neue Forsch. Geo. Austra.* B. Hofmeister and F. Voss, eds. 1–20. Berlin: Universitaetsbibliothek der TU.

Grainger, A. 1989a. Modelling future carbon dioxide emissions in the humid tropics. Contribution to IPCC Tropical Forestry Response Options Workshop, São Paulo, January 9–11, 1990.

———.1989b. Modelling the impact of alternative afforestation strategies to reduce carbon dioxide emissions. Contribution to IPCC Tropical Forestry Response Options Workshop, São Paulo, January 9–11, 1990.

———.1989c. Strategies to control deforestation and associated carbon dioxide emissions in the humid tropics. Contribution to IPCC Tropical Forestry Response Options Workshop, São Paulo, January 9–11, 1990.

Grayson, A.J. 1989. *Carbon dioxide, global warming and forestry.* Farnham: Forestry Commission Research Division, Research Information Note 146.

Groebl, W. 1989. Schutz der tropischen Regenwaelder als umwelt-politische Herausforderung. In *Bulletin,* Ed. Presse-und Informationsamt der Bundesregierung (Bonn) 91:797–799.

Houghton, R., R.D. Boone, J.M. Melillo, D.A. Palm, G.M. Woodwell, N. Myers, B. Moore, and D.L. Skole. 1985. Net flux of carbon dioxide from tropical forests in 1980. *Nature* 316 Aug. 15, 1985:617–620.

Houghton, R., R.D. Boone, J.R. Fruci, J.E. Hobbie, J.M. Melillo, D.A. Palm, B.J. Peterson, G.R. Shaver, G.M. Woodwell, B. Moore, D.L. Skole, and N. Myers. 1987. The flux of carbon from terrestrial ecosystems to the atmosphere in 1980 due to changes in land use: Geographic distribution of the global flux, *Tellus* 39B:122–139.

Houghton, R. 1989. Background information prepared for the Tropical Forestry Response Options Workshop. Contribution to IPCC Tropical Forestry Response Options Workshop, São Paulo, January 9–11, 1990.

———.1990a. Projections of future deforestation and reforestation in the tropics. Contribution to IPCC Tropical Forestry Response Options Workshop, São Paulo, January 9–11, 1990.

———.1990b. Unpublished results of model runs performed for USEPA (1989) and for USEPA in support of the IPCC Tropical Forestry Response Options Workshop, São Paulo, January 9–11, 1990.

IPCC-AFOS Agriculture Workshop, Washington, D.C. 1990.

IPCC AFOS Tropical Forestry Workshop, São Paulo, Brazil, 1990.

IPCC Working Group I Final Report, Summer 1990.

International Steering Committee Meeting. 1989. Background paper on afforestation.

Kienast, F., and N. Kuhn. 1989. Computer-gestuetzte Simulation von Waldentwicklungen. *Schweizerische Zeit. für das Forstwesen* 140(3):189–201.

Kirchoff, V.W.J.H., A.W. Setzer, and M.C. Pereira. 1990. Biomass burning in Amazonia: Seasonal effects on atmospheric 03 and CO. Contribution to IPCC Tropical Forestry Response Options Workshop, São Paulo, January 9–11, 1990.

Moosmayer, H.U. 1988. Stand der Forschung über das "Waldsterben." *Allgemeine Forst Zeit.* 43:1365–1373.

Mueller-Dombois, D. 1988. Canopy dieback and ecosystem processes in the Pacific area. In W. Greuter and B. Zimmer, eds.: *Proc. XIV. Int. Bot. Congress*: 445–465.

Myers, N. 1980. *Conversion of tropical moist forests.* Washington, D.C.: National Academy of Sciences Press.

———.1984. *The primary source.* New York: W.W. Norton.

———.1989. *Deforestation rates in tropical forests and their climatic implications.* London: Friends of the Earth Report.

Neiland, B., and L.A. Viereck. 1977. Forest types and ecosystems. In *North American forest lands at latitudes north of 60 degrees.* Proc. Symp. University of Alaska.

Oberle, B.M., U. Bosshart, and T. Kohler. 1989. Waldzerstörung und Klimaveränderung. Zurich: Schweizerisches Bundesamt für Umwelt, Wald und Landschaft (BUWAL).

Parry, M. 1989. The greenhouse effect and agriculture in the future. The Ninth Asher Winegarden Memorial Lecture given at Agriculture House on May 17, 1989.

Pohjonen, V. 1989. *Metsityksen suuri linja-kommenttiteokseen: Malinen.* 1989. Kehitysapu taeysremonttiin.

Postel, S., and L. Heise. 1988. Reforesting the Earth. *Worldwatch Paper* 83:1–67.

Ramanathan, V., R.J. Cicerone, H.B. Singh, and J.T. Kiehl. 1985. Trace gas trends and their potential role in climate change. *J. Geophys. Rev.* 90D3:5547–5566.

Sanchez, P. 1988. *Deforestation reduction initiative: An imperative for world sustainability in the twenty-first century.* Paper presented at the Bureau of Science and Technology, U.S. Agency for International Development, Washington, D.C., July 22.

SanMartin, R.K. 1989. *Environmental emissions from energy technology systems: The total fuel cycle.* U.S. Department of Energy Report.

Scholz, F. 1989. *Schriftliche Stellungnahme zu dem Fragenkatalog für die öffentliche Anhörung der Enquete Kommission des Deutschen Bundestages "Vorsorge zum Schutz der Erdatmosphäre" zum Thema Schutz der Wälder in mittleren und nördlichen Breiten,* 149–161. Bonn: Kommissionsdrucksache 11/83.

Scholz, F., H.-R. Gregorius, and D. Rudin. 1989. Genetic of air pollutants in forest tree populations. Berlin: Springer-Verlag.

Sedjo. R.A. 1989. Forests: A tool to moderate global warming? *Environ.* 31(1):14–20.

Serrao, A. 1990. Pasture development and carbon emissions accumulation in the Amazon: Topics for discussion. Contribution to IPCC Tropical Forestry Response Options Workshop. São Paulo, January 9–11, 1990.

Shands, W.E., and J.S. Hoffmann. eds. 1987. The greenhouse effect, climate change, and U.S. forests. Washington, D.C.: The Conservation Foundation.

Shcherbakov, I.P. 1977. Forest vegetation in burned and logged areas of Yakutsk. In *North American forest lands at latitudes north of 60 degrees.* Proc. Symp. University of Alaska.

Shula, R.G. 1989. The upper limits of radiata pine stem-volume production in New Zealand. *NZ Forestry.*

Smith, J.B., and D.A. Tirpak, ed. 1988. The potential effects of global climate change on the United States. Draft Report to Congress, Executive Summary, in United States Environmental Protection Agency.

Strasburger. 1971. Lehrbuch der Botanik, 30th ed. Stuttgart: Gustav Fischer. 842 pp.

Tiedemann, H. 1988. Berge, Wasser, Katastrophen. In *Schweizer Rueck*, 81–83. Zurich.

Trabalka, J.R. 1985. Atmosphere carbon dioxide and the global carbon cycle. USDOE/ER-0239, In *Der Treibhauseffekt und die Gemeinschaft*, ed.: Kommission der europäischen Gemeinschaften, Brussels, 16 November 1988.

Trexler, M.C., P.E. Faeth, and J.M. Kramer. 1989. Forestry as a response to global warming: An analysis of the Guatemala agroforestry and carbon fixation project. Washington, D.C.: World Resources Institute.

Tucker, G.B. 1981. *The carbon dioxide climate connection: A global problem from an Australian perspective.* Canberra: Australian Academy of Sciences.

Ulrich B., K.J. Meiwes, N. Koenig, and P.K. Khanna. 1984. Untersuchung-sverfahren und Kriterien zur Bewertung der Versauerung und ihre Folgen in Waldböden. *Forst- und Holzwirt* 39:278–286.

USEPA. 1989. *Policy options for stabilizing global climate.* Washington, D.C.: U.S. Environmental Protection Agency, Office of Policy, Planning, and Evaluation, vol. 1.

Walter, H. 1971. Vegetationszonen und Klima. Stuttgart: UTB, Ulmer. 244 pp.

Walter, H. 1985. Vegetation of the Earth and ecological systems of the geo-biosphere. Heidelberg Science Library. 318 pp.

WRI et al. (World Resources Institute, International Institute for Environment and Development, and United Nations Environment Programme). 1988. *World Resources 1988–89.* New York: Basic Books.

Yearbook of forest statistics. 1988. The Finnish Forest Research Institute. *Agriculture and Forestry* 1989:1.

Zasuda, J.C., K. van Cleve, R.A. Werner, and J.A. McQueen. 1977. Forest biology and management in high-latitude North American forests. In *North American forest lands at latitudes north of 60 degrees.* Proc. Symp. University of Alaska.

APPENDIX 4.1

DEFORESTATION OF ALL TROPICAL FORESTS, 1980–2050: TWO BASE CASE PROJECTIONS OF INCREASING CARBON EMISSIONS AND TWO POLICY OPTION CASES FOR DECREASING CARBON EMISSIONS

YEAR	POLICY OPTION CASES				BASE CASES		
	LOW		HIGH		LOW		HIGH
1980	.544		.797		.85		2.52
Latin America	.223	(41%)	.320	(43%)	.409	(48%)	1.037 (41%)
Africa	.174	(32%)	.235	(29%)	.204	(24%)	.584 (23%)
Asia	.147	(27%)	.220	(27%)	.237	(28%)	.898 (36%)
2000	.387		.668		.96		2.91
Latin America	.150	(39%)	.294	(44%)	.463	(48%)	1.432 (49%)
Africa	.133	(34%)	.201	(30%)	.229	(23%)	.803 (28%)
Asia	.104	(27%)	.173	(26%)	.268	(28%)	.674 (23%)
2020	.221		.530		1.05		3.89
Latin America	.053	(24%)	.202	(38%)	.509	(48%)	1.887 (48%)
Africa	.104	(47%)	.168	(32%)	.245	(23%)	1.054 (27%)
Asia	.065	(29%)	.160	(30%)	.295	(28%)	.948 (24%)
2050					1.18		4.99
Latin America					.576	(49%)	2.289 (46%)
Africa					.269	(23%)	1.389 (28%)
Asia					.335	(28%)	1.311 (26%)

Policy Option Cases (Grainger, 1989a): Decreased forestation and emissions are expected to result from increases in agricultural productivity that relieve pressure to clear new land. Increases in productivity are adjusted for changes in per capita consumption for a net gain. In the high case, net agricultural productivity per ha rises by 0.5%, 0.0% and 0.5% per year for Africa, Asia-Pacific and Latin America respectively; and by 1.0%, 0.0% and 1.0% per year respectively, in low case scenario. Biomass estimates are the same in both cases, based on "volume over bark" estimates FAO (1981) and procedures of Brown and Lugo (1984).

Base Cases (Houghton, 1990b): Increased deforestation is forecast. The low base case assumes low forest biomass estimates (roughly) comparable to Grainger's biomass assumptions) and linear increases in the rate of deforestation. The high base case assumes high forest biomass and population based deforestation rates. The biomass estimates are outlined in Houghton et al., 1985.

APPENDIX 4.2

LOW AND HIGH BASE CASE CARBON EMISSIONS PROJECTIONS ON A COUNTRY-BY-COUNTRY BASIS FOR DEFORESTATION OF ALL TROPICAL FOREST FOR 1980–2050
(*in Billion Tonnes of Carbon per Year*)

LOW BASE CASE
NET FLUX FROM LINEAR RATES OF FOREST CONVERSION AND LOW FOREST BIOMASS ESTIMATES

COUNTRIES BY GLOBAL REGION	1980	2000	2025	2050
TROPICAL AMERICA				
Belize	0.3	0.3	0.4	0.4
Bolivia	3.3	3.8	4.3	4.7
Brazil	174.7	198.0	223.6	249.2
Colombia	61.2	69.4	78.4	87.4
Costa Rica	4.6	5.2	5.9	6.6
Cuba	0.1	0.1	0.1	0.1
Dominican Republic	0.1	0.1	0.1	0.1
Ecuador	19.7	22.3	25.2	28.1
El Salvador	0.1	0.2	0.2	0.2
French Guiana	0.2	0.2	0.3	0.3
Guatemala	4.9	5.5	6.3	7.0
Guyana	0.3	0.3	0.4	0.4
Haiti	0.1	0.1	0.1	0.1
Honduras	5.0	5.7	6.4	7.2
Jamaica	0.1	0.1	0.1	0.1
Mexico	23.7	26.9	30.4	33.8
Nicaragua	8.4	9.5	10.7	11.9
Panama (Not Canal Zone)	2.7	3.0	3.4	3.8
Paraguay	5.9	6.7	7.6	8.5
Peru	22.2	25.2	28.5	31.7

APPENDIX 4.2 (*continued*): LOW BASE CASE

COUNTRIES BY GLOBAL REGION	1980	2000	2025	2050
Suriname	0.3	0.3	0.4	0.4
Trinidad & Tobago	0.1	0.1	0.1	0.1
Venezuela	12.7	14.4	16.2	18.1
FAR EAST				
Bangladesh	0.5	0.6	0.6	0.7
Bhutan	0.2	0.2	0.3	0.3
Brunei	1.9	2.2	2.4	2.7
Burma	13.4	15.2	17.1	19.1
Democratic Kampuchea	3.4	3.9	4.4	4.8
Sri Lanka	1.2	1.4	1.5	1.7
India	8.7	9.9	11.1	12.4
Indonesia (& East Timor)	51.1	57.9	65.4	72.9
Lao People's Democ. Rep.	23.6	26.8	30.3	33.7
Malaysia	35.4	40.0	45.3	50.3
Pakistan	0.8	0.9	1.0	1.1
Thailand	27.2	30.8	34.8	38.8
Nepal	4.9	5.5	6.3	7.0
Philippines	14.9	16.9	19.0	21.2
Viet Nam	9.4	10.7	12.1	13.4
Papua New Guinea	2.5	2.8	3.2	3.5
AFRICA				
Algeria				
Angola	3.9	4.4	5.0	5.6
Burundi				
United Rep. of Cameroon	11.0	12.5	14.1	15.8
Central African Rep.	2.5	2.9	3.3	3.6
Chad	3.0	3.4	3.8	4.2
Congo	2.4	2.7	3.1	3.4
Equatorial Guinea	0.2	0.2	0.3	0.3
Benin	1.7	1.9	2.2	2.4
Mali	1.5	1.7	1.9	2.1
United Rep. of Tanzania	3.5	3.9	4.4	4.9
Burkina Faso	3.0	3.4	3.8	4.2
Mozambique	5.0	5.6	6.3	7.1
Niger	1.1	1.3	1.5	1.6
Nigeria	23.4	26.5	29.9	33.3

APPENDIX 4.2 (*continued*) : LOW BASE CASE

COUNTRIES BY GLOBAL REGION	1980	2000	2025	2050
Guinea Bissau	1.8	2.1	2.4	2.6
Rwanda	0.2	0.2	0.3	0.3
Uganda	1.6	1.8	2.0	2.2
Zambia	3.0	3.4	3.8	4.2
Botswana	0.5	0.6	0.6	0.7
Liberia	2.9	3.3	3.7	4.1
Namibia	0.7	0.8	0.9	1.0
Ethiopia	5.5	6.3	7.1	7.9
Gambia	0.1	0.2	0.2	0.2
Ghana	3.8	4.3	4.8	5.4
Guinea	3.9	4.4	5.0	5.6
Kenya	1.2	1.4	1.5	1.7
Malawi	11.1	12.6	14.2	15.9
Sierra Leone	0.4	0.4	0.5	0.5
Côte d'Ivoire (Ivory Coast)	41.4	46.9	53.0	59.1
Madagascar	8.7	9.9	11.1	12.4
Togo	0.4	0.4	0.5	0.5
Senegal	2.1	2.3	2.6	2.9
Somalia	0.8	0.9	1.0	1.1
Zimbabwe	3.0	3.4	3.8	4.2
Gabon	1.6	1.8	2.0	2.2
Sudan	18.8	21.3	24.1	26.9
Zaire	24.8	28.1	31.7	35.4
TROPICS TOTAL	**750.0**	**850.0**	**960.0**	**1070.0**
TROPICAL AMERICA	**350.6**	**397.4**	**448.8**	**500.2**
TROPICAL ASIA	**199.0**	**225.6**	**254.8**	**284.0**
TROPICAL AFRICA	**200.3**	**227.0**	**256.4**	**285.8**

Note: Totals reflect rounding.

HIGH BASE CASE
NET FLUX FROM POPULATION-BASED RATES OF FOREST CONVERSION AND HIGH FOREST BIOMASS ESTIMATES

COUNTRIES BY GLOBAL REGION	1980	2000	2025	2050
TROPICAL AMERICA				
Belize	0.9	1.1	1.5	1.8
Bolivia	10.3	11.9	17.1	20.5
Brazil	510.5	589.5	844.8	1010.9
Colombia	187.3	216.3	309.9	370.9

APPENDIX 4.2 (*continued*): HIGH BASE CASE

COUNTRIES BY GLOBAL REGION	1980	2000	2025	2050
Costa Rica	14.3	16.5	23.6	28.3
Cuba	0.2	0.2	0.3	0.3
Dominican Republic	0.2	0.2	0.3	0.3
Ecuador	60.6	70.0	100.3	120.0
El Salvador	0.3	0.4	0.5	0.6
French Guiana	0.5	0.5	0.8	0.9
Guatemala	15.2	17.5	25.1	30.1
Guyana	0.6	0.7	1.0	1.2
Haiti	0.2	0.2	0.3	0.3
Honduras	15.3	17.7	25.4	30.4
Jamaica	0.2	0.2	0.3	0.3
Mexico	50.7	58.6	84.0	100.5
Nicaragua	25.7	29.6	42.5	50.8
Panama (Not Canal Zone)	8.4	9.6	13.8	16.5
Paraguay	12.8	14.7	21.1	25.3
Peru	68.4	78.9	113.1	135.4
Suriname	0.6	0.7	1.0	1.2
Trinidad & Tobago	0.2	0.2	0.3	0.3
Venezuela	27.2	31.4	45.0	53.8

FAR EAST

	1980	2000	2025	2050
Bangladesh	3.2	3.7	5.3	6.3
Bhutan	0.5	0.5	0.8	0.9
Brunei	4.1	4.7	6.8	8.1
Burma	77.8	89.8	128.7	154.0
Democratic Kampuchea	7.3	8.4	12.1	14.4
Sri Lanka	2.6	3.0	4.3	5.1
India	50.1	57.9	82.9	99.3
Indonesia (& East Timor)	291.5	336.6	482.4	577.2
Lao People's Democ. Rep.	128.7	148.6	212.9	254.8
Malaysia	75.8	87.5	125.3	150.1
Pakistan	1.7	1.9	2.8	3.3
Thailand	143.5	165.8	237.5	284.2
Nepal	10.5	12.1	17.3	20.8
Philippines	86.1	99.5	142.5	170.5
Viet Nam	54.8	63.3	90.7	108.6
Papua New Guinea	5.3	6.1	8.8	10.5

APPENDIX 4.2 *(continued)*: HIGH BASE CASE

COUNTRIES BY GLOBAL REGION	1980	2000	2025	2050
AFRICA				
Algeria				
Angola	8.4	9.6	13.8	16.5
Burundi				
United Rep. of Cameroon	23.7	27.4	39.2	46.9
Central African Rep.	5.5	6.3	9.0	10.8
Chad	6.4	7.4	10.6	12.6
Congo	5.2	6.0	8.5	10.2
Equatorial Guinea	0.5	0.5	0.8	0.9
Benin	3.8	4.4	6.3	7.5
Mali	3.2	3.7	5.3	6.3
United Rep. of Tanzania	47.4	8.6	12.3	14.7
Burkina Faso	6.4	7.4	10.6	12.6
Mozambique	10.6	12.3	17.6	21.1
Niger	2.4	2.8	4.0	4.8
Nigeria	90.4	104.4	149.6	179.0
Guinea Bissau	4.6	5.3	7.5	9.0
Rwanda	0.5	0.5	0.8	0.9
Uganda	3.3	3.9	5.5	6.6
Zambia	6.4	7.4	10.6	12.6
Botswana	1.1	1.2	1.8	2.1
Liberia	11.8	13.7	19.6	23.5
Namibia	1.5	1.8	2.5	3.0
Ethiopia	11.8	13.7	19.6	23.5
Gambia	0.3	0.4	0.5	0.6
Ghana	11.7	13.5	19.4	23.2
Guinea	13.4	15.4	22.1	26.5
Kenya	2.6	3.0	4.3	5.1
Malawi	23.8	27.5	39.5	47.2
Sierra Leone	1.5	1.8	2.5	3.0
Côte d'Ivoire (Ivory Coast)	152.7	176.3	252.6	302.3
Madagascar	35.2	40.7	58.3	69.8
Togo	1.1	1.2	1.8	2.1
Senegal	4.4	5.1	7.3	8.7
Somalia	1.5	1.8	2.5	3.0
Zimbabwe	6.4	7.4	10.6	12.6

APPENDIX 4.2 *(continued)*: HIGH BASE CASE

COUNTRIES BY GLOBAL REGION	1980	2000	2025	2050
Gabon	3.3	3.9	5.5	6.6
Sudan	40.4	46.7	66.9	80.0
Zaire	53.2	61.4	88.0	105.3
TROPICS TOTAL	2520.0	2910.0	4170.0	4990.0
TROPICAL AMERICA	1010.3	1166.6	1671.8	2000.5
TROPICAL ASIA	943.4	1089.5	1561.2	1868.2
TROPICAL AFRICA	566.3	653.9	937.1	1121.3

Note: Totals reflect rounding.

These low and high base case projections represent the upper and lower bounds for the range in emissions possible in models of linear, exponential, and population based deforestation rates, run with both high and low forest biomass estimates. The low base case uses linear deforestation rates reported by FAO/UNEP (1981) for the years 1975–85, coupled with low forest biomass estimates. The high base case uses rates of deforestation that are a function of rates of population growth, coupled with high forest biomass estimates.

Estimates of flux are given for 1980, 2000, 2025, and 2050. After 2050, country-by-country estimates are especially dangerous for the reasons given below. Estimates for the year 2025 in this Appendix (instead of 2020) were selected so they would be directly comparable with the EIS data. They are, however, basically interchangeable with the estimates for 2020 in Appendix 4.1 and Figure 4.2 of this report given the error factor involved in making such projections. The low base case emissions estimates are slightly lower in this Appendix than in Appendix 4.1 and Figure 4.2 due to subsequent refinements in the accuracy of the model.

Two important qualifications should be recognized for these projections:

1) The projections were based on rates of deforestation for *entire regions*, not countries.

Thus, projected fluxes drop to zero only when the forests of an entire region are gone. In fact, some countries will eliminate their forests before others, and fluxes there will drop to zero before all forests in the region are gone. If the projections were carried out country-by-country, the estimates of flux, for countries as well as regions, would probably be lower.

2) All estimates of flux are based on extrapolations or functions of FAO/UNEP data that are now more than ten years old. It is, therefore, interesting to compare these projections with estimates of flux from Myers' new rates of deforestation. Such a comparison can be done for closed forests only, but if Myers' recent estimates are correct, all the projections are low.

1990 Fluxes of Carbon from Deforestation of Closed Forests Only

LINEAR		POPULATION		MYERS' RECENT ESTIMATE
Low	HIGH	Low	HIGH	
0.80	1.58	0.85	1.68	0.9 to 2.3

Source: Houghton, 1990b

5

Coastal Zone Management

CO-CHAIRS
J. Gilbert (New Zealand)
P. Vellinga (Netherlands)

CONTENTS

COASTAL ZONE MANAGEMENT

PREFACE

ACKNOWLEDGMENTS

The focus of the Coastal Zone Management Subgroup (CZMS) is on options for adapting to sea level rise and other impacts of global change on coastal areas. The CZMS held workshops in Miami, Florida, and in Perth, Western Australia, to generate information on available adaptive response options and their environmental, economic, social, cultural, legal, institutional, and financial implications. The countries that contributed to the work of the CZMS by sending experts to these workshops are listed below. Special acknowledgments are addressed to the United States and Australia for hosting these workshops.

The writing of this report was entrusted to Messrs. J. Dronkers, R. Misdorp, P.C. Schröder (the Netherlands), J.J. Carey, J.R. Spradley, L. Vallianos, J.G. Titus, L.W. Butler, Ms. K.L. Ries (United States), J. T. E. Gilbert, J. Campbell, Ms. J. von Dadelszen (New Zealand), Mr. N. Quin, and Ms. C. McKenzie and Ms. E. James (Australia).

We wish to recognize and thank all of the participants and reviewers who contributed their time, energy, and knowledge to the preparation of this report. We hope that the report will help nations in beginning to prepare for the potential impacts of global climate change on their coastal areas.

PARTICIPATING COUNTRIES

Algeria, Antigua and Barbuda, Argentina, Australia, Bahamas, Bangladesh, Barbados, Benin, Brazil, Brunei, Canada, Chile, China, Colombia, Costa Rica, Denmark, Egypt, Fiji, France, Fed. Rep. of Germany, Ghana, Greece, Guyana, India, Indonesia, Iran, Italy, Ivory Coast, Jamaica, Japan, Kenya, Kiribati, Liberia, Maldives, Mauritius, Mexico, Micronesia, the Netherlands, New Caledonia, New Zealand, Nigeria, Pakistan, Papua New Guinea, Philippines, Poland, Portugal, Senegal, Seychelles, South Korea, Spain, Sri Lanka, St. Pierre and Miquelon, St. Vincent and the Grenadines, Thailand, Tonga, Trinidad and Tobago, Tunisia, Turkey, Tuvalu, United Kingdom, United States, USSR, Vanuatu, Venezuela, Vietnam, Western Samoa, Yugoslavia.

PARTICIPATING INTERNATIONAL ORGANIZATIONS

Greenpeace, International Oceanographic Commission (IOC), Organization for Economic Coordination and Development (OECD), South Pacific Regional Seas Programme (SPREP), United Nations Environment Programme (UNEP), United Nations Educational, Scientific and Cultural Organization (UNESCO), World Meteorological Organization (WMO).

—J.T.E. GILBERT & P. VELLINGA
The Hague, April 1990

EXECUTIVE SUMMARY

REASONS FOR CONCERN

Global climate change may raise sea level as much as one meter over the next century and, in some areas, increase the frequency and severity of storms. Hundreds of thousands of square kilometers of coastal wetlands and other lowlands could be inundated. Beaches could retreat as much as a few hundred meters and protective structures may be breached. Flooding would threaten lives, agriculture, livestock, buildings, and infrastructures. Saltwater would advance landward into aquifers and up estuaries, threatening water supplies, ecosystems, and agriculture in some areas.

Some nations are particularly vulnerable. Eight to ten million people live within one meter of high tide in each of the unprotected river deltas of Bangladesh, Egypt, and Vietnam. Half a million people live in archipelagoes and coral atoll nations that lie almost entirely within three meters of sea level, such as the Maldives, the Marshall Islands, Tuvalu, Kiribati, and Tokelau. Other archipelagoes and island nations in the Pacific, Indian Ocean, and Caribbean could lose much of their beaches and arable lands, which would cause severe economic and social disruption.

Even in nations that are not, on the whole, particularly vulnerable to sea level rise, some areas could be seriously threatened. Examples include Sydney, Shanghai, coastal Louisiana, and other areas economically dependent on fisheries or sensitive to changes in estuarine habitats.

As a result of present population growth and development, coastal areas worldwide are under increasing stress. In addition, increased exploitation of non-renewable resources is degrading the functions and values of coastal zones in many parts of the world. Consequently, populated coastal areas are becoming more and more vulnerable to sea level rise and other impacts of climate change. Even a small rise in sea level could have serious adverse effects.

The Coastal Zone Management Subgroup has examined the physical and institutional strategies for adapting to the potential consequences of global climate change. Particular attention was focused on sea level rise, where most research on impacts has been conducted. The Subgroup also has reviewed the various responses and has recommended actions to reduce vulnerability to sea level rise and other impacts of climate change.

RESPONSES

The responses required to protect human life and property fall broadly into three categories: retreat, accommodation, and protection. *Retreat* involves no effort to protect the land from the sea. The coastal zone is abandoned and ecosystems shift landward. This choice can be motivated by excessive economic or environmental impacts of protection. In the extreme case, an entire area may be abandoned. *Accommodation* implies that people continue to use the land at risk but do not attempt to prevent the land from being flooded. This option includes erecting emergency flood shelters, elevating buildings on piles, converting agriculture to fish farming, or growing flood- or salt-tolerant crops.

Protection involves hard structures such as seawalls and dikes, as well as soft solutions such as dunes and vegetation, to protect the land from the sea so that existing land uses can continue.

The appropriate mechanism for implementation depends on the particular response. Assuming that land for settlement is available, retreat can be implemented through anticipatory land-use regulations, building codes, or economic incentives. Accommodation may evolve without governmental action, but could be assisted by strengthening flood preparation and flood insurance programmes. Protection can be implemented by the authorities currently responsible for water resources and coastal protection.

Improving scientific and public understanding of the problem is also a critical component of any response strategy. The highest priorities for basic research are better projections of changes in the rate of sea level rise, precipitation, and the frequency and intensity of storms. Equally important, but more often overlooked, is the need for applied research to determine which options are warranted, given current information. Finally, the available information on coastal land elevation is poor. Maps for most nations only show contours of five meters or greater, making it difficult to determine the areas and resources vulnerable to impacts of a one-meter rise in sea level. Except for a few countries, there are no reliable data from which to determine how many people and how much development are at risk. There are many uncertainties, and they increase as we look further into the future.

ENVIRONMENTAL IMPLICATIONS

Two-thirds of the world's fish catch, and many marine species, depend on coastal wetlands for their survival. Without human interference, (the retreat option), ecosystems could migrate landward as sea level rises, and thus could remain largely intact, although the total area of wetlands would decline. Under the protection option, a much larger proportion of these ecosystems would be lost, especially if hard structures block their landward migration.

Along marine coasts hard structures can have a greater impact than soft solutions. Hard structures influence banks, channels, beach profiles, sediment deposits and morphology of the coastal zone.

Protective structures should be designed—as much as possible—to avoid adverse environmental impacts. Artificial reefs can create new habitats for marine species, and dams can mitigate saltwater intrusion, though sometimes at the cost of adverse environmental impacts elsewhere. Soft solutions such as beach nourishment retain natural shorelines; but the necessary sand mining can disrupt habitats.

ECONOMIC IMPLICATIONS

No response strategy can completely eliminate the economic impacts of climate change. In the retreat option, coastal landowners and communities would suffer from loss of property, resettlement costs, and the costs for rebuilding infrastructure. Under accommodation, there would be changing property values, increasing damage from storms, and costs for modifying infrastructure. Under the protection option, nations and communities would face the costs of the necessary structures. The structures would protect economic development, but could adversely affect economic interests that depend on recreation and fisheries.

An annex of the Coastal Zone Management Subgroup Report shows that if sea level rises by one meter, about 360,000 kilometers of coastal defenses would be required at a total cost of U.S.$500 billion by the year 2100. (This sum only reflects the marginal or added costs and is not discounted.) This value does not include costs necessary to meet present coastal defense needs. The estimate does not include the value of the unprotected dry land or ecosystems that would be lost, nor does it consider the costs of responding to saltwater intrusion or the impacts of increased storm frequency. *Therefore the overall cost will be considerably higher.* Although some nations could bear all or part of these costs, other nations—including many small island states—could not.

To ensure that coastal development is sustainable, decisions on response strategies should be based on long-term as well as short-term costs and benefits.

SOCIAL IMPLICATIONS

Under the retreat option, resettlement could create major problems. Resettled people are not always well received; they often face language problems, racial and religious discrimination, and difficulties in obtaining employment. Even when they feel welcome, the disruption of families, friendships, and traditions can be stressful.

Although the impacts of accommodation and protection would be less, they may still be important. The loss of traditional environments—which normally sustain economies and cultures and provide for recreational needs—could disrupt family life and create social instability. Regardless of the response eventually chosen, community participation in the decision making process is the best way to ensure that these implications are recognized.

LEGAL AND INSTITUTIONAL IMPLICATIONS

Existing institutions and legal frameworks may be inadequate to implement a response. Issues such as compensation for use of private property and liability for failure of coastal protection structures require national adjudication. For some options, such as resettlement (retreat option) and structures that block sediments (protection option), there are transboundary implications that must be addressed on a regional basis. International action may be required through existing conventions if inundation of land results in disputes over national borders and maritime boundaries, such as exclusive economic zones or archipelagic waters. New authorities may be required, both to implement options and to manage them over long periods of time in the face of pressures for development. National coastal management plans and other new laws and institutions are needed to plan, implement, and maintain the necessary adaptive options.

CONCLUSIONS

Scientists and officials from some 70 nations have expressed their views on the implications of sea level rise and other coastal impacts of global climate change at Coastal Zone Management Subgroup workshops in Miami and Perth. They indicated that, in several noteworthy cases, the impacts could be disastrous; that in a few cases impacts would be trivial; but that for most coastal nations, at least for the foreseeable future, the impacts of sea level rise would be serious but manageable if appropriate actions are taken.

It is urgent for coastal nations to begin the process of adapting to sea level rise not because there is an impending catastrophe, but because *there are opportunities to avoid adverse impacts by acting now*—opportunities that may be lost if the process is delayed. This is also consistent with good coastal zone management practice irrespective of whether climate change occurs or not. Accordingly, the following actions are appropriate:

National Coastal Planning

1) *By the year 2000, coastal nations should implement comprehensive coastal zone management plans.* These plans should deal with both sea level rise and other impacts of global climate change. They should ensure that risks to populations are minimized, while recognizing the need to protect and maintain important coastal ecosystems.

2) *Coastal areas at risk should be identified.* National efforts should be undertaken to (a) identify functions and resources at risk from a one-meter rise in sea level and (b) assess the implications of adaptive response measures on them. Improved mapping will be vital for completing this task.

3) *Nations should ensure that coastal development does not increase vulnerability to sea level rise.* Structural measures to prepare for sea level rise may not yet be warranted. Nevertheless, the design and location of coastal

infrastructure and coastal defenses should include consideration of sea level rise and other impacts of climate change. It is sometimes less expensive to incorporate these factors into the initial design of a structure than to rebuild it later. Actions in particular need of review include river levees and dams, conversions of mangroves and other wetlands for agriculture and human habitation, harvesting of coral and increased settlement in low-lying areas.

4) *Emergency preparedness and coastal zone response mechanisms need to be reviewed and strengthened.* Efforts should be undertaken to develop emergency preparedness plans for reducing vulnerability to coastal storms, through better evacuation planning and the development of coastal defense mechanisms that recognize the impact of sea level rise.

INTERNATIONAL COOPERATION

5) *A continuing international focus on the impacts of sea level rise needs to be maintained.* Existing international organizations should be augmented with new mechanisms to focus awareness and attention on sea level change and to encourage nations of the world to develop appropriate responses.

6) *Technical assistance for developing nations should be provided and cooperation stimulated.* Institutions offering financial support should recognize the need for technical assistance in developing coastal management plans, assessing coastal resources at risk, and increasing a nation's ability—through education, training, and technology transfer—to address sea level rise.

7) *International organizations should support national efforts to limit population growth in coastal areas.* In the final analysis, rapid population growth is the underlying problem with the greatest impact on both the efficacy of coastal zone management and the success of adaptive response options.

RESEARCH, DATA, AND INFORMATION

8) *Research on the impacts of global climate change on sea level rise should be strengthened.* International and national climate research programmes need to be directed at understanding and predicting changes in sea level, extreme events, precipitation, and other impacts of global climate change on coastal areas.

9) *A global ocean-observing network should be developed and implemented.* Member nations are strongly encouraged to support the efforts of the IOC, WMO, and UNEP to establish a coordinated international ocean-observing network that will allow for accurate assessments and continuous monitoring of changes in the world's oceans and coastal areas, particularly sea level change.

10) *Data and information on sea level change and adaptive options should be made widely available.* An international mechanism should be identified with the participation of the parties concerned for collecting and exchanging data and information on climate change and its impact on sea level and the coastal zone, and on various adaptive options. Sharing this information with developing countries is critically important for preparation of coastal management plans.

PROPOSAL OF THE CZM CHAIRMEN FOR FUTURE ACTIVITIES

Based on the views of the delegates and the recommendations of the Miami and Perth IPCC-CZMS workshops, the chairmen of the CZM Subgroup and their advisers have undertaken the task to facilitate the implementation of the CZM actions. They suggest that three parallel efforts be undertaken:

1) *Data Collection.* Efforts to build a current global data base on coastal resources at risk due to sea level rise need to be vigorously pursued. The IPCC-CZM Subgroup has developed a questionnaire that can serve as a first step in the collection of this information and in identifying the countries where additional work needs to be done. It is also suggested that a data base or monitoring system be set up which would provide access to and information on adaption techniques, and which could be maintained in an international or regional "clearing house."

2) *International Protocol.* Efforts should commence immediately on the development of an international protocol to provide a framework for international and multinational cooperation in dealing with the full range of concerns related to impacts of sea level rise and climate change impacts on the coastal zone. A protocol is needed to both establish the international frames of reference as well as to establish a clear set of goals and objectives.

Possible elements contained in such a protocol are outlined in Table 5.1.

3) *Organizational Requirements.* A process should be set in motion to guide and assist countries, particularly developing countries in carrying out the IPCC-CZM actions. For this purpose IPCC could consider the formation of a small advisory group to assist in the development of more specific guidelines. Such an advisory group could be formalized at a later stage to support the secretariat for the parties to a future protocol on CZM and sea level rise.

The goals and actions presented in this report are based on problems common to all coastal nations; their achievement can benefit significantly from coordination at the international level.

The three activities described above are considered crucial steps in realizing the full potential of the IPCC process. The Miami and Perth workshops demonstrated very clearly that many developing nations will not be able to respond effectively to the needs that have been identified without some form of assistance.

Additionally, and in accordance with the primary action for the development of comprehensive coastal zone management plans, a timeline (Table 5.2) of essential actions for the formulation of such plans is suggested. Countries that do not currently have coastal management plans could use this timeline as a basis for their own planning process over the next decade.

TABLE 5.1: Possible Elements to Be Included in a Protocol on Coastal Zone Management and Sea Level Rise

Signatories endeavor to develop before the year 2000 a comprehensive coastal management programme. Giving priority to the most vulnerable areas, they agree to:

- *provide* support to institutions conducting research on sea level rise and other impacts of climate change on the coastal zone;

- *cooperate* in international efforts to monitor sea level rise and other impacts of climate change on the coastal zone;

- *contribute* to systematic mapping and resource assessment of coastal zones to identify functions and critical areas at risk;

- *support* international initiatives to provide information and technical assistance to cooperating countries for the preparation of coastal management programmes;

- *contribute* to the exchange of information, expertise and technology between countries pertaining to the response to sea level rise and other impacts of climate change on the coastal zone;

- *promote* public and political awareness of the implications of sea level rise and other impacts of climate change on the coastal zone;

- *manage* the coastal zone so that environmental values are preserved whenever possible;

- *avoid* taking measures that are detrimental to the coastal zones of adjoining states;

- *provide* emergency relief to coastal nations struck by storm surge disasters;

- *establish* a secretariat supported by a small advisory group to facilitate the implementation of the protocol agreements.

COMPOSITION AND FUNCTIONS OF A GROUP OF ADVISERS

ADVISORY GROUP: COMPOSITION AND FUNCTIONS

In order to facilitate the development of responses to the threat of sea level rise and other impacts of climate change on the world's coastal zones, a functional nucleus of experts is required. Its task should be limited to requests by coastal states for assistance in achieving the goal of having a comprehensive coastal zone management programme in place by the year 2000.

Upon receipt of a request for assistance, the IPCC may send an investigative mission to the requesting country or encourage multilateral or bilateral aid organizations to do so. The mission should assess the country's institutional, technical, and financial needs and means, i.e., its requirements in these three areas. The advisory group could prepare guidelines for such missions or provide other support if asked for.

Countries should have the institutional capability to develop their own coastal management programmes and to establish a regulatory framework and the means for enforcement. The required technical capability should be brought to an adequate level by training programmes, expert advice, and appropriate equipment. An estimate of the costs involved (excluding equipment) is presented in Table 5.3.

It should further be determined to what extent the necessary funding can be generated within the country itself and what part could be requested from outside financing institutions.

The mission report referred to above should then be considered against and in the light of worldwide data, synthesized from information supplied by countries with a marine coast. These data should initially be compiled on the basis of the responses to a comprehensive questionnaire sent to all coastal countries, and augmented as required.

Finally, the group of advisers would report to the IPCC panel on country assessments and priorities in terms of vulnerability to the coastal impacts of climate change and on related institutional needs.

TABLE 5.2: Suggested Ten-Year Timeline for the Implementation of Comprehensive Coastal Zone Management Plans

1991	Designate (a) national coastal coordinating bodies, (b) national coastal work teams, and (c) an international coastal management advisory group to support the IPCC-CZM Subgroup and assist national work teams
1991–1993	Develop preliminary national coastal management plans; begin public education and involvement
1991–1993	Begin data collection and survey studies of key physical, social, and economic parameters assisted by international advisory group. For example: • Topographic information • Tidal and wave range • Land use • Population statistics • Natural resources at risk
1992	Adoption of a "Coastal Zone Management and Sea Level Rise" protocol, with a secretariat of the parties, supported by the international coastal management advisory group
1992–1995	Begin development of coastal management capabilities, including training programmes; strengthening of institutional mechanisms
1995	Completion of survey studies, including identification of problems requiring immediate solution and of possible impacts of sea level rise and climate change impacts on the coastal zone
1996	Assessment of the economic, social, cultural, environmental, legal, and financial implications of response options
1997	Presentation to and the reaction from public and policymakers on response options and response selection
1998	Full preparation of coastal management plans and modifications of plans as required
1999	Adoption of comprehensive coastal management plans and development of legislation and regulations necessary for implementation
2000	Staffing and funding of coastal management activities
2001	Implementation of comprehensive coastal zone management plans

TABLE 5.3: Operational Costs for Implementation of CZM-Actions 1, 6, 10

Estimated funding to provide the necessary support to meet the year 2000 coastal zone management plan proposal:

1	120 consultant-months @ U.S.$10,000 per month	= U.S.$ 1,200,000
	Expenses and travel	= U.S.$ 800,000
		= U.S.$ 2,000,000
2	Training of 100 in-countries personnel to strengthen coastal zone technical and planning capabilities 100 people @ U.S.$30,000 each	= U.S.$ 3,000,000
3	Expenses for secretariat and advisory group	= U.S.$ 3,000,000
4	Conferences & workshops	= U.S.$ 1,000,000
5	Contingency	= U.S.$ 1,000,000
	Total for 5 years, 1992–1997	= U.S.$10,000,000

5.1 INTRODUCTION

5.1.1 IMPORTANCE OF THE COASTAL ZONE

A large portion of the world's population has always inhabited coastal areas. Fertile coastal lowlands, abundant marine resources, water transportation, aesthetic beauty, and intrinsic values have long motivated coastal habitation.

The coastal zone includes both the area of land subject to marine influence and the area of sea subject to land influence. Coastal economies include commercial, recreational, and subsistence fisheries; ports and industrial facilities that rely on shipping; and tourism, agriculture and forestry dependent on the coastal climate. Coastal areas are a critical part of the economies of virtually all nations bordering the sea, particularly subsistence economies. Coastal habitats provide important areas for fish and wildlife, including many endangered species. They filter and process agricultural and industrial wastes, and buffer inland areas against storm and wave damage.

FIGURE 5.1: Schematic World Map of Population Densities in Coastal Areas

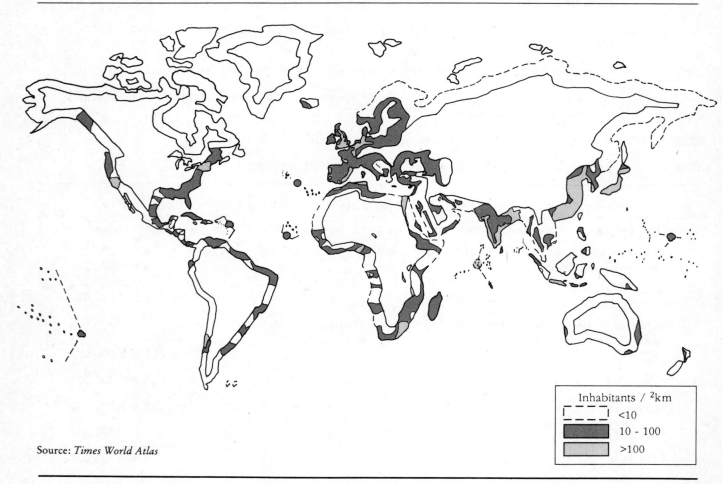

Inhabitants / km^2

(dashed)	<10
(dark)	10 - 100
(gray)	>100

Source: *Times World Atlas*

142

5.1.2 Existing Problems

Throughout the world, nations are facing a growing number of coastal problems as a result of development and increased population pressures. In many areas the functions and values normally associated with coastal areas are being degraded.[1] Flooding, erosion, habitat loss and modification, structural damage, silting and shoaling, pollution, and over-exploitation of living resources, all have major public safety and economic consequences. Yet while these risks are substantial and commonly recognized, the local benefits of using coastal resources outweigh the risks—sometimes significantly—and they continue to attract human activity and development to the coastal zone.

Shoreline alterations, mangrove and coral harvesting, dredge and fill activities, sand and gravel extraction, and disposal of wastes in the marine environment all result in changes to the natural character of the coast. Inland activities—particularly upstream of river deltas—can also have a significant impact on the coast. Construction of dams, diversion of river flows, and removal of ground water or hydrocarbons can result in coastal erosion, subsidence, and shifts in the fresh and salt water interface[2]—so critical to the maintenance of coastal habitats and fisheries.

Obvious examples of the consequences of human activities include (1) the accelerated retreat of two Nile subdeltas following construction of the Aswan High Dam and loss of the sardine fishery;[3] (2) the rapid loss of land in the Mississippi River delta due to subsidence, river levees, canals, and navigation channels;[4] and (3) the exposure of valuable agricultural land in Malaysia to ocean waves as a result of uncontrolled mangrove harvesting.

If populations in coastal areas continue to grow, balancing environmental and development concerns will be increasingly difficult. Changes in climate and sea level will exacerbate many of these problems, particularly for small islands, deltas, and low coastal plains. High population density is therefore the most fundamental problem faced by coastal areas (Figure 5.1).

5.1.3 Global Climate Change

An accelerated rise in global sea level is generally considered to be the most important impact of global climate change on coastal areas. The IPCC-Working Group I projects a rise in global sea level of 30 to 110 cm by the year 2100 (Figure 5.2), due principally to thermal expansion of the ocean and melting of small mountain glaciers. Such a rate of rise would be 3 to 10 times faster than the current rate. Even with actions to limit emissions, the IPCC-Working Group I concludes that there appears to be enough momentum in the global climate system for a rate of accelerated rise in sea level to be inevitable.[5]

Sea level rise could increase flood-related deaths, damage to property and the environment, and cause some nations to lose territorial seas, and hence change the relative values of the coastal zone to society. This will inevitably lead to decisions regarding response options, for example, to retreat, accommodate or protect.[6] A number of researchers

FIGURE 5.2: Global Sea Level Rise, 1985–2100, for Policy Scenario A (No Limitation of Greenhouse Gases) IPCC-WGI, 1990

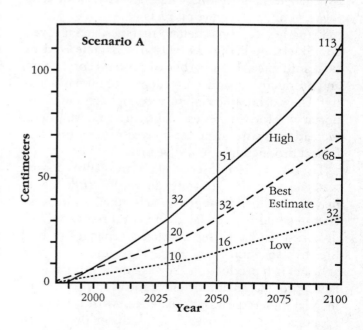

have further suggested that extreme events may become more frequent as a result of climate change.[7] For example, increased ocean temperatures may result in changes in the frequency, duration and intensity of tropical storms. Moreover, the effect of storm surges could be intensified by higher sea levels. Inundation of coastal areas is already common during tropical storms and any increases in the extent or frequency of inundation may render numerous heavily populated areas marginal or uninhabitable.

Because the global climate system is complex, our understanding of it may progress slowly. The existing system for monitoring global sea level cannot yet detect significant changes. Considerable uncertainties remain about the nature, timing, and magnitude of future sea level rise, and the local, national, and regional impacts of human-induced global climate changes.

5.1.4 ECOLOGICAL IMPACTS OF SEA LEVEL RISE

Working Group II[8] suggests that a rise in sea level could: (1) increase shoreline erosion; (2) exacerbate coastal flooding; (3) inundate coastal wetlands and other lowlands; (4) increase the salinity of estuaries and aquifers; (5) alter tidal ranges in rivers and bays; (6) change the locations where rivers deposit sediment; and (7) drown coral reefs.

Estuaries, lagoons, deltas, marshes, mangroves, coral reefs, and seagrass beds are characterized by tidal influence, high turbidity (except coral reefs) and productivity, and a high degree of human activity. Their economic significance includes their importance for fisheries, agriculture, shipping, recreation, waste disposal, coastal protection, biological productivity, and diversity.

The direct effect of sea level rise in shallow coastal waters is an increase in water depth. Intertidal zones may be modified; mangroves and other coastal vegetation could be inundated and coral reefs could be drowned. In turn, this may cause changes in bird life, fish spawning and nursery grounds, and fish and shellfish production. For example, coastal wetlands provide an important contribution to commercial and recreational fisheries, with an annual economic value of over $10 billion in the United States alone.[9] Equally important is the contribution of wetlands to commercial and subsistence fisheries in many coastal and island states. Table 5.4 lists the areas of coastal wetlands of "international importance" for major regions of the world.

In general, the effects on shallow coastal ecosystems are strongly determined by local conditions. A good understanding of the physical and biological processes and topography is required to forecast local impacts. But if the accumulation of sediments cannot keep pace with rising waters, or if inland expansion of wetlands and intertidal areas is not possible[10] (because of infrastructure or a steeply rising coast), major impacts could occur.

The estuarine response to rising sea level is likely to be characterized by a slow but continually adjusting environment. With a change in estuarine vegetation there could be an adjustment in the animal species living in and around the wetlands. Climate change may also provoke shifts in the hydrological regimes of coastal rivers and lead to increased discharge and sediment yields and, consequently, to increased turbidity. These changes, together with a rise in sea level, could modify the shape and location of banks and channels. If no protective structures are built, wetlands can migrate inland; however, a net loss of wetlands would still result.

5.1.5 SOCIAL AND ECONOMIC IMPACTS OF SEA LEVEL RISE

Many developing countries have rapid rates of population growth, with large proportions of their populations inhabiting low-lying coastal areas. A one-meter rise in sea level could inundate 15 percent of Bangladesh,[11] destroy rice fields and mariculture of the Mekong delta, and flood many populated atolls, including the Republic of Maldives, Cocos Island, Tokelau, Tuvalu, Kiribati, the Marshall Islands, and Torres Strait Islands.[12] Shanghai and Lagos, the largest cities of China and Nigeria, lie less than two meters above sea level, as does 20 percent of the population and farmland of Egypt.[13]

Four highly populated developing countries, India, Bangladesh, Vietnam, and Egypt, are especially vulnerable to sea level rise because their low-lying coastal plains are already suffering the effects of flooding and coastal storms. Since 1960, India and Bangladesh have been struck by at least eight tropical cyclones, each of which killed more than 10,000

TABLE 5.4: Areas of Coastal Wetlands* of International Importance in Sq km and As Percentage of Country Areas

	REGION	AREA OF WETLANDS IN KM²	WETLAND AS % OF TOTAL COUNTRY AREA
1.	North America, excluding Canada and USA	32330	1.639
2.	Central America	25319	0.882
3.	Caribbean Islands	24452	9.431
4.	South America Atlantic Ocean Coast	158260	1.132
5.	South America Pacific Ocean Coast	12413	0.534
6.	Atlantic Ocean Small Islands	400	3.287
7.	North and West Europe	31515	0.713
8.	Baltic Sea Coast	2123	0.176
9.	Northern Mediterranean	6497	0.609
10.	Southern Mediterranean	3941	0.095
11.	Africa Atlantic Ocean Coast	44369	0.559
12.	Africa Indian Ocean Coast	11755	0.161
13.	Gulf States	1657	0.079
14.	Asia Indian Ocean Coast	59530	1.196
15.	Indian Ocean Small Islands		
16.	South-East Asia	122595	3.424
17.	East Asia	102074	0.999
18.	Pacific Ocean Large Islands, excluding Australia and New Zealand	89500	19.385
19.	Pacific Ocean Small Islands	—	—
20.	USSR	4191	0.019
	TOTALS	732921	0.846

Source: "A Global Survey of Coastal Wetlands, Their Functions and Threats in Relation to Adaptive Responses to Sea Level Rise." Paper by Dutch Delegation to IPCC-CZM Workshop, Perth, Australia, February 1990.

* Based on: Directories of Wetlands, issued by IUCN/UNEP (1980–90), 120 countries, excluding among others Australia, Canada, New Zealand, USA.

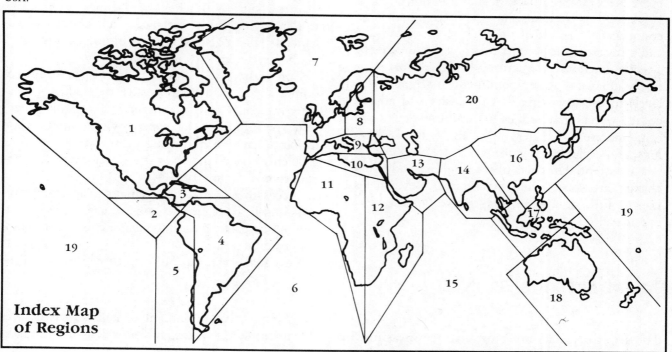

Index Map of Regions

people. In late 1970, storm surges killed approximately 300,000 people in Bangladesh and reached over 150 kilometers inland. Eight to ten million people live within one meter of high tide in each of the unprotected river deltas of Bangladesh, Egypt, and Vietnam.[14] Even more people in these countries would be threatened by increased intensity and frequency of storms.

Sea level rise could increase the severity of storm related flooding. The higher base for storm surges would be an important additional threat in areas where hurricanes, tropical cyclones and typhoons are frequent, particularly for islands in the Caribbean Sea, the southeastern United States, the tropical Pacific, and the Indian subcontinent. Had flood defenses not already been constructed, London, Hamburg, and much of the Netherlands would already be threatened by winter storms.

Many small island states are also particularly vulnerable.[15] This is reflected in their very high ratios of coastline length to land area. The most seriously threatened island states would be those consisting solely, or mostly, of atolls with little or no land more than a few meters above sea level. Tropical storms further increase their vulnerability and, while less in magnitude than those experienced by some of the world's densely populated deltas, on a proportional basis such storms can have a much more devastating impact on island nations.

Disruption could also be severe in industrialized countries as a result of the high value of buildings and infrastructure. River water levels could rise and affect related infrastructure, bridges, port structures, quays, and embankments. Higher water levels in the lower reaches of rivers and adjacent coastal waters may reduce natural drainage of adjacent land areas, which would damage roads, buildings, and agricultural land.

The potential impacts of sea level rise and climate change are varied and uncertain. Nevertheless, there is little doubt that adaptive responses will be necessary.

5.2 ADAPTIVE RESPONSES

The selection and timing of adaptive measures in response to sea level rise would depend on the physical, social, economic, political, and environmental characteristics of the affected areas. Although such measures could be implemented on case by case bases, growing population pressures and conflicting demands in many of the world's coastal areas favor implementation of comprehensive and systematic coastal management programs.

5.2.1 COASTAL MANAGEMENT

The three principal objectives of coastal management are to: (1) Avoid development in areas that are vulnerable to inundation; (2) Ensure that critical natural systems continue to function; and (3) Protect human lives, essential properties, and economic activities against the ravages of the seas. Accordingly, such programmes should give full consideration to ecological, cultural, historic, and aesthetic values, and to the needs for human safety and economic development.[16]

Coastal management programmes usually include governmental controls and private-sector incentives. Vulnerable areas are managed to minimize loss of life and property through such means as setback lines, limits on population densities, minimum building elevations, and coastal hazard insurance requirements. Resilient natural protective features, such as beaches, sand dunes, mangroves, wetlands, and coral reefs, are conserved and enhanced, which also maintains biological diversity, aesthetic values, and recreation.

Comprehensive plans for protecting existing economic activities help to ensure that defense measures are consistent with other coastal management objectives. Policies that specify which activities and development are permitted in new areas promote efficient private land use with the least risk of exposure to coastal hazards.

Successful coastal management programmes require public education to gain broad-based support, and public participation to ensure equal representation of interests.

Response strategies fall into three broad categories:

- *Retreat*: Abandonment of land and structures in vulnerable areas, and resettlement of inhabitants.

- *Accommodation*: Continued occupancy and use of vulnerable areas.
- *Protection*: Defense of vulnerable areas, especially population centers, economic activities, and natural resources.

5.2.2 RETREAT

Options for retreat include:

1) *Preventing development* in areas near the coast.

2) *Allowing development to take place on the condition* that it will be abandoned if necessary (planned phase out).

3) *No direct government role* other than through withdrawal of subsidies and provision of information about associated risks.

Governmental efforts to limit development generally involve land acquisition, land-use restrictions, prohibited reconstruction of property damaged by storms, and reductions of subsidies and incentives for development in vulnerable areas. Many nations have purchased large areas on the coast and designated them as nature reserves. Preventing development can reduce future expenditures for adaptation.

India, Sri Lanka, Tonga, Fiji, Mauritius, Australia, and the United States already require new buildings be set back from the sea. These regulations could be modified to consider the future impacts from a rising sea level, but most nations would require compensation for coastal property owners.[17]

The second option gives the government a more limited role, in that it lays out the "rules of the game"—the eventual transgression of the sea.[18] Investors are accustomed to evaluating uncertainty and can determine whether development should proceed, given the constraint. This approach can be implemented through (a) regulations that prohibit private construction of protective structures, or (b) conversion of land ownership to long-term or conditional leases that expire when the sea reaches a particular level or when the property owner dies.

The third option would be to depend on the workings of the private market. Productive crop and timber lands may be left to slowly and progressively deteriorate as a result of salt intrusion into the groundwater or by surface flooding. Wells and surface water exposed to saltwater intrusion would gradually be abandoned. Natural resources, such as mangroves, marshes, and coral reefs, would be left to their natural processes as sea level rises.

Under this option, governments could take the more limited role of ensuring that all participants in potentially vulnerable areas have full knowledge about the expected sea level rise and its associated uncertainties. Development would presumably not occur if developers, lenders, and insurers were not willing to accept the risks. However, if people continue to build in vulnerable areas, governments must be prepared to take the necessary actions to ensure public safety.

For small island states, retreat does not offer a broadly applicable alternative. There would be little or no land for resettlement, in addition to loss of heritage and cultural upheaval.

5.2.3 ACCOMMODATION

The strategy of accommodation, like that of retreat, requires advanced planning and acceptance that some coastal zone values could be lost. Many coastal structures, particularly residential and small commercial buildings, could be elevated on pilings for protection from floods. To counter surging water and high winds, building codes could specify minimum floor elevations and piling depths, as well as structural bracing. Drainage could be modified. Storm warning and preparedness plans could be instituted to protect the affected population from extreme events. Where saltwater damages agricultural lands and traditional crops, salt-tolerant crops may be a feasible alternative. Fundamental changes in land use may be desirable, such as the conversion of some agricultural lands to aquacultural uses.

Human activities that destroy the natural protection values of coastal resources can be prohibited. Perhaps the most important controls would be to prohibit filling wetlands, damming rivers, mining coral and beach sands, and cutting mangroves. Undeveloped land with sufficient elevation and slope can be set aside to accommodate natural reestablishment of wetlands and mangroves. Within deltaic areas, natural processes can be maintained by diverting water and sediment. In response to salinity intrusion into groundwater aquifers, management controls can be implemented to regulate pumping and withdrawal practices.

Requiring private insurance coverage in vulnerable areas is an important method to compensate injuries and damages caused by natural disasters. It forces people to consider whether risks are worth taking and provides the necessary funds to repair damages and compensate victims.

5.2.4 PROTECTION

This strategy involves defensive measures and other activities to protect areas against inundation, tidal flooding, effects of waves on infrastructure, shore erosion, salinity intrusion and the loss of natural resources.[19] The measures may be drawn from an array of "hard" and "soft" structural solutions.[20] They can be applied alone or in combination, depending on the specific conditions of the site.

There is no single or generic "best solution," as each situation must be evaluated and treated on its particular merits. However, there are some basic steps in the selection of measures likely to produce the highest economic returns. First, those charged with planning, design or management responsibilities in the coastal zones should be cognizant of the potential for future sea level rise. Moreover, proposed plans should leave options open for the most appropriate future response. For example, many protection structures can be planned and designed with features that allow for future incremental additions that, if needed, could accommodate increased water levels and wave action. This can often be done without significant additional costs in the initial investment.

It should be noted that the capital costs associated with the "hard" set of options may prove a barrier to consideration of this option by developing countries and small island states.

5.2.4.1 Hard Structural Options

Dikes, Levees, and Floodwalls are raised embankments or walls constructed for flood protection purposes. Depending on circumstances, internal drainage may be accomplished by gravity flow, tide gates, or pumping systems.

Seawalls, Revetments, and Bulkheads protect inland properties from the direct effects of waves and storm tides. Seawalls and heavy revetments (sloping armored surfaces) are constructed along open coast areas to defend areas against severe wave attack. Lighter revetments and bulkheads usually serve as secondary lines of defense along open coast areas, or as first lines of defense along more sheltered interior shores with low to moderate wave exposure.

Groins are structures placed perpendicular to the shoreline. They generally extend from the land into the near shore zone and trap sediment moving along the shore in order to widen the beach or prevent it from eroding.

Detached Breakwaters are robust structures placed offshore, usually parallel to the shoreline, for the purpose of dissipating the energy of incoming waves to reduce both erosion and damage from storms.

Raising Existing Defensive Structures may be facilitated through the incorporation of such a possibility in the initial design. Some dikes, levees, floodwalls, seawalls, revetments, and breakwaters can be easily raised and strengthened in the event of sea level rise or increased storm exposure.

Infrastructure Modifications may involve the elevation of piers, wharves, bridges, and road and rail beds; modifications to drainage systems; relocations of various facilities and the institution of flood-proofing measures.

Floodgates or Tidal Barriers, which are adjustable, dam-like structures, can be placed across estuaries to prevent the upstream flooding from storm tides. Such barriers are usually left open to avoid interfering with existing flows.

Saltwater Intrusion Barriers in surface water streams can consist of locks or dams that directly block upstream penetration of saline water. Dams upstream of a salt penetration zone may be operated so that water released from the reservoirs at appropriate times can act to minimize the upstream movement of salt water. Under certain conditions, underground barriers can be placed by open-cut or injection methods to prevent saline water intrusion in groundwater aquifers. Fresh groundwater lenses in coastal areas can be maintained by fresh water recharging techniques.

5.2.4.2 *Soft Structural Options*

Beach Filling and Subsequent Renourishment involves the placement of sandy material along the shore to establish and subsequently maintain a desired beach width and shoreline position to dissipate wave energy and enhance beaches, particularly for recreational and aesthetic purposes.[21]

Dune Building, and/or the maintenance and preservation of existing dunes, in combination with adequate beach strands, provides an effective measure of protection to upland properties against the effects of storm tides and wave action.

Wetland/Mangrove Creation can be accomplished through the placement of fill material to appropriate elevations with subsequent plantations.

Other Possible Solutions may be found through increasing resilience and reducing vulnerability of coastal zone features that are under threat of degradation. Options include continued field research in the use of artificial seaweed, artificial reef creation, the rehabilitation of natural coral enhance growth, increasing coastal protection; instituting pollution controls, and preventing the harvesting of mangroves.

5.3 ENVIRONMENTAL IMPLICATIONS OF ADAPTIVE RESPONSES

5.3.1 INTRODUCTION

Most coastal areas contain habitats that are important to fish, shellfish, sea turtles, sea birds, and marine mammals. These areas also have high recreational, cultural, and aesthetic values for many people. Working Group II concluded that a large net

wetland loss would result as sea level rises, because the area onto which new wetlands might expand is less than the area of wetlands at risk.

5.3.2 RETREAT

Enabling wetlands to migrate inland is one possible motivation for a retreat strategy. Coastal wetlands can be found along most of the world's coastal margins, notably in the tropical and subtropical regions. From a global perspective, there is presently a large scale de facto retreat in process, given that most of the world's coastal wetlands border on land areas with low population densities and little major development (Fig. 5.3).[22] Nevertheless, these areas

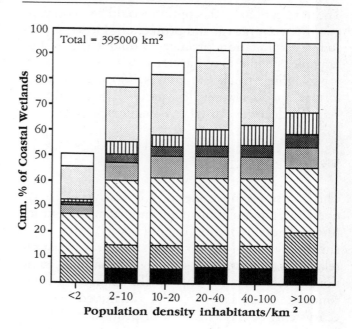

FIGURE 5.3: Cumulative Percentages of Major Coastal Wetlands and Their Population Densities

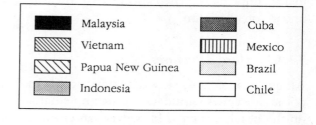

may be developed in the future. Governments should focus attention on their wetland areas and, where appropriate, establish zones to which wetlands will be allowed to retreat using the measures outlined in the previous section. However, even a retreat cannot prevent a large net loss of coastal wetlands.

Developed nations with large land masses, such as the United States and Australia, have implemented retreat strategies along sections of their coasts in the interest of allowing coastal ecosystems, particularly tidal wetlands, to adjust to increased levels of the sea through a slow landward migration. By contrast, on small islands the lack of land for inland migration would restrict the applicability of this option; in the case of atolls, many ecosystems could be completely lost.

5.3.3   ACCOMMODATION

The implications of this option would be a compromise between retreat and protection. However, resource exploitation practices would change. For example, people may harvest mangrove wood for use as pilings to elevate houses. Flood control efforts might alter water flow patterns that could adversely affect the coastal environment.

5.3.4   PROTECTION

This strategy is most relevant for areas having relatively large populations and important infrastructure. These conditions inherently alter the environments. However, the structural measures related to a protection strategy can impose additional alterations not only to the immediate environment but also to the unaltered coastal ecosystems beyond the area of protection. Therefore, environmental impact assessments are particularly important when protective measures are under consideration.

5.3.4.1   *Hard Structures*

Along ocean coasts, seawalls constructed landward of the shorelines would have little immediate impact on the beach systems. However, an eroding shore would eventually reach the seawalls and result in a loss of the natural beach.[23] This impact can be avoided by means of beach nourishment. Similarly, structures could block the inland migration of coastal wetlands. For example, in the United States, the loss from a one-meter rise would be 29–66 percent under the retreat option, but 50–82 percent if shores are protected with bulkheads.[24]

Groins trap sediment moving along the shore. However, protection of one area is generally at the expense of increased erosion downdrift from the protected area. Because these structures do not increase the total sediment available to beaches and barrier islands, their long-term impact is primarily a geographic shift of the erosion. Detached breakwaters often have similar effects, although they allow for nearshore habitat shifts in some cases, and often provide desirable fish habitats in much the same fashion as natural reefs.

Dams and saltwater intrusion barriers can protect water supplies and freshwater habitats. On the negative side, these structures can retain sediments that in turn can increase erosion of coastal headlands and impair the ability of deltaic wetlands to keep pace with sea level rise.

In deltaic areas, levees might be constructed along rivers to prevent flooding due to sea level rise. The resulting "channeling" of rivers could, in some cases, prevent annual river floods from providing sediment and nutrients necessary to enable deltas to keep pace with sea level rise and maintain the fertility of agricultural lands.[25]

5.3.4.2   *Soft Structures*

Soft structures have a less severe impact than hard structures, since they usually consist of simulated natural features, such as beaches and wetlands.[26] The most common "soft engineering" approach is beach nourishment, which involves dredging sand from back bays, navigation channels, or offshore, or excavating material from a land-based source and placing it on the beach. Because beach ecosystems are already adapted to annual erosion/accretion cycles, the placement of sand on the beach generally has little impact on beach ecosystems. By contrast, the dredging itself can seriously disrupt shallow-water ecosystems and wetland habitats, both due to the direct effects of removing material and the resulting increase in turbidity.

5.4 ECONOMIC IMPLICATIONS

5.4.1 INTRODUCTION

The potential economic implications of responses to sea level and temperature rise over periods of fifty to one hundred years are extremely difficult to quantify. The variables to be considered include both the cost of the strategies themselves, and the effects of those costs on national economies. Thus far, only the cost of protecting against inundation and erosion has been estimated worldwide. Much more research needs to be done.

The cost of an adaptive response is site-specific. The nationwide impact of such costs will be greater on rural, subsistence economies, often found in coastal areas in developing countries. Losses of resources such as biologically productive wetland areas, and important mangrove stands and their products would compound such hardship. Reduction in the productivity of fisheries and the loss of land, resources, and jobs are a further consideration. Significant costs can also be associated with the establishment and operation of the institutional mechanisms necessary to implement retreat or accommodation strategies. Finally, and especially if structural response options are exercised, operation and maintenance costs are a factor.

A fundamental element in the decision-making process is a cost-benefit assessment to weigh the life-cycle costs and economic returns of the various alternatives.[27] Not all of the important factors are totally quantifiable in monetary terms, however. This is particularly so for cultural, environmental, and social factors. Nevertheless, these nonquantifiable aspects must be evaluated and given due consideration in an equitable trade-off analysis in order to formulate and implement an acceptable adaptive response.

5.4.2 RETREAT

In densely populated and productive areas, retreat may prove to be the least economically viable response option because of nearly irretrievable losses involved, or, in the case of small islands, the lack of land on which people can resettle. Inundation of fertile coastal agricultural land and frequent flooding of industrial sites and urban centers would threaten the value of past investments and drastically limit future growth. In such cases, it is highly unlikely that the economic benefits of retreat would exceed the costs.

Large-scale resettlements could severely tax the planning, infrastructural, and distributive capabilities of most countries, especially for developing countries. In particular, small island nations would face the most serious economic implications of retreat. At its most extreme, it would involve resettlement of the populations of entire nations.

The slow (albeit increasing) rate of sea level rise permits appropriate planning and incremental implementation of retreat options, and this may reduce costs.[28] However, in the case of arable lands, the inability to produce an adequate food supply may cause further national hardship through both unemployment and loss of exports.

5.4.3 ACCOMMODATION

Accommodation provides opportunities for inundated land to be used for new purposes. Thus, some compensatory economic benefits could be derived from accommodation or adaptation to inundation and flooding. For example, agricultural land may, in some instances, be found suitable for aquaculture; salt-resistant crops may be grown in areas previously dependent on freshwater. Nonetheless, considerable costs may be involved in the planning and restructuring of land use. The necessary expenditure may place significant stress on national budgets, especially in developing countries. In the case of an increase in extreme events induced by climate change, such as tropical storms, altered wave regimes, and storm-surge frequencies, significant expenditures would be involved in disaster planning and preventing loss of life. Responding to such events would require considerable national planning and might involve compensation.

5.4.4 PROTECTION

The economic benefits accruing from protection depend on the values of the land being protected. Benefit categories, as measured against taking no action include (a) prevention of physical damage to property as a result of waves and flooding; (b) prevention of loss of (economic) production and income; (c) prevention of land loss through erosion; and (d) the prevention of loss of natural resources (environmental and recreational).

Costs include capital, operation, and maintenance of the protective measures, as well as any cultural, environmental, and social changes that may result. For example, some hard structural protection works may cause beaches to disappear. For economies heavily dependent on tourism (e.g., Caribbean Islands) this may have serious adverse consequences. As previously stated, the nonquantifiable aspects of cultural, environmental, and social impacts must be considered when selecting any response strategy. Options may be restricted for some developing countries because of costs or lack of technology.

5.4.4.1 *Cost Estimates for Protection*

Although the potential economic developments in the next few decades are difficult to predict, an approximation of basic implementation costs is possible. Although any such calculations are only rough approximations, they provide a useful first estimate and a guide for future data-collection efforts. Table 5.5 illustrates estimates based on a sea level rise scenario of one meter in 100 years, for 181 countries and territories with a marine coast. These estimates show that preventing inundation alone would cost, at a minimum, some U.S.$500 billion (not discounted over time).

This value reflects only the marginal or added costs of protecting against the effects of a one meter rise in water level by the year 2100. It does not include any costs associated with basic coastal protection already in place or necessary to meet present coastal defense needs. The estimate does not include the value of the unprotected dry land or ecosystems that would be lost, nor does it consider the costs of responding to saltwater intrusion nor the impacts of increased storm frequency.[29]

The annual cost of protection amounts to 0.037 percent of total Gross National Product (GNP). It is important to note, however, that the cost burden in terms of GNP is not uniform within the community of nations. For example, the small low-lying island nations of the world would have to commit a relatively high proportion of their GNPs to protect against a one-meter rise in sea level. Specifically, the small island states in the Indian and Pacific Oceans would, respectively, have to commit 0.91 of 1 percent and 0.75 of 1 percent of GNP in the one-meter rise scenario. For some atoll islands the annual cost may be as much as 10–20 percent of their GNP.

5.5 SOCIAL AND CULTURAL IMPLICATIONS

5.5.1 INTRODUCTION

The social and cultural implications of adaptive response measures may affect hundreds of millions of people living in coastal zones, which have an average width of 50 kilometers.[30] In some coastal areas, inhabitants are highly concentrated in a narrow coastal belt (e.g., Java, India, and China).

The lifestyles of many people are tied directly to the coast and its predominant local features. The coast also features strongly in the mythology of many cultures. Numerous places of particular cultural significance are situated on the coast, and many people in developed and developing nations view the sea, coasts, reefs, and beaches as central to their lives.

Social and cultural implications of adaptive options are likely to vary considerably from country to country and from site to site. Options that are socially and culturally beneficial in some situations may be less desirable in others. It is particularly important that the affected communities are consulted and participate in the decisions to adopt particular options. This is probably one of the best means available to identify the social and cultural implications for particular cases.

5.5.2 RETREAT

Retreat, as an option, may imply a partial, incremental process or a sudden large-scale event. In some circumstances, there may be a need to relocate inhabitants, or even entire communities, which could have major financial and social implications in developing countries. The loss of the traditional

TABLE 5.5: Estimate of Marginal Costs Involved in Protecting Countries Worldwide Against the Effects of a 1-Meter Sea Level Rise in 100 Years

REGION	TOTAL PROTECTION COSTS (BILLIONS U.S.$)	STANDARD DEVIATION TOTAL COSTS (IN %)	TOTAL COSTS PER CAPITA (U.S.$)	ANNUAL PROTECTION COSTS AS % OF GNP
1. North America	106.2	43	306	0.03
2. Central America	3.0	19	117	0.12
3. Caribbean Islands	11.1	17	360	0.20
4. South America Atlantic Ocean Coast	37.6	36	173	0.09
5. South America Pacific Ocean Coast	1.7	30	41	0.04
6. Atlantic Ocean Small Islands	0.2	28	333	0.12
7. North and West Europe	49.8	14	190	0.02
8. Baltic Sea Coast	28.9	23	429	0.07
9. Northern Mediterranean	21.0	23	167	0.04
10. Southern Mediterranean	13.5	16	87	0.06
11. Africa Atlantic Ocean Coast	22.8	11	99	0.17
12. Africa Indian Ocean Coast	17.4	27	98	0.17
13. Gulf States	9.1	17	115	0.02
14. Asia Indian Ocean Coast	35.9	26	34	0.14
15. Indian Ocean Small Islands	3.1	27	1333	0.91
16. South-East Asia	25.3	18	69	0.11
17. East Asia	37.6	56	38	0.02
18. Pacific Ocean Large Islands	35.0	23	1550	0.17
19. Pacific Ocean Small Islands	3.9	22	1809	0.75
20. USSR	25.0	36	89	0.01
TOTALS	488.1	12	103	0.04

Source: "Sea Level Rise: A Worldwide Cost Estimate of Basic Coastal Defense Measures." Paper by Dutch Delegation (Rijkswaterstaat/Delft, Hydraulics, Note No. H1068) to IPCC-CZM Workshop, Perth, Australia, February 1990.

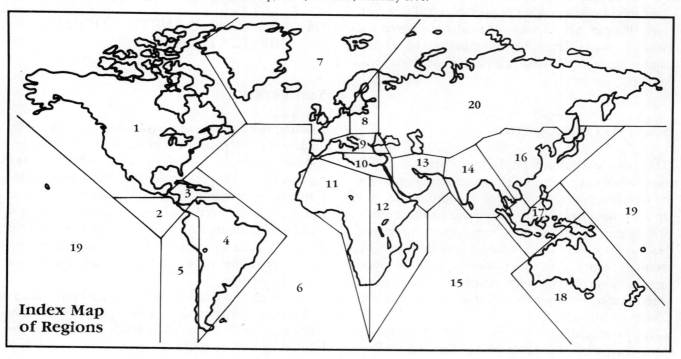

Index Map of Regions

environment that normally sustains economies and cultures and provides for recreational needs, could severely disrupt family life and create social instability, with a resulting adverse impact on the entire community, especially on the young and the elderly. In addition, places of great cultural significance, for example, burial grounds, historic places, or religious centers, could also be lost if retreat occurs.

All retreat options have been identified as having potentially significant implications both socially and culturally. This is particularly the case with abandonment and the resulting need to resettle whole populations. Even though migration is relatively common in some areas, for example, the South Pacific, there remains the need for social adjustment. Situations where an individual's or a community's identity is closely associated with a particular piece of land or access to particular resources, as in most subsistence economies, can have implications that are difficult to resolve.

The greatest implications of retreat may lie in being denied access to the original coast. A well-planned retreat that provides for access to alternative resources could minimize some of these impacts.[31] An associated issue is that of the social implications for the host people at the place of relocation. There exists a potential for conflict, and existing social services may be heavily taxed in the host area if relocation is not well-planned and managed. People may choose not to abandon even vulnerable coastal areas in anticipation of climate change impacts, if there is strong population pressure in adjacent areas.

5.5.3 ACCOMMODATION

The social and cultural implications of accommodation, while not as severe as those of retreat, may still be significant. A change in the economic activity of an area, for instance, from farming to aquaculture, will change lifestyles. Accommodating change may lead to living conditions being less desirable, for example, if properties are subject to periodic flooding, or if problems with sewage disposal occur. Public safety and health will thus be adversely affected by this option.

Accommodation is a more socially desirable option when applied in areas where there is a tradition

of adapting to water, for instance, if people live in houses on stilts or in houseboats.

5.5.4 PROTECTION

Protection has fewer identified social and cultural implications. However, hard structures are likely to have less aesthetic value than the original environment, and access to the shore may also be restricted by some protective options. Beach losses could impair recreation, while loss of wetlands may affect fish stocks. If protective options involve non-local labor, there may be social and cultural friction which could lead to community disruption. Options that can be implemented by communities themselves are less likely to have social and cultural implications than those which require outside labor.

If the protection structures cause alteration to places of cultural significance, there could be opposition to their construction. The loss of any biological resources resulting from protection activities could also be of cultural significance. In some areas, for example, if a significant species is seriously threatened it may no longer be available for ritualistic or economic purposes.

5.6 LEGAL AND INSTITUTIONAL IMPLICATIONS

5.6.1 INTRODUCTION

Existing institutions and legal frameworks may be inadequate to plan and implement adaptive responses. New institutions and legal authorities may be needed in many coastal states. National legislation and institutions for coastal zone management can provide the needed planning. In addition, legal structures to require advance consideration of likely impacts—such as environmental impact review by those planning new projects—can encourage needed foresight.

One matter to consider is that virtually any adaptive option involves the use of "private" land. In some nations such use by individuals may be prohibited by law; in other states the government may not have authority to use the land without the consent of the landowner; and, in others, the government may have the authority to use private land, but only upon providing compensation to the landowner.[32]

An accelerated rate of increase in the global sea level also raises the possibility of legal issues pertaining to maritime boundaries and jurisdiction, and transboundary matters. These issues may require a review of existing international arbitration procedures. An example of the first issue would be if a nation loses maritime boundary base-points and therefore a legal claim to sea territory, or if beach nourishment measures are required in the vicinity of national borders. An example of the second issue would be if protective measures interrupt or impede the longshore sediment transport benefiting an adjoining coastal state. In the worst case, sea level rise may result in the total land loss of an island nation; the resulting legal implications are difficult to predict.[33]

5.6.2 RETREAT AND ACCOMMODATION

The resettlement option could raise significant transboundary implications. The legal authority and institutional capability to manage or direct a relocation on a temporary or permanent basis must be clearly established. Authorities to facilitate and encourage relocation from vulnerable areas, and to subsequently deal with the use of abandoned lands may be needed. In extreme cases when individuals will not leave areas subject to great risk, authorities for condemnation of land and facilities may also be necessary.

Whether relocation is on a temporary or a permanent basis, accommodations for displaced inhabitants must be provided. In some coastal states, relocation could involve tens of millions of inhabitants. The relocation may be further complicated by the lack of land within small coastal or island states. If relocation outside such states is required, then the assistance of regional or international institutions may be needed.

The first option for both measures is to discourage growth in population, or additional development in vulnerable areas that would increase either the risks of losses or would increase the costs of later retreat to unacceptable levels. In order to imple-

ment this option, the coastal state must have the institutional facility for identifying vulnerable coastal zones. Many developing countries do not have the institutional structure and will require assistance to develop a national plan for management of coastal resources and coastal development. Alternatively, a state might choose to encourage private retreat and accommodation actions through non-regulatory measures, such as providing information to the affected population.

In addition to an institutional structure to plan and manage coastal development, legal authorities are needed to enforce restrictions or conditions on coastal development if a coastal or island state chooses to take an active regulatory role for implementing those strategies. Legal authorities may also be needed both to ensure the integrity of natural coastal protection systems and to avoid placing coastal populations and developments in jeopardy from sea level rise. For example, a coastal or island state may need new authority to restrict access or activities to certain areas in order to protect natural systems (such as from the use of mangrove for firewood) as well as have the authority to restrict residential and commercial development (such as new settlements on deltas).

5.6.3 PROTECTION

An important implication of selecting an option to protect against sea level rise is liability for the failure of public protection structures. Structures to protect against sea level rise enable commercial, agricultural, and residential activities to continue in protected areas. Therefore, people and economic resources will be attracted to and concentrate in areas so protected. Should the structure fail, significant loss of life and property could result. Each type of structure is different, and each requires some type of maintenance in order to perform as designed. Where the entity responsible for maintenance is different from the entity that designed and/or constructed the structure, it may be difficult to assess any liability for damage resulting from a failure of the structure. Some public or private entity within each nation, therefore, must have responsibility for maintenance of the physical integrity of these structures.

5.7 PRIORITIES FOR ADAPTIVE RESPONSES

5.7.1 INTRODUCTION

The projected rise in sea level warrants urgent policy responses in many coastal states, particularly those with populated coral atolls and deltas, or those with estuary-dependent fisheries. It is imperative that such actions focus on human safety and on sustainable development of coastal resources.

Even though sea level rise is predicted to be a relatively gradual phenomenon, adaptive strategies may require lead times in the order of 50 to 100 years, to tailor them to the unique physical, social, economic, environmental, and cultural considerations of a particular coastal area. Moreover, even though there may be no need to begin building dikes that are not needed for 50 years, it is appropriate to begin planning now to avoid actions that could increase vulnerability to the impacts of sea level rise. It will take 10 years to implement plans, in view of the time required for the necessary analyses, training the people, developing the plans and mobilizing public and political awareness and support. Therefore, the process should begin today.

Protection from coastal impacts of sea level rise and other impacts of global climate change include both capital investment in defense structures and maintenance costs. Moreover, if the sea continues to rise these structures may have to be augmented or replaced. Similarly, non-structural options to reduce vulnerability to impacts of sea level rise, such as land use planning, may require actions to implement and enforce them.

It is important to recognize that decisions today on planning for coastal development will greatly influence costs for later adaptation to impacts of sea level rise. Venice, Shanghai, New Orleans, and Lagos are all vulnerable because of decisions made 200–2,000 years ago. It is therefore necessary to establish some immediate priorities for planning and management of coastal resources, and for technical and financial assistance to developing countries to facilitate their planning.

There is a need to provide developing countries with the technical and financial assistance required to plan for coastal development in order to reduce vulnerability to impacts of sea level rise. There is also a need to estimate the future long-term funding requirements for developing countries that may be required if protection options are needed.[34]

Finally, the success of strategies to limit climate change is a factor to be considered. Limitation measures will be likely to reduce the costs of adaptation to the coastal impacts of sea level rise; however, it is likely that some adaptation to sea level rise will be required regardless of the limitation strategies eventually implemented.[35]

5.7.2 PRIORITIES FOR ADAPTATION

5.7.2.1 Science/Monitoring

There is still considerable uncertainty regarding sea level rise and other impacts of global climate change. This makes the selection of adaptation options extremely difficult. In particular, there is a lack of regional, national, and site-specific data that is needed to make decisions on adaptive options.

For example, a system to monitor, detect and predict sea level rise is needed to assist in determining the need for construction of protective structures or relocation of coastal inhabitants. There is also a need for information on other impacts, such as changes in tropical storms, in order to plan for natural emergencies.[36]

5.7.2.2 Information

There is a great need to identify those areas that are most vulnerable to the impacts of sea level rise. The identification should concentrate on densely populated low-lying areas, deltas, and small atoll islands.

The need for clearinghouse arrangements to facilitate exchange of information and international data bases accessible to all nations has also been identified.

Development of models and assessment techniques to support coastal planning needs to be undertaken in order to provide decision makers insight into the complicated interactions and conflicting interests that are involved in coastal zone management. Equally important is the transfer to developing countries of existing coastal adaptation technologies and the provision of training in coastal zone management, engineering, and environmental monitoring. Such training might also include technology research centers, extension services, technology advisory committees, technology research and development, technology conferences, and pilot projects to enhance technology transfer.

5.7.2.3　*Planning*

Many priorities have been identified within the broad area of planning. These include:

- *Emergency management planning* to reduce vulnerability of inhabitants in areas exposed to extreme weather events.
- *Coastal management planning* to reduce impacts on development structures and on natural resources of the highest priority. Technical and financial assistance to developing countries may be required to develop and implement national plans for management of coastal development.

5.7.2.4　*Education and Community Participation*

Public education and education of decision makers regarding the impacts of sea level rise and the impacts of ongoing activities is essential, so that everyone understands the risks of development in coastal areas.[37] The involvement of members of the local communities in selecting and implementing response options is also essential for the success of adaptive responses.

5.7.3　PRIORITIES FOR IMPLEMENTING ADAPTIVE OPTIONS IN DEVELOPING COUNTRIES

5.7.3.1　*Retreat*

Technical assistance to developing countries is required for timely planning for resettlement and emergency management pending resettlement. Financial assistance also may be needed to facilitate the resettlement. Assessments of potential relocation sites should be made to minimize dislocation difficulties such as linguistic diversity, cultural differences, and long-term viability.

5.7.3.2　*Accommodation*

Education, technical assistance, and training are required for developing countries so that their populations can understand the risks of development in coastal areas in order to reduce vulnerability to impacts of possible sea level rise.

Technical assistance on alternative economic activities—for example, mariculture instead of agriculture—is required to mitigate the social, cultural, and economic implications of various options. Experience in this field exists in several developing countries and should be shared. The same holds true for alternatives to current coastal development activities. For example, using coastal areas for tourism rather than for industrial or residential activities, may also be a solution.

5.7.3.3　*Protection*

Protection options involving structures in most developing countries are likely to require external assistance. For example, the building of hard structures could require assistance such as transfer of skills and/or capital. There may also be a need for transfer of planning skills to support the choice of appropriate options.

5.7.4　CRITERIA FOR ALLOCATION OF RESOURCES

In addition to identifying priorities for adaptive responses it is also appropriate to identify the priorities that might be used to allocate resources. As the necessary resources vary considerably, depending on the adaptive option and the coastal area, allocation criteria must include consideration of both the options and the area.

A list of sample criteria are provided in Table 5.6. There is no intention to suggest that any one criteria should be pre-eminent. Some may be more significant in some situations while other criteria may be more important than others.

TABLE 5.6: Criteria for Allocation of Resources

RELATED TO THE COASTAL AREA

1. The contribution of current activities within the coastal area that contribute to its vulnerability to sea level rise;

2. The importance of the coastal area in terms of:
   - urgency of risk;
   - proportion of national land area;
   - population affected;
   - environmental importance;
   - economic importance;
   - social and cultural importance; and
   - regional importance.

3. The national ability to finance the response option;

4. The institutional and political ability to realize implementation.

RELATED TO THE ADAPTIVE RESPONSE OPTION

1. The cost of the option;

2. The effectiveness of the option;

3. Cost effectiveness;

4. The economic, environmental, social, cultural, legal, and institutional implications of the adaptive option;

5. The vulnerability of the option to the impacts of an accelerated sea level rise;

6. Performance under uncertainty;

7. Equity.

REFERENCES

1. Misdorp, R. 1990. Existing problems in the coastal zones: A concern for the IPCC? In *Changing climate and the coast: Report to the IPCC from the Miami conference on adaptive responses to sea level rise and other impacts of global climate change*, Proceedings of the Miami Workshop.

2. Commonwealth Secretariat. 1989. *Global climate change: Meeting the challenge*, p. 131, London.

3. Halim, Y. 1974. The Nile and the East Levantine Sea, past and present. In *Recent researches in estuarine biology*, ed. R. Natarujan, p. 76–84. Hindustan Publishing Cooperation, Delhi, India.

4. Day, J.W., and P.H. Templet. Consequences of sea level rise: Implications from the Mississippi Delta. In *Expected effects of climatic change on marine coastal ecosystems*, ed. J.J. Beukema, W.J. Wolff, and J.J.W.M. Bronns, p. 155–165, Boston: Kluwer Academic Publishers.

5. Warrick, R.A., and J. Oerlemans. 1990. IPCC Working Group I: Chapter 9: Sea level rise.

6. Barth, M.C., and J.G. Titus, ed. 1984. *Greenhouse effect and sea level rise: A challenge for this generation*. New York: Van Nostrand Reinhold. Dean, R.G., et al. 1987. *Responding to changes in sea level*. Washington, D.C.: National Academy Press.

7. Emmanuel, K.A. 1988. The dependence of hurricane intensity on climate. *Nature* 326:483–85.

8. IPCC-Working Group II. 1990. Chapter 5: World ocean and coastal zones.

9. U.S. National Marine Fisheries Service, May 1989. Fisheries of the United States 1988. NOAA/NMFS, 1335 East-West Highway, Silver Spring, MD 20910, U.S.A.

10. Misdorp, R., F. Steyaert, F. Hallie, and J. De Ronde. 1990. Climate change, sea level rise and morphological developments in the Dutch Wadden Sea, a marine wetland. In *Expected effects of climatic change on marine coastal ecosystems*, ed. J.J. Beukema, W.J. Wolff, and J.J.W.M. Bronns, p. 123–133. Dordrecht: Kluwer Academic Publishers.

11. Broadus, J.M., J.D. Milliman, S.F. Edwards, D.G. Aubrey, and F. Gable. 1986. Rising sea level and damming of rivers: Possible effects in Egypt and Bangladesh. In *Effects of changes in stratospheric ozone and global climate*. Washington, D.C.: United Nations Environment Programme and Environmental Protection Agency.

12. Hulm, Peter. 1989. A climate of crisis: Global warming and the South Pacific islands. Port Moresby, Papua New Guinea: The Associations of South Pacific Environmental Institutions. Lewis, James. 1989. Sea level rise: Some implications for Tuvalu. *Ambio*, vol. 18, no. 8.

13. Broadus, J., J. Milliman, and F. Gable. 1986. Sea level rise and damming of rivers. In *Effects of changes in stratospheric ozone and global climate*. UNEP.

14. Personal communication with Dr. Nguyen Ngoc Thuy, Marine Hydrometeorological Center, 4 Dang thai Than Street, HANOI, Vietnam, at the Perth CZMS Workshop.

15. Jacobson, J.L. 1990. Holding back the sea. In *Changing climate and the coast: Report to the IPCC from the Miami conference on adaptive responses to sea level rise and other impacts of global climate change*, Proceedings of the Miami Workshop.

16. Jansen, M. 1990. The role of coastal zone management in sea level rise response. In *Changing climate and the coast*, ibid.

17. Leatherman, S.P. 1990. Environmental implications of shore protection strategies along open coasts (with a focus on the United States). In *Changing climate and the coast*, op cit.

18. Titus, J.G. 1990. Strategies for adapting to the greenhouse effect. *J. of the Am. Planning Association*. J.G. Titus, *Greenhouse effect and coastal wetland policy*. Environmental Management. In press.

19. Pope, J.J., and T.A. Chisholm. 1990. Coastal engineering options by which a hypothetical community might adapt to changing climate. In *Changing climate and the coast*. Sorensen, R.M., R.N. Weisman, and G.P. Lennon, 1984. Control of erosion, inundation, and salinity intrusion. In Barth and Titus (eds.), op cit.

20. U.S. Army Corps of Engineers, Coastal Engineering Research Center. 1977. *Shore Protection Manual*. Coastal Engineering Research Center, Fort Belvoir, Virginia, U.S.A.

21. Misdorp, R., and R. Boeije. 1990. A world-wide overview of near-future dredging projects planned in the coastal zone. In *Changing climate and the coast*. Op cit. Titus, J.G. Greenhouse effect, sea level rise, and barrier islands. *Coastal Management* 18:1.

22. Misdorp, R. 1990. Strategies for adapting to the greenhouse effect: A global survey of coastal wetlands. The Netherlands, Rijkswater-staat, Tidal Waters Division, Note no. GWWS-90.008.

23. Howard, J.D., O.J. Pilkey, and A. Kaufman. 1985. Strategy for beach preservation proposed. *Geotimes* 30:12:15–19.

24. Titus, J.G., R. Part, and S. Leatherman. 1990. The cost of holding back the sea. *Coastal management*. In press.

25. Park, R.A. 1990. Implications of response strategies for water quality. In *Changing climate and the coast*, op cit.

26. Leatherman, S.P. 1990. Environmental impacts of sea level response strategies. In *Changing climate and the coast*, op cit.

27. Moser, D.A., E.Z. Stakhiv, and L. Vallianos. 1990. Risk-cost aspects of sea level rise and climate change in the evaluation of coastal protection projects. In *Changing climate and the coast*, op cit.

28. Yohe, G.W. 1990. Toward an analysis of policy, timing, and the value of information in the face of uncertain greenhouse-induced sea level rise. In *Changing climate and the coast*, op cit.

29. The calculations underlying these estimates also assume a one-meter sea level rise in 100 years; that externalities such as other effects of climate change are nil; that present boundary conditions (geomorphological, economic, social) are maintained; and that costs are based on present conditions. The estimates assume that current flood risks remain constant; e.g., areas flooded once every ten years today would still be flooded every ten years when sea level has risen one meter. The complete study can be found in an annex of the CZMS Report.

30. Charlier, R.H. 1987. Planning for coastal areas. In *Ecology for environmental planning*, ed. F.C. Wollf. Norges Geologiske Underskolse. Trondheim, Norway.

31. Yohe, G.W. 1990. Toward an analysis of policy, timing, and the value of information in the face of uncertain greenhouse-induced sea level rise. In *Changing climate and the coast*, op cit.

32. Fishman, R.L., and L. St. Amand. 1990. Preserving coastal wetlands and sea level rises: Legal opportunities and constraints. In *Changing climate and the coast*, op cit.

33. Shihab, H., ed. Proceedings of the Small States Conference on Sea Level Rise, Environment Section. Male, Republic of the Maldives.

34. Campbell, J. 1990. Funding implications for coastal adaptations to climate change: Some preliminary considerations. In *Changing climate and the coast*, op cit.

35. Warrick, R.A., and J. Oerlemans. 1990. IPCC-Working Group I: Chapter 9: Sea Level Rise.

36. Intergovernmental Oceanographic Commission of UNESCO. Global Sea Level Observing System (GLOSS) Implementation Plan. UNESCO/IOC Secretariat, 7, Place de Fontenoy, Paris, France, 75700.

37. Maroukian, K. 1990. Implications of sea level rise for Greece; Erol, O. Impacts of sea level rise on Turkey; Muehe, D., and C.F. Neves. Potential impacts of sea level rise on the coast of Brazil; Andrade B., and C. Castro. Impacts of and responses to sea level rise in Chile; Adam, K.S. Implications of sea level rise for Togo and Benin; and Ibe, A.C. Adjustments to the impact of sea level rise along the west and central African coast. In *Changing climate and the coast*, op cit.

6

Resource Use and Management

CO-CHAIRS
R. Pentland (Canada)
J. Theys (France)
I. Abrol (India)

CONTENTS

RESOURCE USE AND MANAGEMENT

PREFACE

The First Report of the Resource Use and Management Subgroup (RUMS) focuses on that subset of natural resources which are of common interest to most nations, i.e., food, water, land, and biological resources. The topics, which were selected at the RUMS initial meetings, were: agriculture, animal husbandry, fisheries, desertification and salinization, forestry, land use, unmanaged ecosystems (including biological diversity), and water resources. Various nations produced papers addressing one or more of these topics. Based upon these contributions, theme papers addressing water resources, biological diversity and food security were commissioned from individuals associated with the American Association for the Advancement of Science's Water Resources and Climate Change Panel, United Nations Environment Programme, and the Food and Agricultural Organization, respectively. In addition, the several papers on each of the above-mentioned topics were consolidated.

On October 30–November 1, 1989, RUMS held a workshop in Geneva, Switzerland to discuss the theme and consolidated topic papers, and to solicit additional input from experts from developing nations, academia, and other non-governmental entities. This workshop was scheduled immediately prior to a meeting of the Agriculture, Forestry, and Other Human Activities Subgroup (AFOS) to assure broad participation from that Subgroup because of the potential overlaps on food and forests. As this report was produced in parallel with the IPCC Science and Impacts Working Groups' first report, rather than the more logical series approach, special efforts were made to obtain these groups' interim findings. Accordingly, the workshop also heard reports from Working Groups I and II on their efforts, particularly by participants in those Groups concerned with agriculture and food security. This First Report is a synthesis of all of the above efforts.

Appendix 6.1 lists all the papers contributed to the Subgroup, which form the basis of this report. The Subgroup is grateful for these contributions, and to the IPCC Secretariat for making this report possible.

EXECUTIVE SUMMARY

This report addresses adaptation responses pertinent to resource use and management in the event of human-induced climate change. These will help societies anticipate and reduce any negative impacts as well as capitalize on any positive consequences. Topics addressed here include food security (i.e., crops, animal husbandry, fisheries, desertification, and salinization), water resources, land use, and managed and unmanaged ecosystems (including forests and consideration of biological diversity).

The impacts of, and responses to, human-induced climate changes on resource use and management must be considered against a backdrop of (a) population growth, which increases the pressures and demands on all resources, (b) technological change, which may alleviate some pressures while creating others, (c) the hoped-for advancements in economic progress that could, on the one hand, increase the per capita demand for some resources but, on the other hand, allow for greater support for resource conservation and environmental protection, and (d) natural climatic variability.

In many instances, the impacts of human-induced climate change could be a relatively small perturbation on the larger effects resulting from increasing populations and other socio-economic factors. In other instances, they could be quite significant. Human-induced climate change may, in some areas, add to these pressures; in others, it may relieve them.

There are several reasons for focusing on adaptation. First, even in the absence of human-induced climate change, we need to deal with the climate's natural variability. Second, if human-induced climate change is significant, because of the time lags between increases in greenhouse gases (GHG) con-

centration and climate change, and climate change and impacts on natural resources, some degree of adaptation will be necessary even if all anthropogenic GHG emissions were halted. Third, the overall approach toward dealing with climate change must consider both limitations and adaptations as one package. Finally, the potential cost of adaptation will help to assess the costs associated with no action and, therefore, inform decision makers on processes for setting limits on greenhouse gases.

There already exists a large reservoir of experience and knowledge to help formulate and implement adaptation response strategies in the event of climate change:

- Society and living things have a built-in ability to adapt to some degree of climate change because climate is inherently variable at all time scales. Through the ages they have developed the capability, and a suite of responses, to adapt—in many instances, successfully—to extreme events (e.g., floods, droughts).
- As noted in the Scientific Assessment, human activities will alter the climate; however, economic and technological progress will also make it easier to cope with climatic variability and extreme events through earlier warning systems, better infrastructure, and greater financial resources.
- Several climatic zones span the globe, and resource use and management is an ongoing challenge in each of these zones. Therefore, one area could draw upon experiences and practices in those zones that most closely approximate future expected conditions in their area.

This is especially true for activities that are managed intensely (e.g., agriculture, water use, and plantation forests). However, even if there are similarities in the climates, other conditions may be sufficiently different to preclude use of analogues without modifications. In addition, there may be legal, economic, institutional or cultural barriers to translocating adaptation measures from one locale to another. More importantly, the ability to adapt is not uniform: developing nations would have greater difficulty in adapting, especially if that requires substantial financial expenditures and institutional changes, for example those which could occur as a result of loss of cultural and social heritage.

Natural ecosystems, while having a degree of inherent adaptability, are less able to adjust if the disturbance is large or rapid. Increasing populations and associated demand for land for a variety of human activities including agriculture, human settlements, grazing and plantation forests, and for forest products (e.g., fuelwood, timber); and the use of inland and coastal waters for waste disposal have radically altered some natural ecosystems and made others more vulnerable in many parts of the world. Thus, adaptive policies may be necessary to deal with the impacts on such ecosystems: the faster the rate of climate change—or more importantly, the rate of occurrence of the impacts of climate change—the greater the need for developing, evaluating, and implementing such policies.

A major obstacle to development and analysis of adaptive policy options for resource use is the substantial uncertainty associated with each link in the chain of calculations necessary to undertake such analysis. First, credible *regional* estimates of changes in critical climatic factors (e.g., temperature; soil moisture; diurnal, annual, and seasonal variability; frequencies and magnitudes of extreme events such as droughts, floods, and storms) are simply not available. While temperature, if anything, is likely to increase, the rate and magnitude of temperature change is uncertain. For some of the other critical climatic factors, even the direction of change is uncertain. Second, many processes linking changes in climate and in greenhouse gas concentrations to the effects on the biosphere and other natural resources are not well understood or characterized. Third, there are uncertainties regarding the

linkages between natural resource effects and socioeconomic consequences.

Nevertheless, these uncertainties do not preclude planning and taking anticipatory actions, where appropriate. Such action may be indicated in circumstances where the impacts or costs of reacting to climate change could be very high or irreversible; where the lifetime of decisions is long enough to be adversely affected by a change in the climate; or where there are significant technological, informational, cultural, legal, or other barriers to efficient adaptation. Most importantly, many actions that would help societies adapt to any climate change would be worthwhile for other reasons—i.e., they need to be undertaken for wise and sustainable resource use and management whether or not climate changes. As elaborated below, the focus should be on such actions as we further develop scientific and economic understanding.

Temperature increase attributable to atmospheric composition change is expected to be larger at higher latitudes, particularly in the Northern Hemisphere. In the Southern Hemisphere, there may be impacts on agriculture in areas already suffering from environmental problems, e.g., soil degradation. For small island states, sea level rise will expose vulnerabilities. While some semiarid areas may be vulnerable, some in the north may, in fact, benefit, e.g., due to increases in agricultural productivity. Moreover, some sectors are more able to adapt.

In agriculture, and to some extent in forestry, the rate at which technology is introduced is, on the whole, likely to accelerate with the advent of biotechnology and genetic engineering. The efficiency with which many countries have adopted new technology in these sectors, albeit often without adequate environmental assessment, suggests a considerable ability to adapt to new circumstances whatever their cause; however, the closer farming is to subsistence farming the less the likelihood that it would adapt without assistance and "appropriate design."

In recognition of the uncertainties regarding resource use and management, the Subgroup provides a menu of options rather than more specific recommendations. These options need to be analyzed by each nation taking into consideration its specific social, environmental, and economic context. Based on this, a nation can decide on the precise mix of response options that would maximize its net socio-economic (including environmental) well-being. This inevitably must involve achieving the necessary balance between various competing societal objectives (of which dealing with climate change is one) and allocating limited financial, technical and human resources between them. Accordingly, evaluation of each option should be based on the criteria outlined below. An option should be:

- *Flexible*, i.e., adjustable in light of new knowledge, reductions in uncertainties and a variety of climatic conditions, and add to the resilience of resource use and management, i.e., the option should be successful whether or not climate changes.
- *Timely*, considering the time it takes to formulate effective responses, and for any effects on natural resources to be manifested.
- *Feasible*, based on its compatibility with other climate-related responses and socio-economic objectives, and with institutional, economic, legal, and cultural barriers and the degree of difficulty in overcoming them.
- *Economically justifiable*, on grounds other than climate change, while we further develop our scientific and economic understanding. This includes ensuring cost-effectiveness and economic efficiency, and consideration of opportunity costs—aspects that are likely to be met if it provides other non-climate-related benefits. Such analysis should also consider the broad range of social and environmental factors.

In some instances, options may have to be analyzed at the subnational levels, because they may make sense in one area but not in another. In other instances, such options analysis may be done on a national or even on a bi- or multi-national basis (e.g., for rivers crossing international boundaries). To assist in this, national, regional, and international institutions may have to be strengthened (or, where appropriate, new ones established). Such actions could also help implement several of the options mentioned below, e.g., inventorying, monitoring, assessments and information and technology transfer.

The options in this document have been classified into three categories, as described below.

CATEGORY A

Category A consists of options that would augment our knowledge base to make reasoned judgment on response strategies dealing with climate change, and should be undertaken in advance of accurate predictions of climate change impacts. This category includes:

- Developing inventories and data bases of the current state of resources.
- Documenting, cataloguing, and making more accessible the information on resource use and management practices under the widely divergent, existing climatic conditions since many of the response strategies identified here already exist and are in current use around the world.
- Improving our scientific understanding of, and the methodological tools for predicting changes in, critical climatic factors, their impacts on natural resources and their socio-economic consequences.
- Developing estimates of the costs and benefits for both adaptation and limitation measures to help determine the optimal mix of responses that would maximize social well-being.
- Undertaking studies and assessments to gauge the resilience of resources and their vulnerability to climate change which may help establish priorities regarding which areas—and, within areas, which resources—authorities should focus upon.
- Establishing systems that monitor the status of climate, resources, and rates of change in status to give early warning of further potential changes and trends.
- Encouraging research and development by both public and private enterprises directed toward more efficient forestry, agricultural, land and water use practices, and biotechnological innovation with adequate safeguards for public health and safety, and environmental protection. This may require developing new or modified institutional, legal, and financial measures that (i) would allow innovators to benefit from their R&D, and (ii) encourage individuals and communities to develop an economic stake in conservation and in efficient and sustainable resource use and management.
- Continuing existing research and development

on methods to cope with the potentially worst consequences (e.g., developing more drought- or salinity-resistant cultivars) and, in conjunction with improvements in feed, developing more efficient livestock strains and/or breeds using classical and modern breeding techniques such as genetic engineering, which would help keep farming and forestry options open, and research on agrometeorology or agroclimatology.

Other measures in Category A pertinent to food security and land use include new food products that can be more easily stored and improvement of storage facilities where yields tend to be unstable; development of baseline information on and monitoring changes in climate, crop production, pest and disease incidence, livestock fecundity and quality, and fisheries at local, regional, and global levels; protection of germ plasm resources and biodiversity (including increased research on the preservation of biological resources *in situ* and *ex situ*); investigations into the size and location of protected natural areas and conservation corridors; increased R&D to enable fuller utilization of felled trees, and to increase sustainable yields for traditional and new uses of forests and associated products (e.g., for fuel wood, chemicals, materials, fruits, and nuts).

CATEGORY B

Category B consists of responses that are probably economically justified under present-day conditions and could be undertaken for sound resource management reasons, even in the absence of any climate change, to meet increasing demands on these resources in a sustainable manner. In general, this would mean improving the efficiency of the use of natural resources by increased productivity and fuller utilization of the "harvested" component of resources and by waste reduction. Measures that could be implemented in the short-term include:

- Increased emphasis on the development and adoption of technologies which may increase the productivity or efficiency (per unit of land and water) of crops, forests, livestock, fisheries, and human settlements consistent with the principles of sustainable growth and development. Such efficiencies reduce the demand for land for a variety of human activities including agriculture, plantation forests, grazing and human settlements, which is the major cause of conversion of natural ecosystems and loss of biological diversity. In addition to alleviating pressures on land, such measures would also help reduce emissions of greenhouse gases. On the other hand, in areas where carrying capacities are strained or extended, appropriate measures to expand carrying capacities should be considered, e.g., by implementing pollution control measures, improving access to potable water or transportation infrastructure. Examples of options to increase efficiency include more efficient milk and meat production per unit product; improved food storage and distribution; and better irrigation water management practices and drainage, which would allow water supplies to serve greater areas, and limit salinization.

- Increased promotion and strengthening of resource conservation and sustainable resource use—especially in highly vulnerable areas. Climate change may or may not exacerbate conditions in marginal and endangered ecological and agricultural systems and overused water basins. Assessments of the potential impacts of climate change might help clarify this point. In areas that are likely to be further stressed, various initiatives could be explored for conserving the most sensitive and valuable resources including strengthening conservation measures, managing development of highly vulnerable resources, and promoting reforestation and afforestation. Identification of the most sensitive and valuable resources would necessarily be based upon each society's consideration of relevant social, cultural, environmental, and economic factors.

- Acceleration of economic development efforts in developing countries. Because these countries have largely resource-based economies, efforts improving agriculture and natural resource use would be beneficial. Such efforts would help formation of such capital as may be necessary to adapt to climate change, and generally make sustainable growth and development more feasible.

- Developing methods whereby local populations and resource users gain a stake in conservation and sustainable resource use, e.g., by investing resource users with clear property rights and long-term tenure, and allowing voluntary water transfer or other market mechanisms.

Other options in this category include: maintaining flexibility in resource use to the extent practicable; decentralizing, as practicable, decision making on resource use and management, while assuring coordination with adjacent jurisdictions and incorporating mechanisms whereby interests of the broader society are also considered; strengthening existing flood management measures; continuing and improving national and international agricultural and natural resource research/extension institutions; strengthening mechanisms for technology transfer and development; and reviewing subsidies for natural resource use.

CATEGORY C

Category C are responses that, because they are costly, should be considered in the longer-term once uncertainties regarding climate change impacts are reduced. Options in this category include, as and where appropriate: building large capital structures (e.g., dams); examining the feasibility of and, if appropriate, strengthening and enlarging protected natural areas; establishing conservation corridors; as appropriate, reviewing and eliminating direct and indirect subsidies and incentives for and institutional barriers to inefficient resource use; and developing alternatives to current resource-based livelihoods. Many of these options may have to be taken in anticipation of the effects of climate change.

In recognition of the importance and special situation of developing nations, the issues of technology transfer and financial assistance are deferred to the Response Strategy Working Group's Task B report.

In summary, development and analysis of limitation and adaptation strategies must be harmonized. There is a relationship between the timing and costs of limitation and adaptation: the slower the rate of climate change, the easier it would be to adapt, and vice versa. Limitation measures should be consistent with adaptation goals, where possible. Thus, analysis of monoculture plantations to absorb CO_2 should consider their potential negative impact on biological diversity. Moreover, analysis of the relative merits of various limitations strategies should consider other non-warming consequences of controlling the various gases (either individually or together) including, e.g., direct effects of CO_2 and CFCs on the biosphere. Thus a truly comprehensive approach toward human-induced climate change should recognize that controlling the different gases might have different effects on the adaptive capacity of natural resources.

There is a range of adaptive measures that could be used to implement response strategies tailored to national situations. These strategies should focus primarily on addressing current problems that seem likely to be intensified by climate change. In considering the above three categories of options, particular attention and support should be given to those developing nations which appear to be most vulnerable to the impact of climate change but the least able—in financial and human capital terms—of responding to them.

6.1 INTRODUCTION

6.1.1 THE IMPORTANCE OF CLIMATE FOR SOCIO-ECONOMIC AND ENVIRONMENTAL WELL-BEING

The climate affects all living things, and in turn, they affect the climate. It influences where and how human beings live, the kind of shelter we build, the quantity and quality of food and water we ingest, what we wear each day, as well as how and where we spend our leisure time.

Resources and their use and management depend upon the climate. It is a major determinant of agricultural and biological productivity in managed and natural lands, water supply and demand, and the distribution, types, and composition of the flora and fauna. Clearly, a significant climate change could affect, for better or worse, the very resources upon which life itself depends and, through that, the social, environmental, and economic well-being of this planet.

6.1.2 STRUCTURE AND CONTENT OF REPORT

This report offers a menu of options that could help societies adapt resource use and management to any effects of human-induced climate changes (hereafter referred to as "climate change" unless otherwise qualified). Such adaptation will be necessary whether it is desired to reduce negative impacts or capitalize on any positive aspects. The options presented here could (a) reduce the uncertainties regarding the effects of climate change on resource use and management, and (b) improve the ability of nations/regions to adapt to the effects of climate change in both the short and long term, should that

be necessary. Short-term adaptation options identified here may be economically justified under present-day conditions and might be beneficial even if climate were not to change, whereas long-term options would generally be costlier and should be considered after uncertainties regarding climate change and its impacts are reduced.

It is expected that some options would be desirable for some nations but not others. Furthermore, within nations, an option may be suitable in one area, but not in another. It is hoped that nations will evaluate the desirability and suitability of each option, taking into consideration the particular situation(s) within their jurisdiction.

This report generally avoids narrowing the menu by eschewing recommendations of specific options because such recommendations should be based upon evaluations of their social, environmental, and economic consequences (including cost-effectiveness, economic efficiency and suitability for a given area); however, such evaluations cannot currently be undertaken with sufficient confidence—especially at the regional scale, which is the scale at which the status of resources usually has to be assessed.

6.1.3 POSSIBLE AND LIKELY CHANGES IN CRITICAL CLIMATIC FACTORS

Scientists agree that an increase in greenhouse gas concentrations will eventually be followed by an increase in the globally averaged temperature. However, there are significant scientific uncertainties regarding the magnitude, rate, and timing of such change due to any specified level (e.g., a doubling) of greenhouse gas concentrations.

Temperature, moreover, is only one aspect of the climate. There are other climatic factors that are just as, if not more, critical to sustaining life on this planet and to the status of resource use and management. These include precipitation, soil moisture, wind speed, the frequencies and magnitudes of extreme events such as floods, droughts, hot and cold waves, hurricanes, cyclones, and other storms, and climatic variability on diurnal, seasonal, and annual scales. With respect to many of these critical climatic aspects, scientists are uncertain regarding the direction of change as well as their rates, magnitudes, and timings.

These uncertainties multiply further as one attempts to estimate changes on spatial scales which are less-than-global (e.g., regions or watersheds) or temporal scales that are less than annual (e.g., seasonal, weekly, daily). However, it is at these scales that assessments regarding the status of, and impacts on, resource use and management have to be made to be credible. Yet a wide range of plausible climate change scenarios can be devised to test sensitivities of systems.

6.1.4 POSSIBLE AND LIKELY EFFECTS ON RESOURCES AND SOCIO-ECONOMIC WELL-BEING

Resources will be affected directly because of changes in critical climatic factors and in greenhouse gas concentrations, and indirectly because of sea level rise (see the report of the Coastal Zone Management Subgroup). In turn, these impacts on resources will be translated into socio-economic (including environmental) consequences.

The impacts of human-induced climate changes on resource use and management must be considered against a backdrop of population growth that will increase the pressures and demands on all resources, and technological change which may alleviate some pressures while creating others. Moreover, the hoped-for improvement in each individual's economic status will have a complicated effect on resources. On the one hand, economic growth could increase the per capita demand for some resources and may increase greenhouse gas emissions; on the other, it also allows for greater support for resource conservation and environmental protection.

In many instances, the impacts of climate change could be a relatively small perturbation on the larger effects resulting from increasing populations and other socio-economic factors. Human-induced climate change will, in some areas, add to these pressures; in others, it may actually relieve them. In addition to the scientific uncertainties associated with estimates of various critical climatic factors (noted in Section 6.1.3) which themselves serve as inputs to estimates of resource use and socio-economic consequences, there are other uncertainties regarding the impacts of climate change:

- There are uncertainties regarding the relationships between climate, resources and socio-economic consequences.
- There are interconnections (feedbacks) between estimates of atmospheric greenhouse gas concentrations, the various critical climatic factors, resource use, and socio-economic consequences that are not well understood.

While recognizing the uncertainties in all the above steps, the following general statements can be made regarding resource use and management in the event of climate change:

- The ability to match water supply with demand will vary from area to area; temporal and spatial precipitation patterns will change as well as frequencies and magnitudes of floods and droughts. All this will occur in ways that are not for the most part determinable now, though some aspects, such as earlier snowmelt, would be more likely, as would changes in demand for water. The net effect in some areas would be positive; and in others, negative.
- Food production at the global level could, in the face of estimated climate changes, be sustained at levels sufficient to meet world demand. However, there might be substantial dislocations in some areas in the agricultural sector: some would become more productive, others less. Agroclimatic zones in the mid- and higher-latitudes would shift poleward. In addition, there will probably have to be significant adjustments in types of crops raised and in management practices to optimize farm incomes, and to avoid localized food shortages in some areas where poor economic conditions may not allow food requirements to be otherwise met (via domestic and international markets/trade).

However, because agriculture is a heavily managed system with substantial flexibility, such adjustments (either in absolute terms or in their rate) may be within the bounds of human experience and likely technological progress—at least for the next several decades.

- Whether or not fisheries become more or less productive, the spatial and temporal distribution of commercial marine fisheries may be changed. Some fishing communities (and/or nations) may gain at the expense of others.
- Impacts on unmanaged or natural ecosystems could be critical. If the rate of climate change is more rapid than their rate of adaptation, species composition and biodiversity may be affected. However, established growth may persist for decades or longer, and, in some cases, degenerate. Nevertheless shifts in climatic zones could, over the span of several decades or centuries, shift species poleward or to higher altitudes or lead to cases of extinction. Moreover, if the rate of climate change is sufficiently rapid, ensembles of species may end up trapped in climatic regimes for which they are not generally suitable and some species may not adapt quickly enough to prevent their extinction. The effects on migratory species may be mixed: the longer the migratory route, the greater the likelihood of disruption.
- Climate zones for temperate and boreal forests could shift northward, which would affect productivity and biomass generation.
- There may be changes in the location and extent of deserts and salinized lands.

Such climate change as could occur is expected to be greater as one moves poleward—this is particularly true for the Northern Hemisphere. However, the physical effects of climate change on resource use and management may not follow the same pattern (i.e., increasing effects with latitude). However, socio-economic impacts due to these effects could have a more complicated pattern. Many of the northern countries—being more developed, having stronger economies and greater wealth per capita—are better able to adapt to the effects of climate change than would many developing nations. This would be particularly true for developing nations that have relatively long coastlines.

6.1.5 ADAPTATION OBJECTIVES

The ultimate objective of adaptation is to maximize social well-being (which incorporates environmental and economic well-being) for a given set of climatic conditions (or trends in these conditions). Such well-being has to be maximized over the span of this (and succeeding) generations. To achieve this, society has to take advantage of any positive impacts resulting from climate change, while also reducing negative impacts. Thus, another adaptation objective would be to place society in a position to respond rapidly and efficiently to the impacts of climate change.

In attempting to maximize social well-being, it should be noted that there are several societal goals that are partly in competition with each other. These include the need for agricultural and economic development and growth to enhance both economic and food security even as populations increase, and the need to assure that such economic development and growth is done in a sustainable manner so that the resource base is not degraded and is consistent with the need for a clean and healthy environment. The importance that one society (or segment of society) places on each of these goals is determined by a variety of factors including current socio-economic status, cultural traditions, institutional systems, expectations for the future, and historical factors. Because of these competing considerations, as noted in Section 6.1.2, it is to be expected that some adaptation options could maximize social well-being for one set of social, environmental, and economic circumstances but not for another, and that different societies (or segments of societies) would choose to pursue various options to different degrees.

6.2 THE NEED FOR ADAPTIVE POLICY RESPONSES

The menu provided in this report is designed to help societies adapt to climate change by anticipating and reducing expected negative impacts or capitalizing on any positive aspects. Information on the status of resources and their use and management under a variety of climatic regimes would assist societies learn from existing analogues to conditions that might reasonably be expected in the future.

Adaptive policy responses, which could include both actions taken in anticipation of expected climate change (or its impacts) as well as actions undertaken after the impacts are evident, are desirable for several reasons.

- Whether or not there will be significant human-induced climate change, the climate's inherent variability makes adaptation unavoidable.
- Assuming human-induced climate change is significant, the expected delays, which could be decades long, between increases in greenhouse gas concentrations and changes in climate and the status of resources make some adaptation essential—regardless of how rapidly greenhouse gas concentrations are limited.
- Both limitations and adaptations must be considered as a package. This would assure the cost-effectiveness of the entire set of response strategies that might be necessary to deal with human-induced climate change.

To the extent practicable, limitations should not make adaptation more difficult, and vice versa. Options for limitation strategies should be evaluated carefully to ensure they would not be counterproductive in terms of adaptation (or its ultimate objectives). Analysis of monoculture plantations to absorb CO_2 should consider their potential negative impacts on biological diversity. Similarly, one factor to be considered in the analysis of limitation measures is that an increase in carbon dioxide concentration would make crops and vegetation more resilient by enhancing photosynthesis rates and making them more resistant to drought and salinity, but that the combined effects of CO_2 and climate change are uncertain.

- Moreover, decisions on the degree and rate of limitation(s) that may be necessary should take into consideration the rates of change to which societies and ecological systems can adapt.
- Anticipatory policies may be necessary in circumstances where project lifetimes are so long that the project's usefulness (or benefit/cost ratios) could be severely compromised if climate were to change significantly during its

lifetime. Such circumstances could include construction of new dams, establishment or augmentation of conservation areas, and selection of seedlings for forestation.
- Anticipatory policies may also have to be considered for situations where the impacts or costs of reacting to climate change are expected to be very high, unacceptable or irreversible. This might be the case for some actions that might place unique or critical resources—e.g., some species—at greater risks of extinction.
- Anticipatory policies will have to consider whether there are barriers to adaptation, such as lack of technology and information or cultural or legal barriers that may prevent or inhibit efficient responses.

While noting the need for adaptive response strategies, it should also be recognized that adaptation to climate is at least as old as the human species. Even without man's influence, climate is extremely variable. At any location the timing and amount of precipitation and temperature varies at all time scales (seasonal, annual, or decadal). Moreover, droughts, floods (and other extreme events) occur regularly. These climatic variations indicate that humans and living things have some built-in ability to adapt to climate change.

While it is possible that human activities and technology could change the climate, society's ability to cope with such climatic events is now higher than ever because of technology (including better communications, transportation, and food storage), and increased wealth. The lessons learned from these events can all be applied in formulating and implementing responses to human-induced climate change. Moreover, people in one area can learn from those in other areas. We now grow several crops over a wider range of climate (and temperature) than is "predicted" under scenarios that would double greenhouse gas concentrations. Thus, if southern Canada were to become as warm and dry as, say, Texas then it may be able to learn from the agricultural practices in the latter area to better adapt to and cope with the impact of climate change on food security even though not all other conditions would be identical. As long as the new climate in an area has existing analogues anywhere in the world, there is the potential for such adaptation—especially for activities that are managed intensely (e.g., agriculture, water use, and plantation forests).

However, this ability to adapt is not uniform. It will vary from nation to nation depending upon its economic and institutional capabilities, and the degree of net negative impacts due to climate change. Moreover, as noted, even if there are similarities in the climates, other conditions may be sufficiently different to preclude use of analogues. In addition, there may be legal, institutional or cultural barriers to adaptation.

Increasing populations, and associated demand for land for agriculture, human settlements, grazing and plantation forests, and for forest products (e.g., fuelwood, timber), have reduced the area of some natural ecosystems. Moreover, human activities have made certain natural ecosystems more vulnerable in many parts of the world even though they have a degree of inherent variability. Thus, adaptive policies may be necessary to enhance the resilience of such ecosystems: the faster the rate of climate change, the greater the need of developing, evaluating and implementing such policies.

6.3 EVALUATION AND TIMING OF RESPONSE STRATEGIES

As noted above (Section 6.1.5), the basic rationale for considering and/or adopting response strategies is to maximize social well-being (which includes economic and environmental quality). Thus, that should be the basis for any evaluation of responses. While attempting this, one has to be cognizant that resources (whether they be natural or human and financial) are limited. Opportunity costs must be considered: expenditures on a response strategy will divert resources from other potentially worthwhile social uses. These include public health, environmental and safety needs, and economic growth. Wealth generated by such growth will eventually make response measures more affordable. Poverty is one of the major causes of environmental degradation. It is also one reason why poorer nations employ obsolete technologies that often are more inefficient in terms of both energy use and emissions. Therefore, cost-effectiveness and an assessment that benefits exceed costs are necessary, but

not sufficient, criteria for ensuring maximum social well-being and economic efficiency: sufficiency can only be established if it can be determined that the cost (including monetized *and* unmonetized consequences) incurred in the development and implementation of a response strategy is the best use for society's resources.

Several steps should be taken prior to any evaluation of response strategies. As a start, one should differentiate between the possible and likely impacts of climate change. Inherently, anything that does not violate a law of nature (e.g., the Third Law of Thermodynamics) is possible. Thus, dealing with merely what is possible is not a wise use of resources. Ideally, one would need to know the probability distributions for climate change effects (as a function of space and time). However, the uncertainties associated with climate change and its effects generally preclude such differentiation between potential and likely effects. These uncertainties also restrict the analysis of response strategies in terms of their effectiveness and intercomparisons of the social, economic, and environmental consequences of implementing response strategies versus doing nothing. Nevertheless, adaptive response strategies can be evaluated on the basis of several criteria.

In evaluating options, one needs to keep in mind that the direction, magnitude, and timing of impacts are uncertain and take into consideration differences in the carrying capacities for various resources. Specific criteria—all of which should be met—include:

- *Flexibility.* Since the effects of climate change are uncertain, responses need to be successful under a variety of conditions, including no-climate-change. Thus, flexibility is a matter of keeping options open. For instance, a market mechanism for pricing and allocating resources will work under a variety of conditions and, therefore, is flexible.
- *Economically Justifiable Based on Other Benefits.* This is often referred to as "doing things that make sense anyway." Such policies would be justifiable in their own right, i.e., in the absence of climate change. They would necessarily have to meet other societal goals besides preparing for climate change; thus, they would be beneficial even if climate were not to change.

Such policies include those that would enhance net public well-being by, e.g., environmental quality, or food or economic security. Thus, even if climate does not change, society would reap net benefits from this approach. Factors to consider in assessing whether a policy "makes sense anyway" include economic efficiency (including environmental factors), cost-effectiveness, and opportunity costs.

- *Timing*. Since climate change may not be felt for decades, the benefits of adaptive policies may also not be realized for decades. Thus, expensive anticipatory actions will not be justified unless the expected costs of climate change are very high. For instance, a dam should not be constructed today in anticipation of being needed several decades hence. On the other hand, if a dam is being built now, it may be useful to "design in" the ability for future augmentations—if that changes the costs only marginally. Factors to consider in assessing timeliness include: (a) whether there is a critical point in time before the adaptation strategy needs to be implemented, and (b) how much time does it take to efficiently develop the response (and necessary technology), and educate and disseminate it to users/implementers?
- *Feasibility*. Adaptive strategies must be consistent with legal, institutional, political, social, cultural, and financial arrangements. These are critical aspects of their "do-ability." However, in some instances, policies may specifically be directed at modifying or removing such barriers.
- *Compatibility*. Response strategies for one sector (or activity) should not run counter to adaptive strategies in other sectors or activities. Similarly, adaptive strategies should not defeat or negate limitation strategies (or their objectives), and vice versa. See Section 6.1.5.

As noted in Sections 6.1.2 and 6.1.5, in many instances evaluations of adaptive options may have to be undertaken on relatively small geographical scales (i.e., regions or watersheds) and consider the specific social, economic, and environmental context. This is necessary to achieve a balance between various competing societal objectives, and thereby maximize net social well-being. In some instances responses may have to be evaluated at the regional, national or sub-national level. However, no evaluations were attempted for this report.

Readers, while going through the response strategy options in the subsequent sections, should keep the above evaluation criteria in mind to judge for themselves the suitability of the options in their particular context.

6.4 IDENTIFYING AND CLASSIFYING OPTIONS FOR ADAPTATION

The options identified in this report were culled from topic papers prepared for RUMS by various nations, theme papers contributed by invited speakers or organizations, and papers and comments provided at the Subgroup Workshop in October 1989. This workshop attempted to classify identified response strategies into three categories:

- Those that augment our knowledge base to make reasoned judgment on response strategies dealing with climate change and that should be undertaken in advance of the availability of accurate regional predictions (Category A)—e.g., inventorying, monitoring, assessments, and information and technology transfer.
- Responses that are probably economically justified under present-day conditions (see Section 6.3) and that, therefore, could be implemented in the short term (Category B)—e.g., measures that could improve efficiency of use of the "harvested" resource.
- Responses, that should be considered in the longer term (Category C). Because these are generally more costly, it may be prudent to consider them once uncertainties regarding climate change impacts are reduced. Examples of such measures are: preparing communities for a shift in existing resource-based livelihoods, or building new capital structures (e.g., dams).

This classification scheme is used for each of the response strategies in the following sections.

6.5 RESPONSE OPTIONS APPLICABLE TO RESOURCES IN GENERAL

This section offers a menu of options applicable to several types of resources. Measures dealing generally with technology transfer and possible financial assistance in the context of developing nations are addressed in the Response Strategy Working Group's Task B report. Many of the following options would help reduce critical lead times and planning horizons necessary to design and implement specific actions in the face of climate change or its potential impacts.

6.5.1 RESEARCH, INFORMATION, AND TECHNOLOGY DEVELOPMENT AND TRANSFER

The knowledge base relevant to making policy decisions needs to be expanded (A). Clearly, there needs to be a concerted effort to undertake the requisite research to reduce uncertainties associated with predictions of the status of resources, their use and management at various geographical scales (e.g., for terrestrial resources, the regional or watershed level) and their socio-economic consequences.

This means coordinated research programs designed to (a) significantly improve the understanding and predictions of changes in critical climatic factors, the direct and non-climatic effects of changes in atmospheric greenhouse gas concentrations on the terrestrial and marine biosphere, (b) improve and/or develop methodological tools to predict the impacts of these climatic and non-climatic factors on the supply and demand of resources, and the socio-economic consequences of climate change and alternative adaptive response strategies, and (c) developing costs and benefits for both adaptation and limitation measures to help arrive at the optimal mix that maximizes social well-being.

Resource use and management practices under the widely divergent, existing climatic conditions need to be documented, catalogued, and made more accessible (A). Thus, if the future climatic regime

could be predicted with sufficient confidence, such catalogs would allow one area to more easily locate analogues for its future climatic regime. There could be a variety of such catalogs. For instance, one could contain information on the performance of crops, trees, and other species under a variety of climatic (and other) conditions. This would help farmers and foresters select species for cultivation based upon their expectations of both the future climate and the resource base. There could be other catalogs on management practices for agriculture, forestry, livestock husbandry, etc.

Inventories of the current state of resources are needed by resource managers whether or not climate changes (A). Inventories which accurately describe the condition and use of resources (e.g., land and water uses, distribution, and diversity of species) would be of value. Such inventories should also describe the future state of resources, as practicable, taking into consideration different scenarios of population growth.

Studies and assessments to gauge the resilience of resources and their vulnerability to climate change may help establish priorities regarding which areas—and, within areas, which resources—authorities should focus upon (A). Such assessments would help determine the present adaptive capability of localities, nations or even systems. Moreover, it would help provide information regarding adaptability to various rates of climate change. However, given the uncertainties at the regional level, such studies should be used with caution. To help in such assessments considerable effort should be expended on researching, improving, and/or developing appropriate methodological tools to estimate the impacts of climate change on resource use and management, and their socio-economic impacts. Studies of the interrelationships between population growth, changes in greenhouse gas emissions, status and use of natural resources and any responses to climate change would be useful.

Systems to monitor the status of resources need to be established to give early warning of any potential changes and trends (A). Such systems should be designed to detect changes in resources in different locations that may be indications of the effects of climatic perturbation. Of course, such systems, while indicating changes may be in the offing, would generally be unable to indicate whether observed changes are short-term trends due to natural variability or irreversible consequences of human-induced climate change.

Improve existing institutions (or, where appropriate, establish new institutions) to assist in rational use and management of natural resources, and to help localities, nations, and regions better cope with any climate change. Technology development and transfer mechanisms also may need to be supported and strengthened (A, B, C). As far as possible, this should be done via existing institutions such as FAO, UNEP, WMO, UNRRO, UNDP, and other multi- and bilateral-aid agencies as well as institutions within nations. These institutions could also assist in the above mentioned efforts to inventory, assess, and monitor natural resource use and management.

Efforts to educate and inform the public and decision makers on the scientific, policy, and economic aspects of issues surrounding climate change need to be strengthened (B). This should be facilitated by the strengthening of the above-mentioned agencies.

Research and development on more efficient resource use needs to be stimulated (A). Such efforts could help cope with new stresses from climate change. Both public and private enterprises should engage in research and development directed to more efficient forestry, agricultural, and water use practices, and biotechnological innovation. Governments could take measures encouraging such R&D while ensuring that such work is conducted in a manner consistent with public health and safety. Nations may consider developing new or modified institutional, legal, and financial measures that (a) allow innovators to profit from their R&D, and (b) encourage individuals and communities to develop an economic stake in conservation and in efficient and sustainable resource use and management.

While there already are methods of reducing the negative impacts on resources, further research and technology development may be necessary to cope with the potentially worst consequences (A). Thus, for instance, development of more drought- or salinity-resistant cultivars using classical and modern breeding techniques (e.g., genetic engineering) would help keep farming and forestry options open. Such development is already being undertaken and is expected to be beneficial even if there were no climate change. However, given these efforts, the time taken to develop and disseminate such technology, and the fact that adaptation along these lines (for reasons of human-induced climate change) is not imminently necessary, additional emphasis is not warranted at this time.

6.5.2 MAXIMIZING SUSTAINABLE YIELDS

It is necessary to increase the efficiency, productivity and intensity of resource use consistent with the principles of sustainable growth and development to relieve pressures on resources, which will inevitably arise due to population growth, whether or not there is human-induced climate change. There are several facets to this, as elaborated below.

There needs to be increased emphasis on the research, development, and adoption of technologies for increasing the productivity or efficiency (per unit of land and water) of crops, forests, livestock, and fisheries (A, B). Some of the research and techniques developed for increasing productivity have been viewed by many with suspicion because they involve genetic engineering and food and chemical additives. Governments and consumers, in dealing with such research and/or products (e.g., in risk analysis/management) derived from such techniques, should also give due weight to the benefits of increased efficiencies in conserving land and water resources.

Such efficiencies reduce the demand for land for agriculture, plantation forests, and grazing, which is the major cause of conversion of natural ecosystems and loss of biological diversity. In turn, these contribute significantly to increased atmospheric concentrations of greenhouse gases, though plantation forests may be a useful sink for CO_2. Moreover, increased productivity and more efficient use for fuelwood and timber would help reduce atmospheric CO_2 concentrations.

With respect to livestock, increased meat or milk productivity per unit weight of animal also offers a possible means of reducing emissions of methane, which has a greenhouse warming potential twenty to thirty times that of carbon dioxide.

In addition, some areas may be able to change zoning to increase allowable limits on densities of population or development without exceeding carrying capacities. Such increases in the intensity of human settlements would help stem the loss of agricultural land to human settlements and generally reduce pressures on land. Moreover, higher population densities make more viable many energy and resource conservation measures, including mass transit, district heating, reductions in heating or cooling requirements, and waste recycling. This would help reduce greenhouse gas emissions. On the other hand, in areas where carrying capacities are strained or extended, appropriate measures to expand carrying capacities should be considered— e.g., implementing pollution-control measures, improving access to potable water or transportation infrastructure.

Another option to improve the efficiency of resource use would be to *identify and review subsidies for resource use* (B-C). Subsidies encourage use of marginal resources. For instance, crop subsidies can often result in cultivation of more land than is economically justifiable. This removes land from its natural unmanaged state, which leads to deforestation and potentially a loss of biological diversity. Moreover, it contributes further to greenhouse warming. Extension of cultivation through subsidies also reduces water available for other purposes. Grazing subsidies have similar consequences. In addition, these help enlarge livestock populations, which, from the point of view of climate change, contribute to increased methane emissions. However, it should be noted that societies may rationally elect to subsidize various activities for reasons of equity and other social benefits that may not be easily amenable to monetization. In such circumstances, it may be worthwhile to examine if subsidies could be re-formulated so that they still achieve their social goals while diminishing their environmental impacts.

Promoting resource conservation and sustainability of resource use (A, B-C). Conservation practices could assist resources to withstand climatic stresses by helping moderate local climates, water use, and soil erosion, increase genetic variability, and reduce

other stresses from environmental degradation. Particular attention may be given to reducing deforestation, promoting reforestation and afforestation, improving water use efficiency, and increasing the use of sustainable agricultural practices.

Strengthening conservation and protection of highly vulnerable areas (A, B-C). Climate change may or may not exacerbate conditions in marginal and endangered ecological and agricultural systems and overused water basins. Assessments of the potential impacts of climate change (see Category A) might help sort this out.

In areas that could be further stressed, various initiatives could be taken for conserving the most sensitive and valuable resources, including strengthening conservation measures and managing development of highly vulnerable resources. Identification of the most sensitive and valuable resources would necessarily be based upon each society's consideration of relevant social, cultural, environmental, and economic factors.

6.5.3　Increasing the Flexibility of Resource Use and Management

Flexibility for resource management needs to be maintained and/or enhanced (A, B-C). Greater flexibility will increase opportunities to adjust land and water uses to a wide range of possible climatic conditions. Climate change could affect the suitability of a tract of land for various purposes. Its productivity and potential will change in ways that cannot currently be predicted. If today it is used for production of a specific crop, that crop may not be viable in the future; land devoted to conservation of a particular species may be unable to support the appropriate habitat for that species in the future. Thus flexibility is a necessary condition for successful adaptation. There are several implications to this.

First, *consideration should be given to decentralizing, to the extent practicable, decisions on resource use and management* (B-C) i.e., they may be left to individuals and local authorities. They are more likely to have a better understanding of the local context and therefore less likely to err in their evaluations. Moreover, decentralization assures that any errors in judgment—and some are inevitable—are not universal. The other side of the decentralization coin is that there should nevertheless be coordination between adjacent jurisdictions. Moreover, local concerns often override the broader good, which leads to a "not in my backyard" mentality. To deal with this, methods need to be explored on how smaller segments of society may accept taking actions benefiting the larger society even at some additional risk or burden to themselves. Second, *quick and accurate information and technology transfer is critical to maintaining such flexibility* (A). Third, *there is a need to research methods of increasing the flexibility of land and water use for various purposes* (A).

6.6   WATER RESOURCES

6.6.1   SPECIAL CONSIDERATIONS FOR WATER RESOURCES

Water is essential for human civilization, living organisms, and natural habitat. It is used for drinking, cleaning, agriculture, transportation, industry, recreation, animal husbandry, and producing electricity for domestic, industrial, and commercial use.

Even in the absence of human-induced climate change, there is a great deal of climatic variability. In many areas it is an ongoing challenge to match water supply with demand. The problems that do occur with respect to water are usually on a regional (multi-basin), basin, or smaller scale. Even in "normal" times, problems may occur in one season but not in another. To achieve the goal of matching supply and demand, societies have established elaborate structures and institutions to store, treat, and distribute water; have mined groundwater; and have used demand-side management such as rationing or pricing. In spite of such measures, most societies expect to be forced to cope with various incidents of floods, droughts, and degraded water quality.

Human beings have faced other circumstances with lessons that could be valuable in adapting to

any adverse impacts of climate change on water resources. These circumstances include increased pressure on limited surface and ground water resources due to population growth, migration into arid or flood-prone areas, periods of short-term and prolonged drought, and degraded water quality. Detailed institutional and legal mechanisms and arrangements have been established to make water more available and dependable, or of better quality. There are hundreds of international compacts, treaties, and agreements dealing with water. In addition, numerous other arrangements dealing with water resources exist within nations. Thus, there is a fund of knowledge that can be drawn upon to help devise response strategies that would mitigate adverse impacts or capitalize upon positive impacts that may result from a greenhouse warming.

As noted in Sections 6.1.3 and 6.1.4, there are significant uncertainties regarding the effects of increased greenhouse gas concentrations on resources. The spatial and temporal distribution of precipitation, soil moisture, and run-off, and the frequencies and magnitudes of droughts and floods will change in a manner that is not currently predictable with confidence. While the world may receive more precipitation on a globally averaged basis, some areas will get more, and others less, precipitation. Precipitation, though, is only one factor determining water availability and run-off. Other critical factors include temperature, wind speed, humidity, the nature and extent of vegetation, and the duration of accumulated snowpack. Each of these factors would also change in the event of climate change: higher temperatures would result in greater evaporation and transpiration and earlier spring melting of snowpack; higher wind speeds and changes in humidity would change the frequencies, magnitudes, and patterns of storms; higher carbon dioxide concentrations could result in more efficient water use by vegetation and crops, thus modifying evapotranspiration; annual and seasonal variability of precipitation, temperature, and other climatic factors would change.

The ability to predict the spatial and temporal distribution of precipitation is quite limited (see Section 6.1.4). This predictive ability declines as one goes from global to regional or watershed scales, and from annual to shorter time periods (e.g., seasonal and weekly periods). Moreover, current estimates for run-off and water availability in the event of climate change so far have omitted consideration of many critical factors, such as the effects of changes in vegetation on evapotranspiration and run-off, humidity, and wind speed. Thus, the present ability to predict the direction, magnitude, extent, and timing of changes in water availability, run-off, and other parameters relevant to water resources management for specific areas and basins is limited at best. Although there is some agreement among current models on the likely direction (but not the magnitude or timing) of change in certain geographic regions (e.g., a likely increase in temperature in the arid sections of the western United States), such agreement does not imply accuracy: it could be a result of similar assumptions and simplifications in current models. In addition, the demand for water supplies could also be modified because of changes in rainfall, cropping patterns, and water-use efficiency and in managed and natural ecosystems.

While site-specific effects on water resources cannot generally be identified at this time, the adaptive measures available today to manage water resources would in all probability be valid for future conditions.

The precise options selected by each area need not, of course, be the same as those it employs today. Nevertheless, many responses would be appropriate today as well as in the event of climate change.

Water-resource management in the face of climate change will face somewhat different challenges compared with past water planning. In the past, when climate was assumed to be constant, one could estimate new demands for water by more easily measurable or observable factors—the rate of population growth, the rate of decline of groundwater supplies, or the degree of aridity in areas being newly settled. However, for many years hence, the ability to model future atmospheric changes and their interaction with the hydrologic cycle is unlikely to provide as great a degree of certainty about the degree, or (in many cases) even the direction, of climate change on a regional level. This increases the degree of uncertainty concerning which responses are prudent. Clearly, the more expensive the response strategy, or the greater its adverse social, economic, or environmental conse-

quences, the greater the caution regarding its adoption.

6.6.2 RESPONSE STRATEGIES

Timing of strategies. Although climate change may occur over many decades, it is uncertain whether these changes—should they come—would be gradual or sudden. Many water supply systems are designed to operate under extreme conditions (greatly increased or decreased run-off), but water resources managers will have to consider that the frequency and magnitude of extreme events may be altered. Given the uncertainties regarding the extent, magnitude, and timing of climate change and its effects on water resources, it may be prudent to delay consideration of more costly adaptation measures (Category C) until after these uncertainties are reduced. By the same token, many of the less costly response strategies (Categories A and B), especially those with other benefits, may be appropriate today, as well as in the event of climate change. Some response strategies and programs can be implemented effectively during the short term; e.g., flood warning, evacuation, disaster relief loans or subsidies, and emergency operations. Other strategies may require a longer lead time to respond to climate change; e.g. conducting studies of modifying reservoir operations to meet shifting demands under climatic uncertainty, and incorporating considerations of climatic uncertainty into the design of new water resource structures. Fortunately, many responses to climate change are already embedded in current planning, design, and management practices, and their general application in industrial and developing countries should be promoted.

Determining the flexibility and vulnerability of current water supply systems (A,B). Given the uncertainty over the nature of hydrologic changes to be expected in any particular region and the cost of making any significant changes in existing water supply structures, a logical first step would be to evaluate the flexibility of current water supply systems to the type of changes that might be expected under climate change. Models could be used to estimate the sensitivity of water systems to increased aridity and increased run-off (such as might occur from a shortening of the run-off season). Such models could utilize altered run-off data from a number of possible sources: arbitrary increases or decreases, the use of proxy data on seasonal temperature and precipitation obtained through paleo-climatological methods, or global climate models used in conjunction with hydrologic models.

The greater the vulnerability and/or inflexibility in a particular water system, and the greater the impact on human population and on ecosystems, the more important it is to monitor relevant parameters with a view toward determining trends, to strive to reduce uncertainties regarding the effects of climate changes on water resources, and to consider measures to enhance the flexibility of the water supply system. Such system models would also necessitate assimilating data on current facilities, streamflow, and other statistics—data that would be required for many of the additional response strategies.

System optimization (A,B). Water supply facilities are often built by one particular jurisdiction or agency to service its needs, and the reservoir operating rules are developed to serve the needs of that jurisdiction or agency only. Hence, system operation may not be optimized across existing jurisdictions or agencies. Significant increases in system yield can often be obtained by joint use and revised operating rules if different jurisdictions or agencies are willing to execute agreements to do so. These agreements involve exchange of storage and flood control capacity between reservoirs at different times of year, as well as specifying rules for joint operation of facilities. The increases obtainable from such measures can be enhanced by more up-to-date data on meteorological and soil moisture conditions, as well as the application of more sophisticated computer models. In the long-term, once the flexibility and vulnerability of a water supply system to respond to a variety of hydrologic changes is better understood, the next step would be an attempt to optimize the water yield, hydropower production, flood control, recreational use, maintenance of fish and wildlife habitat, and other outputs available from existing facilities under various climate change scenarios, as well as under current climate. Optimization of international water-resource systems may require intensified international cooperation among countries sharing river basins.

Enhancement of scientific measurement, monitoring, knowledge, and forecasting (A). Given the natural variability in meteorological and hydrologic conditions, one of the initial challenges of planners may be to determine whether long-term changes are, in fact, occurring or are expected to occur in a particular region. Such assessments are based on comprehensive and accurate monitoring of hydrological and meteorological factors. However, the relevant observational networks are far from satisfactory in most of the developing world. As climate change is a global phenomenon, there is a need for a global approach in monitoring. There should be continued study of the interaction of the hydrologic system with the rest of the climatic system with the eventual goal of enabling area or basin-specific predictions, or detection of trends, with respect to changes in water availability and other parameters useful for water resources management. This could eventually enable planners, designers, and managers to incorporate predicted climate trends in their use of stream flow and other time-dependent data series.

Water conservation (A,B). Water conservation measures have been widely discussed over recent years for a variety of reasons, including the increased demands for water and the high financial, social, and environmental costs of construction of additional storage facilities. Under conditions of increased aridity, conservation measures may become even more important. Large savings in water are possible in agriculture. Irrigation is the largest consumer of freshwater in many areas and relatively small percentage reductions in irrigation water use can make large amounts of water available for new uses.

Agricultural water conservation measures include irrigation management scheduling (monitoring of soil moisture and atmospheric conditions to more precisely schedule the amount and timing of water deliveries), lining of canals to prevent seepage, tail-water recovery (recycling water that reaches the end of field rows), drip irrigation, using more drought-resistant crops and/or cultivars, and tillage practices that retain soil moisture. Conservation of municipal and industrial water supplies can be achieved through education, better measurement and metering, technological improvements, specifying the use of more efficient water-using appliances in building codes, and, in arid climates, use of low-water-use landscaping, rather than grass lawns. Under extreme drought conditions over one or two dry years, voluntary rationing of domestic use and mandatory restrictions (allowing water use during just certain hours, restricting lawn watering, etc.) have also proven effective. In addition to these measures, pricing has more potential as an incentive for water conservation.

Demand management through pricing (B). Water prices provide signals and incentives to conserve water, develop new supplies, and allocate limited water supplies among competing uses. Since water use is sensitive to price, water users facing higher prices will generally conserve water and modify technologies and crop selection. Therefore, pricing by water supply authorities to reflect real or replacement costs promotes efficient use. However, the cost of water supply facilities constructed by government entities is often recovered partially through property taxes or means other than through commodity charges levied on final consumers. In other cases, the facilities are subsidized, with the cost being financed through general government revenues or income taxes. The result is that water is often priced inefficiently, below its cost of delivery or its long-run marginal cost. This leads to overuse of water and the other resources needed to construct water supply facilities. There is substantial additional opportunity for cities and irrigation districts to utilize pricing as a means of conserving water by employing marginal-cost pricing (charging for the cost of the last-added and most expensive increment of supply) or progressive-rate pricing (charging more per unit to users of large amounts). A first step would be to perform studies of the effect of higher prices and different water-rate structures on water use in the particular area under consideration.

It would also be possible to extend these concepts to employ pricing as a means of allocating water use during drought episodes. To some extent this is already done: some areas have a two-tier rate structure, where lower rates are charged for interruptible supplies of water. As discussed below, allowing water entitlements to be traded at market value may achieve results similar to raising prices. Institutional arrangements for pricing and for trading water would need to take into account potential adverse

(or beneficial) impacts on public uses of water, such as recreation and preservation of wildlife, whose value is often not incorporated into current pricing practices.

Voluntary water transfers or markets (B). One response to more arid conditions is to establish institutional arrangements to assure that water is directed to where it is most needed and where it will be the most productive. One means for doing this is to establish a system of property rights in water that can be traded as economic or hydrologic conditions vary.

For example, a growing city or a new industrial user desiring a senior water right can enter the market place to purchase a senior right from an existing irrigation water user.

The amount of water a rights holder could transfer is normally limited to the consumptive-use portion of his entitlement rather than his full diversion entitlement from the stream, in order to assure that other water users are not injured in the transfer process. Government also has a role in protecting instream, public uses of water for water quality, fish and wildlife, and recreational and other uses. Water transfers can be annual rentals, short-term leases, permanent sales of water rights, payments for conservation investments in exchange for the conserved water, or dry-year option agreements under which the water is transferred only under specified drought conditions. The viability of such approaches would vary, depending upon the extent to which property rights and markets have been relied upon in the past.

Modification of cropping systems (C). Any long-term changes in temperature, evaporation, the length of the growing season, the amount and temporal distribution of precipitation, or other climate-related parameters may lead to the modification of cropping practices. Modifications in response to climate change can include: shifts to more (or less) drought- or heat-tolerant species or varieties, changes in planting and harvesting dates, selection of varieties with shorter or longer growing seasons, and adjustments in the number of crops grown per season (e.g., conversion from double cropping to single cropping or vice versa). Therefore, changes in climatic factors may also result in (1) migration of current crops to new areas to take advantage of changed conditions or (2) changes in the demand for irrigation. These changes could, in turn, affect water demand and supply in the new as well as old areas, as well as having impacts on the volume of contaminants in agricultural run-off.

Modifications of tillage systems (B, C). Although historically the primary functions of soil tillage have been to prepare a seedbed and control weeds, certain tillage systems are effective measures for conserving soil and/or water. Practices that leave crop residue on the soil surface tend to increase surface roughness and organic matter, thereby increasing infiltration and reducing the potential for soil erosion. Surface residues also help to reduce soil loss due to wind erosion. Any tillage system that avoids exposing subsurface soil moisture to evaporative loss or that creates a surface soil barrier to evaporation can contribute significantly to agricultural water conservation, especially in rain-fed areas. Thus, tillage systems can be used to make more efficient use of precipitation in more arid areas, to reduce erosion resulting from excess precipitation and to reduce the off-farm impacts of soil erosion and farm chemicals transported by run-off. These modified tillage practices can be combined with other land management practices designed to reduce water use and soil erosion, such as terracing, laser-leveling of fields, and water harvesting systems that recover water run-off.

Natural resources management (B, C). Natural resources management programs are implemented in many regions of the world to address deforestation and desertification and to promote the sustained yield and conservation of natural resources. By including considerations of the potential impacts and risks of climate change, such programs may mitigate the impacts of climate change on water resources.

Examples of such programs include integrated river basin or watershed management programs; integration with forestry practices and reforestation in upland areas; soil conservation, forage selection, livestock grazing practices, land management, and other agricultural practices in the plains; and coastal estuarine, marsh, and mangrove management.

Flood management. Flood management strategies are now based on the computed magnitude and frequency of flood events based largely on historic

data. The potential effects of climate change are changes in the magnitudes and frequencies of storm events and in the magnitude, rate, and timing of the melting of snowpack. Therefore, some areas can expect increased run-off over shorter time periods. In addition to systems operation studies designed to accommodate a wider range of future climatic conditions, potential response strategies include the following:

- *Improvement of flood forecasting* (A). Collection of hydrometeorological data by GEOS satellites and other advanced systems provides real-time information on rainfall, streamflow/stage, and reservoir levels. Broader collection and use of similar real-time data in concert with improved quantitative precipitation forecasting techniques could enable water managers to respond more rapidly and effectively to potential flooding.
- *Evacuation plans* (B). Comprehensive flood preparedness plans may include provisions for temporary evacuation of flood plain occupants during flood events. Improved flood warning and forecasting abilities could enable additional actions, such as removing or raising building contents to reduce flood losses.
- *Flood warning* (B). Implementation of flood warning systems can often be relatively inexpensive, quick to design and build, and easy to modify to changing conditions. Generally such systems can be operated by local people in a decentralized, independent mode.
- *Floodplain zoning* (B). Zoning flood plain areas to prevent construction of structures and activities likely to suffer from floods is another means of avoiding losses.
- *Flood insurance* (B). Flood insurance can serve a double function. The flood insurance premiums can be price signals to the insured to discourage locating in flood-prone areas. Second, once flooding has occurred, insurance can be an effective means of reducing the economic impact of losses. Climatic change could necessitate more frequent review and revision of flood insurance programs.

Disaster relief and emergency preparedness (B, C). Because of the uncertainty concerning the impacts of climate change, improving disaster relief programs may be an effective response. Flood insurance or financial assistance in the form of emergency loans and subsidies could reduce the impact of economic losses and social disruption. Grants or loans could be used to fund such measures as construction of emergency levees (e.g., sand bagging) and emergency debris removal to impede imminent flooding, as well as for post-flood rehabilitation efforts.

Advance planning of these programs would reduce conflicts over their implementation and make potentially affected parties aware of their availability.

Design modifications (B, C). In situations where it is cost-effective, designing more capacity into spillways at the time of project construction and other design modifications, such as increased capacity for levees and dikes, can assist in handling larger flows of water.

Adjustments in river transportation (C). Increased precipitation could help inland navigation systems by providing a more constant depth of water in the free-flowing reaches of inland systems. If climate changes result in less run-off, there is the possibility that interruptions could occur in the free-flowing reaches. If greenhouse warming results in a more extreme hydrological regime, river navigation will tend to be adversely affected by more frequent floods and droughts. Response strategies include dredging of shoals or sand bars in the major river systems to maintain adequate depths; lessening the likelihood of groundings by lightening barge loads; placing greater reliance on other forms of transportation during drought episodes; and augmenting low river flows.

Education, technology transfer, and financial assistance and special considerations for developing countries (A). In many developed countries a comprehensive system of physical structures is already in place to deal with excess or inadequacy of water supply and to manage distribution. Similarly, such countries have well-established institutions with a long record of dealing with water resource problems and making adjustments. Furthermore the costs of making the necessary adjustments can be accommodated largely within the financial and human resources of the countries concerned.

In other, less well developed countries, the converse may be true. In many cases, developing countries are not able to cope with adverse water resource conditions under existing climate. Thus, in such countries, additional efforts may first be needed to raise standards of water resource management. Education, training, and technical assistance efforts directed at water managers and water users could play a role in making water use more efficient and in responding to climate change. These programs could include national, regional, and international efforts such as joint scientific research; exchange of research results on new crops, products, and technology; and assistance to developing nations for training and technical assistance. The United Nations, the World Bank, and other international and bilateral agencies provide a framework through which technical and financial assistance is provided to developing countries and regions.

Where it does not already exist, nations could develop an infrastructure (e.g., extension services similar to that existing in the United States) to assist in rapid dissemination of new and appropriate technology, management techniques, and practices to help assure sustainable use of forest and agricultural resources. The same infrastructure could also be used to educate farmers and the local public about the role of vegetation in controlling erosion and in modifying the hydrological cycle.

In order to enhance the robustness and resilience of water systems, it would be important to identify appropriate technologies, depending in part on the economic base and level of economic development in a region; cultural and institutional factors (for example, market-based strategies may be more difficult to implement in some settings); international and bilateral trade and debt policies; and guidelines for development projects.

Countries where water resource systems have not been developed may first want to concentrate on the provision of adequate storage and delivery systems, conservation practices, and appropriate water allocation institutions under conditions of current climate.

Modification of storage and other augmentation measures (C). Although nonstructural measures are generally less expensive and should be exhausted first, additional storage may become a method for responding to climate change to accommodate changes in the magnitude and timing of precipitation and/or snow melt, either through raising existing dams, construction of new facilities, inter-basin transfers of water from areas of surplus to deficit water areas, or recharge of underground aquifers from available surface supplies. Planning for such measures would need to take into account potential adverse and beneficial environmental and economic impacts. Transportation of emergency water supplies could be provided when drought conditions threaten public health and well-being.

In those coastal areas where water has reached a very high value, desalinization technologies could be used to augment supplies.

For a number of years, research has proceeded on cloud seeding to determine whether this technique has value for augmenting water supplies in certain areas. One of the main problems has been that it is difficult to separate out the amount of additional water that is the result of cloud seeding, and therefore it is difficult to find financial sponsors of cloud-seeding experiments.

Dam safety and other design criteria (B, C). Response strategies to safeguard the integrity of existing deficient impoundment structures generally include some combination of enlargement of spillways, raising of dams, and modification of water-control plans. Increased run-off due to climate change could potentially pose a severe threat to the safety of existing dams with design deficiencies. Design criteria for dams may require re-evaluation to incorporate the effects of climate change.

Adjustments in protecting water quality in rivers and reservoirs (B, C). Climate change could modify the amount of fresh water available. If freshwater quantities are reduced, this could affect the ability to dilute contaminants and salts, to dissipate heat, to leach salts from agricultural soils, and to regulate water temperatures in order to forestall changes in the thermal stratification, aquatic biota, and ecosystems of lakes, rivers, and streams. The potential effects of climate change on water quality also relate to the magnitude and frequency of storm events as well as seasonal changes in temperature. For example, the onset, duration, and characteristics of reservoir and lake stratification would respond to seasonal temperature changes, particularly in temperate regions. Dissolved oxygen concentrations could also be affected, and eutrophication problems could worsen. Climate change could also affect the recharge rates of aquifers, which could affect the quality of underground water supplies. However, if climate change involved increased flows in a region, greater dilution of pollutants and other water-quality benefits may result. These various changes in water quality may affect the usable supplies of fresh water.

The efficient operation of systems to manage water quality may become more critical. Transferable discharge permits are one means to allow ambient water-quality standards to be met at the least cost by trading pollution reduction capabilities among dischargers.

Various in-place technologies, such as aeration and destratification and localized mixing systems, can mitigate adverse changes in water quality. Modifying the operation of reservoirs with multi-level withdrawals, or adding this capability to existing reservoirs, would increase the ability to manage changes in water-quality conditions. Water-quality problems are also affected by the level of discharges into a stream, including non-point source run-off from the watershed. Therefore, watershed-management programs to control non-point sources as well as point sources can help maintain water quality.

Adjustments in protecting estuarine water quality (C). Estuarine water quality will be subject to similar hydrometeorological changes that affect fresh water, as well as changes that may occur in the oceans, such as sea level rise or tidal variations. One of the effects could be saltwater intrusion into surface and ground water, having unpredictable and possibly adverse impacts on fishery resources and wildlife and on water supplies.

In addition to the response strategies listed for maintaining river and reservoir water quality, the following could prove useful: relocating water supply intakes out of areas that may be susceptible to higher saltwater intrusion; and providing saltwater barriers to further saltwater intrusion in estuaries and tidal rivers (such barriers would have to be evaluated against possible adverse effects on migratory fish and shellfish resources passing seasonally through the system).

Utilization of hydropower (B, C). There is considerable potential for developing hydropower in Asia,

Africa, and Latin America, as well as in other parts of the world. Development of these resources could help reduce combustion of fossil and wood fuel, thereby reducing carbon dioxide emissions directly. Moreover, conserving forests would reduce erosion and the frequency and magnitude of flooding, especially in mountainous terrain. In planning for hydropower projects, consideration would also have to be given to potential beneficial and adverse environmental and economic impacts (such as inundation of agricultural lands, forested lands, and wetlands; downstream impacts on navigation, flood control, and water supplies; and effects on aquatic resources and recreation).

6.7 LAND USE AND MANAGED AND NATURAL ECOSYSTEMS

6.7.1 INTRODUCTION

This section deals with adaptive responses to the impacts of climate change on land use and managed and unmanaged ecosystems, including forests and biologically diverse areas.

Biological diversity refers to the variety and variability among living organisms and the ecological complexes in which they occur. Biological diversity is organized at many levels, ranging from complete ecosystems (i.e., systems of plants, animals, and microorganisms together with the non-living components of their environment) to the chemical structures that are the molecular basis of heredity. Thus, the term encompasses different ecosystems, species, and genes, and their relative abundance.

More important, while biodiversity is related, it is not identical to the number or abundance of particular species. There are economic and non-economic reasons for maintaining biological diversity, forests, and other ecosystems. In many areas, food production and livelihood (e.g., tourism, forest products) depend directly upon these functions. Nature contains blueprints for substances that could be of great benefit to agriculture, medicine, and forests under a variety of climatic conditions. If climate were to change, such blueprints could be the source of new cultivars better adapted to future conditions. There are also aesthetic and cultural reasons for protecting biological diversity. Moreover, disruption of forests and land cover can affect availability of quality water, run-off, and soil erosion.

Climate change could alter the physical suitability and economic viability of land for different uses in many areas. The climate has played a significant role in present land-use patterns and on the occurrence and distribution of present-day agricultural and forest lands, human settlements, and biota. Hence, it seems likely that these distributions could be altered if significant changes in the climate occur. In this process, forests and other ecosystems could be altered with change in inter- and intra-species diversity with some species extinctions becoming possible (appropriate actions could reduce these impacts), while some species could benefit.

Adjustments, including changes in land use and improved biodiversity conservation efforts, can be made in response to changing climatic conditions and population growth rates. In open market economies, many of these adjustments will be made by private resource managers who are guided by market incentives (changes in prices and costs). Yet, many decisions on land use and biodiversity conservation have environmental and social consequences that are not considered in private benefit-cost calculations.

Section 6.7 outlines the changing pressures on land use that could result from global climate change and suggests planning and management options for adapting to climate change. It also makes suggestions for helping rapidly growing populations meet the demand for land for various human activities, while assuring conservation of the environment and maintenance of biological diversity.

6.7.2 CURRENT PRESSURES ON LAND AND ECOSYSTEMS

The effects of increasing greenhouse gas concentrations and associated climate change on land use and biodiversity must be considered against a backdrop of rapidly increasing population growth. Such growth alone will result in increasing demand for food, fiber, and forest products and for living and recreational space. These demands increase pressures to (a) remove more land from its unmanaged state, thus increasing stresses on less intensively managed habitats, biodiversity and ecosystems and increasing atmospheric concentrations of carbon dioxide as forests are converted to other land uses, (b) adopt more intensive land uses that could increase soil erosion and further degrade water and other environmental quality, and (c) convert agricultural and other lands to urban and suburban uses.

6.7.3   POSSIBLE ADDITIONAL PRESSURES RESULTING FROM CLIMATE CHANGE

Increasing greenhouse gas concentrations and associated global climate change could affect patterns and intensity of land use in significant ways. Climate change could shift regions of suitable climate for a particular species toward the poles and higher elevations. Over a period of decades or centuries, the species present in plant and animal communities could dissociate as a result of differences in thermal tolerances, habitat requirements, and dispersal and colonization abilities, and assemble to form new communities under the new climatic conditions. At this time, however, neither the precise direction, rate, distance nor success of migration of species can be predicted.

While such changes provide information on possible shifts in ecological systems, they cannot easily be translated into specific shifts in uses of land for agriculture, forestry, or other purposes. Current land-use patterns could change: some areas that today are used for agriculture may change in the intensity of use because of changes in climate or availability of water; for the same reasons, other areas that today are not used for intensive agriculture may be able to support it in the future. Switches in crops or cultivars will likely occur to optimize expected farm income. The demand for lands for human settlement may be affected: some coastal areas may become uninhabitable in the event of sea level rise; areas in the higher latitudes may become more hospitable for human habitation. In some areas, current land uses may be continued only if relatively expensive measures are taken to mitigate the effects of climate change; in other areas, adapting the land to new uses may result in a net social benefit (over current uses).

Changes in the habitat could lead to a species invading the range of another species. In at least some cases this may lead to a reduction of population size or even extinction of the competitively weaker species. (Appropriate actions could reduce these impacts.) Declines in populations and species extinctions could occur due to loss of habitat, lack of new land suitable for colonization, and inability to keep pace with changes in the climate. On the other hand, other species may benefit because certain habitats may become more abundant and productive. Ecosystems restricted by human activities to small isolated areas could face the greatest risks from climate change due to the exacerbation of current stresses on such systems and the fact that colonization by new individuals may be hampered by the fragmented nature of the landscape. Any inter- and intra-specific declines in biodiversity could result in irreversible loss of genes and gene complexes.

Some migratory organisms may also face increased threats. Migratory species are dependent on the quantity and quality of habitat in more than one area and they may be adversely affected if suitable habitat in only one of these areas is reduced. These effects could be mitigated because migratory species have the mobility to locate suitable new habitat. Moreover, suitable habitat may expand or become more productive. In these instances, migratory species may benefit.

6.7.4   ADAPTATION MEASURES

To meet the various demands on land and to conserve natural resources under future climatic regimes, resource users and managers would need to consider management and development efforts to maintain or enhance economically efficient, sustainable land use and biodiversity. However, responses to climate change should optimize socioeconomic well-being and growth subject to environmental constraints.

Moreover, both adaptation and emission limiting strategies should be considered as a package (see Section 6.3). Since the rate of climate change has such a strong influence on the ability of various ecosystems to adapt, efforts to slow the rate of global warming, particularly reducing deforestation and promoting forestation, can go a long way toward limiting adverse impacts. In addition, several adaptive responses should be considered to improve resilience to changes in the climate.

With respect to adaptation, emphasis should be on identifying and considering removal of barriers to rapid and efficient adaptation, identifying decisions with long-term consequences, maintaining flexibility and improving resource use and management where possible, limiting costs and administrative burden, and promoting public input and acceptance. Effective adaptation to climate change is largely dependent on the integration of information on the impacts of climate change with land use planning and biodiversity conservation efforts. It may also require more dynamic nature conservation (i.e., greater human intervention) rather than strict preservation. Adaptive measures that meet the above objectives are described below under the categories of research and short- and long-term actions.

6.7.4.1 *Research, Planning, and Information Dissemination (A)*

This subsection identifies measures that should be considered to increase the knowledge base so that societies can respond rationally and efficiently to possible changes in the climate. These options are classified Category A according to the classification scheme adopted in Section 6.4. Studies of climate change impacts on land use and biodiversity are needed to identify the resources that are the most vulnerable to climate change and to characterize the dynamics of the responses of managed and unmanaged ecosystems. This information, together with assessments of the effectiveness of particular response strategies, will allow for timely and efficient planning for and modification of land use and conservation of biodiversity.

i) *Inventory* (A). Resource managers could use information on the current state of resources to analyze what is vulnerable to climate change and what might be done about it. Inventories that describe the current uses of land, such as now exist in some countries, and the current distribution and diversity of species, would be of value. Developing such inventories where they do not exist would be a

useful step regardless of whether climate were to change.

ii) *Assessment* (A). Based on the above-mentioned inventories, resource managers could examine vulnerabilities of natural resources to climate change, assess how various ecosystems and land-use patterns could be affected by increased greenhouse gas concentrations, and identify potential land-use conflicts resulting from climate change. Significant effort needs to be devoted to the development of biogeographic models to investigate the response of various species, including migratory species, to higher greenhouse gas concentrations and associated climate change. Research is also needed to improve the understanding of behavior of collections of isolated populations of species in a fragmented landscape and rates and constraints to colonization and dispersal. To achieve this, research is also needed on: (a) the size and location of protected natural areas to satisfy various uses and what practices may be necessary, given the existing or expected size limitations; (b) utility, extent, and placement of conservation corridors; (c) rates of population migration, and (d) behavior of species. Monitoring systems also need to be established to detect any changes in disease and pest outbreaks.

iii) *Development and dissemination of new technologies* (A). R&D efforts that result in more efficient and sustainable land use could help in coping with new stresses from climate change by reducing land use demand. Both public and private enterprises should continue research and development directed to more efficient and resilient forestry and agricultural practices (e.g., development of drought- and heat-tolerant species and crop rotation techniques) and biotechnological innovation. Governments should take measures encouraging such R&D while ensuring that such work is conducted in a manner consistent with public health and safety. Nations may consider developing innovative institutional, legal, and financial measures that would take advantage of innovation, including measures that would allow innovators to profit from their R&D. Nations might also consider institutional, legal, and financial measures that would encourage individuals and communities to develop an economic stake in efficient and environmentally sound land use and development.

Research is also needed to improve methods for internalizing the true social costs in land-use decisions to ensure that these decisions reflect such externalities as ecological damages and infrastructure improvement.

Several approaches for limiting adverse impacts on biodiversity need research. If deemed effective, development should be encouraged. These approaches may include (a) techniques for establishing and maintaining conservation corridors between protected areas; (b) *ex situ* conservation techniques (e.g., preservation of species in zoos, botanical gardens, and germ plasm banks); and (c) community restoration and development techniques to introduce species unable to colonize new regions naturally.

Research and development need to be undertaken to promote fuller and more efficient use of forest products especially as sources of biological energy, materials, and chemicals. This includes research and development to assure more complete utilization of felled trees; assure longer life of timber and wood products by improving resistance to fire and pests and diseases; increase efficiency of industrial plantations through genetic selection, breeding, and propagation; increase efficiency of fuel plantations by screening of tree species for charcoal making; develop new uses for felled trees that could displace fossil fuels (e.g., in electrical generation); promote fuller use of forest products other than timber (e.g., resins, oil, fruits, fiber, materials) by researching and enhancing traditional practices, and focusing on "multipurpose" trees (i.e., trees that can provide a variety of benefits—e.g., fruits for human consumption, fodder for livestock, and fuelwood from branches).

Increasing the rate of dissemination of re-

search results and technologies to users (e.g., land-use planners, farmers, foresters, wildlife biologists, and public policymakers) to enable them to quickly adopt new practices, plans, and technologies. The United Nations and its member organizations (FAO, UNESCO, WMO, UNEP), the World Bank, other multi- and bilateral agencies, and several national institutions provide a framework through which technical and other assistance could be provided to developing countries and regions. Information to planners, farmers, foresters, and other land users is needed. Extension services, such as that supported by the U.S. Department of Agriculture (USDA), could be used to disseminate information and educate planners, farmers, foresters, and other land users.

iv) *Research that enhances the stake of local population in preserving biodiversity* (A). Research needs to be undertaken on how to give parties a stake (including an economic stake) in preserving biodiversity—e.g., developing economic incentives to preserve existing (natural) germ plasm and maintain genetic diversity in breeding populations. Increased research is necessary on the economics of biodiversity.

6.7.4.2 *Short- and Long-Term Responses (B or C)*

This subsection describes options that some societies may be able to undertake:

- in the short term, to improve resilience to climate change if they result in little or no additional costs (or consequences). (Category B responses under the classification scheme of Section 6.4.) The options include maintaining flexibility in land use, managing development of highly vulnerable areas, and increasing and maintaining the sustainability and efficiency of land-use management;
- in the long term, as uncertainties regarding climate change are reduced. (Category C, as discussed in Section 6.4.) Possible long-term

initiates include creation of conservation corridors, strengthening protected areas, and modifying economic incentives affecting land use.

i) *Maintaining flexibility in land use* (B). The uncertain, but potentially significant, shifts in land-use suitability associated with climate change argue for enhancing and maintaining flexibility in land-use decisions. Nations could explore methods of making future land use more flexible and adaptable to climate change. Such flexibility might allow for switches in land-use practices to activities that are expected to provide the highest social values under future climatic regimes. For instance, programs designed to acquire and manage recreation or conservation areas should have the capability to adjust to shifts in the needs for—and suitable locations of—these areas.

Greater use of property easements is one way for maintaining flexibility in land use. More flexible methods of acquiring, developing, and managing areas to be set aside for fuelwood, forage production, and expanding human domicile needs (including recreation and other amenities) within or in close proximity to existing human settlements should be explored. However, in many countries ownership of land is perceived as a fundamental right and one not lightly abrogated. Thus, any methods of increasing the flexibility of land use must be consistent with the principle of fair and equitable compensation to land owners. Moreover, if there are uncertainties regarding "property" rights or length of tenure, they will lead to abuse from resource managers (see (iii), below).

In some cases it may be useful to vest or empower local organizations with clear responsibility and authority for coordinating land-use planning that reflects the likely impacts of climate change and rapidly growing human populations. Land-use planning efforts should incorporate climate change concerns and operate with full participation of concerned organizations and interests.

Nations should consider methods such as exchanging land-use rights, "impact" payments and other schemes to satisfy the concerns of local communities with respect to uses that,

while beneficial to the larger society, may not be desirable from the local perspectives.

Ideally, land-use planning should be decentralized and conducted at the local level. Nevertheless, since climate change could exacerbate conflicts between competing resource users, increased coordination between managers of different programs (e.g., biodiversity, agriculture, forests, water supply), between public and private resource managers and between all levels of government could be necessary. In a changing climate, it may become more important for bodies with jurisdiction over different geographic areas and uses to coordinate land-use policies. For example, agricultural and other local interest groups and parks managers should confer where their interests overlap. Coordination might usefully be extended to large areas to incorporate potentially major shifts in land use. Opportunities to strengthen the institutional mechanisms for such coordination could be explored.

ii) Creating an economic stake for local populations in conservation areas (B). Opportunities for increasing the economic stake and social value to surrounding populations in conservation, preservation, and recreation areas should be explored. Without strong ties to the protected area, populations may have a strong incentive to respond negatively to altered land-use demands caused by climate change, resulting, in some cases, in destruction of protected areas.

iii) *Invest resource users with clear "property" rights and long-term tenure so that resource conservation and regeneration are enhanced* (B). Many users of resources do not have an interest in long-term sustainability, since they either have no land tenure rights or have leases with short-term durations.

iv) *Strengthening conservation and protection of highly vulnerable areas* (B). Certain areas that are already under significant stress, such as highly erodible farmland, heavily used water basins, and some natural areas, may be particularly sensitive to climate disturbances. In some currently stressed areas, climate change could relieve existing pressures. Assessments of the potential impacts of climate change would help identify which areas could be subject to increased stress. Future development in areas that may come under increasing stress could consider management tools such as regulation of development, purchase from or payments to the owner, tax incentives, and impact charges. For example, highly erodible or other marginal lands may be protected by offering farmers annual payments in lieu of cultivating such land—with participation in such programs being strictly voluntary. More active management and intervention such as the creation of protected areas, where appropriate, may be necessary to maintain viable populations of certain endangered or threatened species. As for all responses, social, economic, and environmental consequences must be considered in evaluating and selecting options.

v) *Promoting resource conservation* (B). Conservation practices could increase the resilience of resources to climatic stresses by moderating local climates, promoting water retention, decreasing soil erosion, increasing genetic variability, and reducing other stresses from environmental degradation. This may allow for the maintenance of long-term productivity.

Particular attention might be given to reducing deforestation, improving water-use efficiency, increasing the use of sustainable agricultural and forestry practices, and ensuring that the intensity of resource use is consistent with carrying capacity. Also, analysis of land-use options should consider factors such as simultaneously accomplishing several objectives (e.g., floodplain hazard reduction, wetland and fisheries protection, and migration corridor needs, as appropriate).

vi) *Improve storage and food distribution of agricultural products* (B). Improved methods of storing and distributing food and agricultural products, supplemented by methods of minimizing storage losses, would lessen the severity of future food deficits, whether or not caused by climate change, and would reduce pressures for additional land for food production, thus enabling the world to better cope with future supply instabilities, especially as human populations grow substantially.

vii) *Encourage efficient and environmentally safe levels of agricultural and forest practices, and location and densities for human settlements* (B). Government review of practices, products, and technologies that enhance agricultural and forest productivity and efficiency of land use for human settlements should proceed expeditiously, while balancing the potential benefits of such reviews against the costs of delays. Such reviews must carefully evaluate environmental and health impacts, yet still meet the needs of a substantially increasing human population. In some areas, changing zoning to allow higher population densities would slow the amount of land devoted to human settlements and result in less agricultural land taken out of production. Increased human settlement density would also make more efficient energy use possible by, for example, increasing the economic feasibility of mass transit and district heating and cooling. Enhanced research and field trials are necessary to improve identification and dissemination of new production technologies which take into account all externalities of production usually left out of the resource management calculus (e.g., soil erosion rates, net emissions of greenhouse gases, deleterious side effects of pesticide use on wildlife). In addition to focusing on local environmental effects, environmental reviews should take into consideration broader effects such as net efficiency of land use.

viii) *Modifying economic incentives* (B or C). Direct and indirect subsidies to agriculture, forestry, and development of human settlements can influence land-use practices. Incentives should be reviewed to ensure that they are economically efficient and consider the sustainability of land use and conservation of biodiversity. However, some nations may rationally elect to subsidize certain activities for reasons of equity or to meet other social goals that may be hard to monetize.

ix) *Strengthening and enlarging—and establishing conservation corridors between—protected nat-*

ural areas (C). Depending upon the outcome of the research outlined in Section 6.7.4.1, item (ii) above, further strengthening or enlargement of protected areas could be beneficial for the maintenance of biodiversity and recreation opportunities; poleward or up-slope additions in ecotone (i.e., transition) regions could be particularly beneficial under a warmer climate. In areas/regions that are built up, or fenced off, or where very little land is in its natural condition, conservation corridors, such as greenways, river corridors, trails, hedgerows along the edges of fields, and transportation and transmission corridors could serve to facilitate migration of species as well as increase the degree of protection to the species involved.

This, too, would be more beneficial to ecotone areas. In addition, corridors could enhance the capacity for species to shift distributions in response to climate change. On the other hand, in areas where there is sufficient land in its natural state or which is not fenced off or built up, such corridors may not be functional and might even serve as a barrier to efficient migration. Another potential problem with purchasing lands for natural areas or corridors is that, given the uncertainties about regional climate change, the eventual direction and magnitude of dispersal cannot be predicted with certainty. An approach that could be explored would be to set aside protected areas with concentric buffer zones of protection (e.g., as in biosphere reserves). The most sensitive zone could be the most protected, with other zones allowing more human use and occupancy. Such a design should also consider the possibility that species in the highly protected zones could migrate out of the area in response to climate change. In many cases strengthening of existing protected areas through the provision of greater financial or managerial support or increasing the economic stake and social value to local communities in protected areas may be the best management approach. In any case, the social and economic consequences of designating lands as protected areas should be considered in view of the ecological benefits.

6.8 FOOD SECURITY

6.8.1 INTRODUCTION

Over the last 50 years, technological advances in irrigation, mechanization, pesticides, fertilizers, and crop and livestock breeding have stabilized and increased agricultural production in many parts of the world. Our ability to produce food and fiber is greater now than at any time in the past. Yet we have not succeeded in ensuring food security for all the world's population.

It is estimated that between 500 and 700 million people in the developing world do not have access to enough food. Malnutrition contributes to the deaths of 35,000 children each day. Despite great progress in countries such as India, many less-developed countries already have a serious food security problem. Food security is determined by the availability of food and the ability to acquire it by "dependable long-term access to food through local production, or through the power to purchase food via local, national, regional, or international markets." Even in the absence of any climate change, several countries will find it difficult to maintain or enhance food security given expected population growth.

To ensure food security for the world's increasing population, it will be necessary to sustain and enhance the natural resources on which we depend. Economic growth and equity are also musts. But most important, the spatial and temporal uncertainties about the impacts of greenhouse gases and associated climate change may demand the development of flexible policies that allow and encourage local adaptations/solutions, allow course corrections, and take a long-term approach.

6.8.2 FACTORS AFFECTING FOOD SECURITY

Food security at the national level has three main elements: adequacy of supplies, stability of supplies, and access to supplies.

Ensuring *adequacy of supplies* involves determining the appropriate balance between domestic production and trade. Increasing domestic production is not simply a resource or technology problem, although the availability of both are critically important, and this will become increasingly the case in those countries where climate change has negative effects on food and agricultural production. It is a function of four interrelated factors—the four "i's" of agricultural development: *incentives* to encourage farmers to produce surplus food for the market; *inputs* to boost productivity; *institutions* to provide credit, technical advice, marketing services, etc.; and *infrastructure* in the form of roads, storage, facilities, etc., to link the farmer with input supplies and food markets, particularly urban ones.

For some countries and communities the most serious issue is not the long-run adequacy of supplies but year-to-year *instability of supplies*. Such instability arises primarily from dependence on rain-fed agriculture in drought-prone areas; from population pressures that have forced people to cultivate more marginal land and/or to maintain more livestock than the rangeland or pastures can feed, with consequent wide fluctuations in crop yields and high livestock mortality during times of drought; from the unwillingness or inability to maintain sufficient stocks; from cyclical supply/demand patterns in the world food market; and from frequent changes in government policies, support prices, etc.

Access to supplies has two major dimensions: a national dimension reflecting the ability of countries to enter the world market to buy additional food, which in part is a function of the openness of the trade system and the ability of food deficit countries to earn foreign exchange through exports; and a personal income dimension, in that extensive poverty restricts the ability of individuals and households to buy all the food they need for a healthy diet. Thus, local or imported supplies must be matched by effective demand and appropriate mechanisms to ensure access to supplies by those lacking purchasing power. Without such mechanisms the market will maintain supplies at levels that leave many people hungry.

It is apparent then that the solutions to food insecurity are not simply technological, but also involve economics, infrastructure, and governmental policies. Developed countries, with their strong economies, well-established infrastructure, institutions and governments, as well as with less dependence on the agricultural sector (farming contributes approximately 2 percent of the annual GNP in the United States, compared with 40–60 percent of the annual GNP in some less-developed countries), are not as vulnerable to the impacts of climate change. Although developed countries will no doubt undergo adjustments in their economies and agricultural sectors in the event of climate change, national food security should be relatively unaffected. They will buy food and fiber that they cannot produce.

Low-income countries are more seriously affected by food insecurity. In order to reduce existing poverty in developing countries from 50 to 10 percent of the population and thereby improve food security, it has been estimated that a 3 percent annual growth in per capita income would be necessary over the next 20 to 30 years. With expected population increases, this translates into an overall national income growth of 5–6 percent. Such economic growth will be difficult if not impossible even under the best of circumstances, unless the countries receive greater international support.

Without technical assistance from developed nations and significant gains in sustainable economic growth and development, food security problems could be exacerbated in many nations.

6.8.3 FOOD SECURITY IMPLICATIONS AND IMPACTS

Climate change will not affect each of the three elements of food security uniformly, and within each element the effects may differ as to their global, regional, or national importance.

6.8.3.1 *Climate Change and Adequacy of Supplies*

While the impacts of climate change are uncertain, it is possible that the overall impact on global food and fiber supplies could be positive. In general there may be positive impacts due to increased productivity resulting from higher CO_2 levels. Production could be further augmented in some areas because of longer growing seasons and reduced frost damage. On the other hand, production may be reduced in some areas due to higher temperatures, less soil moisture and greater pest infestation, and loss of land due to sea level rise (see below).

At the regional level the picture is less certain, but there could be substantial potential for intra-regional compensation.

Such problems as may occur are likely to be at the national or subnational level, particularly in some of today's food-deficit developing countries. Food security in these countries is already a problem because of poverty and the lack of effective demand, stemming in part from the impact of high population pressures on limited and low-quality natural resources.

The adequacy of supplies could be affected in four principal ways: spatial shifts in the agroclimatic zones suited to the growth of specific food crops; changes in crop yields, livestock output, and fisheries productivity; changes in the water available for irrigation; changes in productivity and use of land because of sea level rise. Other impacts, such as alterations in the protein or starch content of crops seem likely to be of far lesser importance.

Shifts in agroclimatic zones. Climate change could shift agroclimatic zones with both positive and negative effects. In general, zones would shift poleward. In the middle and higher latitudes, higher temperatures may extend the growing season and reduce frost damage, though some of this positive potential may not be realizable (see Section 6.8.3.2). At the same time, there may be a loss of agricultural land if conditions become drier in currently arid or semi-arid regions (though higher carbon dioxide concentrations may increase water-use efficiency of many—especially cool season—crops). Such shifts in agroclimatic zones could cause regional dislocations.

Sea level rise could have a major effect on food security in several areas. Some of the most productive areas are low-lying coastal plains and estuaries with fertile alluvial soils. In Asia, for example, a high proportion of rice production comes from low-lying coastal areas, mainly former swamps and marshes.

Production on these lands could be lost through submergence, longer or deep freshwater flooding in some inland parts of low-lying coastal areas, and increased saltwater intrusions in coastal aquifers

used for irrigation or for livestock drinking water. However, while these lands could be lost to traditional uses, new uses (such as fisheries and aquaculture) could also take their place. In reality it may be a major threat to food security only in those areas already at risk in certain low-income food-deficit developing countries.

Changes in irrigation potential could occur in two main ways. First, through changes in the spatial and temporal patterns of precipitation, surface run-off and recharge rates of aquifers. Second, through salt-water intrusions into coastal aquifers. In some regions these changes, coupled with changes in demand, could exacerbate existing water shortages; in other areas, existing shortages could be mitigated.

Impacts on crop yields. With changes in temperature and rainfall regimes and in CO_2 concentrations, countries will become more or less optimal for the production of certain crops. Thus, yields will rise, or fall—unless there are compensating management actions or developments in technology.

While there are uncertainties associated with model results, especially on the regional and local levels, they project the following.

a) Considerable increases in potential crop yields in northern temperate countries—up to several percent in some instances; however, such increases may have little impact on national food security since these countries are already able to import all of the foods they cannot currently produce by spending a small fraction of their national income. Even so, such gains in potential production may not be fully realizable because of plant pests and diseases (see Section 6.8.3.2).

b) In the middle latitudes, potential declines in yields in the hotter, drier interior of continents because decreased rainfalls and increased evapotranspiration may offset positive CO_2 impacts on photosynthesis and water-use efficiency.

c) Limitation of yields may occur in some semi-arid tropical and sub-tropical areas if the net effect of change in temperature and precipitation is a reduction in crop water availability.

Animal husbandry. The effects of climate change on animal husbandry will largely mirror those on cropping. Any change which reduces or enhances biomass production will likewise reduce or enhance livestock carrying capacity. Whereas climate changes are expected to have little effect on crop quality, they may have a significant impact on forage quality. High temperatures may adversely affect reproduction, milk and meat production, but conversely will reduce maintenance requirements—particularly in temperate areas.

As with crops, warmer temperatures and increased precipitation will provide favorable conditions for parasites, fungi, bacteria, viruses, and insects. This can be expected to contribute to the more rapid deterioration of animal products. Increases in variability or intensity of rainfall resulting in droughts or flooding can cause tremendous losses at the local level.

Fisheries. Fisheries may experience problems similar to cropping and animal husbandry due to changes in temperature and precipitation. Current ranges of important commercial species may shift poleward along with marine habitats of higher productivity. Freshwater ponds which are already near the upper temperature limits of tolerance will be lost as habitats to some species. Changes in evaporative demand, surface run-off and melting in the high latitudes may also affect streams and lakes. Coastal wetlands, marshes and shallows (which are important to most fisheries) and aquaculture facilities may be relocated as existing areas are lost and new areas created as a result of sea level rise. (However, creation of suitable new areas may be hampered because of human barriers in their path.) Again, the impact of these effects on economies and food security will be felt mostly on a local or national level.

6.8.3.2 *Climate Change and Stability of Supplies*

Although it is difficult to determine the potential net effect of climate change on supply stability at the global level, at least four sources of changes need to be considered here:

- Reduction of snap frost damage in high altitude tropics and in northern latitudes. Early and late season frosts currently have irregular, but at times severe, impacts on production, although

the damage is largely confined to fruit and vegetable crops.
- Possible increases in the geographic range and severity of both plant and animal pests and diseases. In the absence of corrective actions the overall net effect could be negative. Higher temperatures will allow pests and diseases to over-winter for the first time, or to over-winter in larger populations, thereby providing the conditions for rapid spread early in the growing season to epidemic levels, consequently resulting in greater production losses. Heavier and more prolonged rainfall could have similar effects. Such outbreaks currently occur on an irregular basis, causing food supply to fall by as much as 50 percent over substantial areas. Climate change could make such outbreaks a much more frequent and widespread event.
- Possible changes in the annual variability of the climate, i.e., reliability of arrival of seasonal rainfall, its frequency and intensity. Erratic climate can lead to delayed planting, mid-season and other droughts, and incomplete plant growth cycles. In some present instances it can reduce cereal production by as much as half, and cattle populations by a quarter, leading to wide annual swings in food supply. How climate change will affect annual climate variability is not known.
- Possible changes in the frequency and magnitude of rainfall may lead to changes in flooding and erosion. The effect of climate change on these factors is also unknown.

In addition, changes in climatic variability could have impacts on grain quality and, therefore, market and nutrition value. However, the direction of the net change on this aspect is currently unknown.

6.8.3.3 *Climate Change and Access to Supplies*

Given that global food supplies may be sufficient to meet world demand (and possibly increase), and export supplies may be available, any problem of access to supplies will likely be more acute in low income food-deficit countries, and in low income low-lying islands. Although climate change may have negative impacts on agricultural production in high- and middle-income countries, the consequent falls in export earnings or food production are unlikely to be critical for national food security since food imports or average per capita food expenditure are low in proportion to the respective totals. Nonetheless, even within these countries there may be particular income groups whose food purchasing power will be reduced, either directly or indirectly, by climate change.

Access to food supplies could be affected by changes outside agriculture, e.g., loss of tourist revenues by low-lying islands and other areas following sea-level rise. The main impacts, however, will come from within agriculture. It is conceivable that a number of countries could suffer from decreased production of their major export crops, e.g., ground-nuts and cotton, which would also have serious impacts on employment and agricultural incomes, thereby reducing the ability of individuals and households to buy sufficient food.

Where domestic food production decreases, significant income effects are also likely to arise.

6.8.4 RESPONSE STRATEGIES AND ADAPTATION MEASURES

The uncertainties concerning the magnitude, timing, and spatial distribution of the impacts on food security point to the need for response strategies with two principal objectives. First, the alleviation or elimination of those current food security problems that might be intensified by climate change—e.g., flooding in Bangladesh and drought in the Sahel. Second, the maintenance for future generations of as wide a range of adaptation measures as possible—e.g., breeding from rare livestock breeds and crop land races that are well adapted to the agroclimatic conditions that could emerge but that are currently dying out through the lack of adequate conservation efforts.

The first objective can be achieved through general improvement in economic conditions, especially in developing nations, and through a range of economically justifiable adaptation measures, most of which already exist or could be developed, given the likely time scale of climate change.

These measures include capital investment, technology development and dissemination, and resource management changes. They are considered in the following sections.

The second objective has essentially three com-

ponents: (a) changes in land-use policy to ensure, for example, that high quality agricultural land is not lost permanently through urban and industrial development and is kept in good condition through appropriate management; (b) enhancement of biodiversity conservation efforts—both *in situ* and *ex situ*; and (c) changes in policies to intensify and sustain agricultural production on the best lands in order to remove pressure from some of the more marginal lands. Response strategies for these are given in Section 6.7 of this document.

The following subsections present options to enhance and/or maintain food security. They employ the classification scheme described in Section 6.4. Under this scheme, Category A consists of options that would improve the knowledge base necessary to fashion rational and efficient responses to increases in greenhouse gas concentrations and associated climate change. Category B consists of options that some societies may be able to undertake in the short term if they result in little or no additional costs (or consequences). Category C identifies long-term options that may be considered once uncertainties regarding greenhouse gas concentrations and climate change are reduced.

6.8.4.1 *General Response Options*

1) Since the developing countries have largely resource-based economies, efforts concentrating on the improvement of agricultural and natural resources would be beneficial (B, C). However, some assistance in improving economic conditions in developing nations comes directly or indirectly from developed nations, some of whom are themselves beset with budget difficulties.

2) Improve agricultural and natural resource research/extension institutions, specifically: (a) national coordination for local level efforts due to the uncertainty of local level effects of climate change; (b) regional networking activities to pool resources and share technology/responses; and (c) support for these activities from developed countries in terms of training researchers, extension agents, and natural resource managers, and technological assistance (A).

3) Gradually reduce international trade barriers and farm subsidies in both developed and developing countries to foster economic growth in less-developed countries and allow the agricultural sector to compete in a more global marketplace (B-C).

4) Intensify agricultural production on a sustainable basis on the best lands in order to remove pressure from some of the more marginal lands. Better water and soil management and more efficient use of water, fertilizer, and pesticide will also be necessary to sustain this intensive production (B, C).

5) Develop new food products that can be more easily stored, and improve storage facilities where yields tend to be unstable (A).

6) Develop baseline information on and monitor changes in climate, crop production, pest and disease incidence, livestock fecundity and quality, and fisheries, at local, regional, and global levels (A). Look for irreversible changes, and changes that can be used as indicators of climate change.

7) Protect germ plasm resources and biodiversity. Increased research on the preservation of plant, animal, insect, nematode, and microorganism resources may be needed. New breeding efforts (including genetic engineering) may be necessary to develop crop and livestock tolerance to both physical and biological stresses. Research on biocontrol and the plant rhizosphere may uncover new, beneficial uses for insects, nematodes, and microorganisms (A).

8) Review the current knowledge base of agricultural technology and identify best bets for further investigation and possible transfer to other locales and regions (A).

9) Continue development and maintenance of data bases for species, technology, and management and land-use practices employed under different climatic conditions, for possible transfer to other locales and regions if, and when, climate shifts become more certain (A).

10) Continue research and development of heat- and drought-tolerant crop and livestock varieties (A).

11) In the long run, as and when uncertainties regarding climate change and its impacts at a regional level are sufficiently reduced, in areas that are likely to be significantly adversely affected agricultural communities may have to shift to new crops and possibly accelerate changes that may have occurred anyway—e.g., by adoption of new technologies, practices, and even new livelihoods (C). Implementing the above set of responses 1–10 would help facilitate any such transitions.

6.8.4.2 Adaptive Measures for Crop Production

Options are available for both investment and management (technical change) and could be used to enhance or maintain food security.

Many of the arid and semi-arid areas vulnerable to climate change have undeveloped irrigation potential. In other similar areas, current irrigation water use is depleting water tables. Solutions to both these situations may be expensive and, therefore, may not be economic for low-value food crops. This cost constraint can be overcome partly by secondary investment in raising water-use efficiency so that a larger area can be irrigated with a given volume of water, and by improving drainage to prevent salinization. Drainage investment will also be required in rain-fed areas where soils are already waterlogged periodically (B-C, depending on whether or not it can be economically justified now), or which could become waterlogged as a result of increased rainfall (C).

In areas where reduced rainfall could lead to lower water tables, it may be economic to consider artificial recharge of aquifers.

Similarly, it may be economic in some areas to consider coast protection, storm surge and tidal barriers, and other forms of flood control. Such measures, however, could be very expensive and result in ecological change. Given that the sea level rise will be gradual, the high-income countries should be able to afford preventive measures like higher coastal defenses and special tidal barriers. For the low-income countries, however, investment capital is already in short supply and in many of them is likely to remain too low for them to be able to afford such coastal defenses. Therefore the building of higher coastal defenses to deal exclusively with climate change should be deferred until uncertainties are reduced (C). On the other hand, augmenting an existing plan to "insure" against climate change may be justified (B).

Finally, there is investment in grain storage to "insure" against greater fluctuation in annual rainfall. When such investments are primarily at the farm level, the costs can be relatively modest and feasible since the stores tend to be built with local materials and household labor (B-C).

Most *technical change* options concern shifts in plant-breeding strategies to improve pest resistance, heat or drought tolerance, etc. Given that plant-breeding cycles are usually less than 10 to 15 years, such improvements could keep pace with slow climate changes provided appropriate germ plasm is available (A).

Adaptation and introduction of existing techniques also has an important role to play. Minimal tillage, for example, has been developed primarily as a technique for highly mechanized farming systems in industrialized countries where improvements in soil erosion control and soil water management are required. If adapted for developing country conditions it could play an important role in areas experiencing greater aridity (A, B-C).

Other techniques to conserve soil and water and increase sustainability of resources which could be more widely used include: tied ridges, conservation bench terracing, mechanical and vegetative bunding, weed control, legume and sod-based rotations, windbreaks, drip irrigation, more efficient fertilizer and pest management, and appropriate cultivars (A, B-C). Areas with adequate water resources can be irrigated, although attention should be given to drainage systems (C). Salinization can be problematic on clay soils with poor drainage and in areas with high evaporative demand. Drainage systems may also be necessary in areas with clay soils that experience precipitation increases.

Other management options include changes in the timing of farm operations—e.g., shifts in planting dates to compensate for changes in rainfall distribution (C). Additional gains could come from changes in cropping patterns through the introduction of "new" crops better suited to the modified agroclimatic conditions (C).

6.8.4.3 *Livestock Sector Options for Adaptation*

The most serious threats to food security for the livestock sector will be confined largely to the arid and semi-arid areas of the middle and lower latitudes, where nomadism or settled extensive or semi-intensive livestock systems are the only way of exploiting vast areas of relatively marginal grassland, particularly in the countries surrounding the Sahara desert.

Other agroecologically marginal areas, notably those in Australia, North America, and the USSR, could also be adversely affected, but without posing a significant risk to national food security, although individual farmers would need support.

In the most severe situations, where pasture production is substantially reduced, the main option may be to assist livestock owners and their families to move out of livestock into crop production or out of agriculture altogether into different areas of the same country (C).

Where impacts are less severe, mitigation of the effects of climate change may be achieved through the installation of water points for livestock (C); by research and breeding to select animals and pasture or fodder plants better adapted to the new environment (A). New breeding techniques, like embryo transfer, increase the probability of being able to achieve improvements on an appropriate time scale. Finally, improvements in livestock management may be possible. These could include: improved rations utilizing supplementary protein, vitamins and minerals (B); reductions in stocking rates (although this is often difficult to achieve in practice where communal grazing land is involved) (B-C); improvements in pasture management (B-C); wider use of feed conservation techniques and fodder banks (B-C).

In higher latitudes where conditions may become warmer and wetter, pasture productivity is likely to rise, but pests and diseases could become more serious. Nonetheless, the net effect on production in such areas is likely to be positive.

6.8.4.4 *Adaptation Measures for Fisheries*

Fish have food, recreational, and cultural value. There will be three main impacts from climate change: changes in the abundance and distribution of fish stocks; overall changes in the variability of individual fish stocks; and changes in inland, coastal, and oceanic habitats.

Options for the first include strengthening or, where appropriate, establishing fisheries institutions that could assist in managing or allocating changing fish stocks due to effects of climate change (B); shifting fishing from coastal to distant waters (C); development of inshore or inland aquaculture (C); and finally, employment creation outside the sector (C).

Responses to the second impact are part of wider measures to improve fisheries management. Fisheries management policies may need to be strengthened or modified, as appropriate, along with associated fish population-and-catch monitoring activities, to prevent overfishing where stocks decline or to ensure sustainable exploitation of improved fisheries stocks (B,C). In some areas, it takes many years for changes in fisheries policies to be developed and become fully effective. The time scale of policy change should be considered in weighing various management options.

Consideration of options regarding changes in inland, coastal, and oceanic habitat is necessary to reduce any potential negative impacts as well as to take advantage of new habitat that may be created. One option is to consider fisheries (wild and aquaculture) habitat needs in planning coastal protection measures (B,C).

Other options are strongly research-based and will need general strengthening of research institutions (A) as well as more specific actions. The latter include restocking with ecologically sound species following biological and commercial evaluation of indigenous and non-indigenous species (A, C). Fish breeding supported by measures to conserve the genetic diversity and inherent variability of existing fish populations could also play an important role (A). Furthermore, increased research into population dynamics and adaptive measures is needed (A). Moreover, the use of aquatic and marine species as indicators of change needs to be explored and, if necessary, enhanced (A).

Given the close interaction between land cover and maintenance of adequate fishery habitat, fishery managers need to cooperate closely with forestry and other resource managers (B). This is critical for policy success in this field.

Marine mammals (e.g., whales, dolphins, seals) are of increasing cultural importance and decreasing significance for food. Most stocks of these animals will likely move as marine ecosystems shift to adapt to changes in temperature and circulation patterns and, as a consequence, may not be significantly affected, at least in abundance. Required adaptation measures are probably limited to scientific monitoring (A) and ensuring that habitat needs are also considered from the standpoints of coastal planning and in ocean pollution control efforts (B-C).

APPENDIX 6.1
LIST OF PAPERS PROVIDED TO SUBGROUP

FOOD

Australia, Department of Primary Industries and Energy. "Potential Impact of Climatic Change on Animal Husbandry—An Australian Perspective," May 5, 1989.

Brazil, EMBRAPA. "Future Scenarios and Agricultural Strategies Against Climatic Changes: The Case of Tropical Savannas," by A. Luchiari, Jr., E.D. Assad, and E. Wagner, October 30–November 1, 1989.

Canada, Fisheries and Oceans Department. "Fisheries: Response Strategies to Climate Change," draft working paper, July 11, 1989.

Food and Agricultural Organization. "Implications of Climate Change for Food Security," Theme paper prepared for IPCC Working Group III, Resource Use and Management Subgroup, by D. Norse, September 1989.

France. "Combatting Desertification and Salinization," 1989.

India, Ministry of Environment and Forestry. "Effect of Climatic Changes Due to Greenhouse Gases on Agriculture: Some Salient Points," 1989.

Jodha, N.S., ICIMOD, Kathmandu, Nepal. "Potential Strategies for Adapting to Greenhouse Warming: Perspectives from the Developing World," October 30–November 1, 1989.

Okigbo, B.N., Michigan State University, East Lansing, United States. "Comments on Response Strategies Related to Food and Agriculture in the Face of Climatic Change with Respect to Tropical Africa," November 1, 1989.

Swaminathan, M.S., President, IUCN. "Agriculture and Food Industry in the 21st Century," October 30, 1989.

FORESTRY

Brazil. "Forestry Report," 1989.

Canada. "Socio-economic and Policy Implications of Climate Change on the World's Managed Forests and the Forest Sector," by Peter N. Duinker, Working Group II (Environmental and Socio-economic Impacts of Climate) of the Intergovernmental Panel on Climate Change, Geneva.

India, Ministry of Environment and Forestry. "Some Salient Points—Agriculture: Effect of Climatic Changes Due to Greenhouse Gases on Forests and the Strategies Needed to Tackle These Effects," 1989.

Japan, Ministry of Agriculture, Forestry and Fisheries. "Japanese Report on Forest and Forestry," 1989.

United Kingdom. "Note on Tropical Forestry and Climate Change," prepared for Intergovernmental Panel on Climate Change, Working Group III, Resource Use Management and Agriculture, Forestry and Other Human Activities Subgroups, August 1989.

United Kingdom, Forestry Commission of Great Britain. "United Kingdom Contribution to Consideration of Adaptation to Possible Climatic Impacts on Forests," by A.J. Grayson, 1989.

United States Interagency Task Force. "Agriculture and Forestry: Adaptive Responses to Climate Change," draft, April 7, 1989.

United States Interagency Task Force. "Land Use and Biological Diversity: Adaptive Responses to Climate Change," September, 1989.

LAND USE/UNMANAGED ECOSYSTEMS

Netherlands. "Climatic Change and Biological Diversity," paper prepared for RUMS workshop, October 30–November 1, 1989.

United Nations Environment Programme. "Biological Diversity: A Unifying Theme Paper," by H. Zedan, prepared for IPCC WG3, Resource Use and Management Subgroup, 1989.

United States Interagency Task Force. "Unmanaged Ecosystems—Biological Diversity: Adaptive Responses to Climate Change," September 13, 1989.

United States Interagency Task Force. "Land Use Management: Adaptive Responses to Climate Change," September 13, 1989.

WATER RESOURCES

da Cunha, L.V., NATO. "Impacts of Climate Change on Water Resources and Potential Response Actions: An Overview of the Situation in the EEC Region," Workshop on Adapting to the Potential Effects of Climate Change, Working Group III of the Intergovernmental Panel on Climate Change, Geneva, Switzerland, October 30–November 1, 1989.

Kaczmarek, Z., NASA, and J. Kindler, Warsaw Technical University. "The Impacts of Climate Variability and Change on Urban and Industrial Water Supply and Wastewater Disposal," prepared for IPCC Working Group III, Resources Use and Management Subgroup, October 30, 1989.

Lins, H., U.S. Geological Survey. "Overview and Status of WG-II Hydrology and Water Resources Subgroup Activities," prepared for Working Group III, RUMS Workshop, October 30–November 1, 1989.

Nepal, Department of Hydrology and Meteorology. "Technology and Practice Summary Report on Water Resources," prepared by S.P. Adhikary, 1989.

Rogers, P., and M. Fiering, Harvard University. "Water Resource Use and Management in the Face of Climate Change Considerations in Establishing Responses," presented at UNEP/WMO Intergovernmental Panel on Climate Change, Resource Use and Management Subgroup, October 30–November 1, 1989.

Stakhiv, E.Z., U.S. Army Corps of Engineers and H. Lins, U.S. Geological Survey. "Impacts of Climate Change on U.S. Water Resources (with reference to the Great Lakes Basin, USA)," presented at the IPCC Resource Uses and Management Strategies Workshop, Geneva, Switzerland, October 30–November 1, 1989.

United States Interagency Task Force. "Water Resources—Adaptive Responses to Climate Change," draft working paper, August 23, 1989.

Waggoner, P., Chairman, AAAS Panel on Climate Change and Water Resources. "Theme Paper—Water Resources Session," prepared for IPCC WG3 Resource Use and Management Subgroup, Geneva, Switzerland, October 30–November 1, 1989.

APPENDIX 6.2
SUBGROUP MEMBERS, PARTICIPANTS, AND OBSERVERS

GOVERNMENTAL MEMBERS AND PARTICIPANTS

Australia (Participant)

Rod Holesgrove
Department of Arts, Sport, Environment, Tourism
and Territories

Nelson Quinn
Conservation Division
Department of Arts, Sport, Environment, Tourism
and Territories

Chris Bee
Department of Arts, Sport, Environment, Tourism
and Territories

Brazil (Member)

Enio Cordeiro
Ministry of External Relations

Antonio Rochas Magalhaes

Frederico S. Duque Estrada Meyer
Brazilian Mission to the U.N.

Canada (Co-chairman)

Ralph Pentland
Environment Canada

China (Member)

Erda Lin
Academy of Agricultural Sciences

Jibin Luo
State Meteorological Agency

France (Co-chairman)

Jacques Theys
Ministere de l'Environnement

India (Co-chairman)

Inder Abrol
Indian Council of Agricultural Research

A.C. Ray

Japan (Member)

T. Inoue
Forestry & Forest Products Research Institute

T. Udagawa
National Institute of Agro-Environmental Sciences

S. Sakaguchi
Ministry of Foreign Affairs

Malta (Participant)

G.N. Busuttil
Ministry of Foreign Affairs

Nepal (Member)

Sharad P. Adhikary
Department of Hydrology & Meteorology

Netherlands (Participant)

R. van Venetie
Ministry of Agriculture and Fisheries

203

Saudi Arabia (Member)

Rihab Massoud
Embassy of Saudi Arabia

Switzerland (Member)

Pascale Morand Francis
Federal Office of Environmental, Forest and Landscape

Tanzania (Participant)

Adolfo Mascarenhas
Institute of Resource Assessment

USSR (Member)

Yuriy Sedunov
USSR State Committee for Hydrometeorology

United Kingdom (Member)

C.A. Robertson
Overseas Development Administration

M. McCausland
Overseas Development Administration

Neil Sanders
Department of the Environment

A. Apling
Department of the Environment

U.S.A. (Rapporteur)

Indur M. Goklany
U.S. Department of the Interior

Norton Strommen
U.S. Department of Agriculture

Gary Evans
U.S. Department of Agriculture

Eugene Stakhiv
U.S. Department of Defense

Joel Smith
U.S. Environmental Protection Agency

Morgan Rees
Department of the Army, Civil Works

Richard Wahl
U.S. Department of the Interior

John T. Everett
National Oceanic & Atmospheric Administration

K. Andrasko
U.S. Environmental Protection Agency

Zimbabwe (Participant)

K. Masamvu
Department of Meteorological Services

PARTICIPATING INTERGOVERNMENTAL ORGANIZATIONS

Commission of the European Communities

Peter Faross
Belgium

Food and Agricultural Organization

David Norse
Italy

UNESCO

Bernd von Droste
France

F. Verhoog
France

UNEP (Participant)

W.V. Kennedy
Kenya

Intergovernmental Panel on Climate Change
   (Participant)

N. Sundararaman
Switzerland

NON-GOVERNMENTAL OBSERVERS

I.J. Graham-Bryce
Coal Industry Advisory Board

Norman Rosenberg
Resources for the Future

R. Sedjo
Resources for the Future

M.C. Trexler
World Resources Institute

M. Markham
WWF International

IMPLEMENTATION MEASURES

7

Public Education and Information Mechanisms

COORDINATORS
G. Evans (U.S.A.)
Ji-Bin Luo (China)

CONTENTS

PUBLIC EDUCATION AND INFORMATION MECHANISMS

7.1 INTRODUCTION

7.1.1 An environmentally informed global population is essential to addressing and coping with climate change. Throughout the world in whatever the sphere of human activity or exchange, there is now ignorance, apprehension, and confusion about climate change. For some individuals, groups, and organizations, the problem is unknown, irrelevant, or remote. For others, it is of overwhelming proportion or misunderstood. Few bring to the world forum a workable grasp of the issue and the potential adverse consequences of climate change. Despite the fact that there are uncertainties and unknown dimensions about the topic, there is little disagreement among experts about the importance of understanding about greenhouse gas emissions and achieving sustainable human development.

Given the importance of a well-informed population, the Working Group developed suggestions and approaches for improving international awareness about the causes and potential impacts of climate change. These suggestions and approaches are summarized in this section. While it was recognized that broad-based understanding is essential, it was also appreciated that no single mechanism or template can work for every group or in every culture or country. The social, economic, and cultural diversity of nations will likely require educational approaches and information tailored to the specific requirements and resources of particular locales, countries, or regions. Nonetheless, international collaboration is critical to taking efficient and effective steps. So, too, is mobilizing and fully integrating the expertise and resources of international organizations. Although education must ultimately be context-specific and reflect what individuals and organizations can best do to make a difference, the commonality of the climate change problem requires cooperative solutions.

Across communities and target audiences, education and information should be geared toward making people more capable of dealing with the problems associated with climate change and better able to make responsible decisions as full global partners. Education is used broadly to encompass formal and informal processes of teaching and learning that occur throughout the life of the individual in the home, the school, the workplace and a range of other settings where instruction, training, and firsthand experience are conveyed. Similarly, information refers to materials and data that are produced and accessible in a variety of formats varying in formality and technical sophistication. Thus, for purposes of IPCC response strategies, both education and information are interactive and dynamic ways of fostering new levels of environmental sensitivity and awareness in public and private lives.

7.1.2 It was recognized that a set of guiding principles must govern the structure for developing and implementing education and public information measures. First, multilevel approaches are important; both education and information measures must be appropriate to international, regional, national, and local formats and to wide-ranging audiences from individuals to private and public sector organizations. Second, education and information strategies must be attentive to the cultural and multi-cultural diversity within and between nations and regions, including language diversity and the capacity to communicate fundamental concepts. Third, there must be sensitivity to other national differences, including those between developed and developing nations in education and information resources, in mass media capacity, and in other regional variations. Fourth, education and information mechanisms must strengthen an appreciation of commonality and global interdependence among people facing the complex issues of climate change. Finally, it is important to convey an understanding that early responses are essential to resolving problems of future generations.

This report examines public education and information on climate change in terms of aims, audiences, measures, and actions. Each is addressed in turn. It must be emphasized, however, that these sections do not stand on their own merit but are closely interrelated. The aims, audiences, measures, and actions identified in this section in reality also fall into three overlapping categories—those of immediate concern, those of the near term, and those of a long-range nature. In devising public education and information mechanisms, a balance between long-term and short-term actions is recommended. Additionally, there must be a balance between top-down and bottom-up target audiences.

7.2 AIMS

7.2.1 The purpose of public education and information is to foster a common understanding about causes and effects of global climate change. There is also a need to change public attitudes and policies to effect responses to climate change.

7.2.2 The core aims may be summarized as follows:

- Promote awareness and knowledge of climate change issues.
- Encourage positive practices to limit and/or adapt to climate change.
- Encourage wide participation by all countries, both developed and developing, in addressing climate change and developing appropriate responses.

Among the aims of education and information is the need to ensure timely flow of authoritative, frank, and dispassionate information about global climate change, exchanging the most up-to-date information as it becomes available. The information must fit into general approaches for information dissemination on world environmental problems.

7.2.3 An aware public is absolutely essential if politicians and other decision makers are to mobilize the necessary means to limit the effects. Most people realize that climate changes exceed the length of human life; yet, unlike the physical Earth where deterioration is visible to the naked eye, deterioration of the air is not a tangible element. Therefore, a priority should be placed on education and information because attitudinal and behavioral changes can bring about structural change in production and consumption processes. Global strategies defined by IPCC and others at all levels can convince leaders of the need for joint or complementary action to achieve common goals—to foster a concept of global common good.

These aims can be met by governments ensuring the understanding, support, knowledge, and expertise of the private sector, including nongovernmental organizations, and the general public and by moving normally long-term measures of public education and information into the mainstream much more rapidly. Recognizing that well-planned and well-organized public information and education are separate activities, both must develop on parallel tracks, both must form an inseparable part of response strategies, and both must be open to public action so that people feel directly concerned and mobilized.

7.3 AUDIENCES

7.3.1 The audience or target groups include everyone. No person or organization should be excluded. For purposes of discussion, audiences are aggregated into common or related groups that might best be served by particular activities.

7.3.2 These audiences include present and future generations of children and youth as well as individuals at household levels. Public and private sector policymakers and leaders at each level (local, national, and international) are a high priority, special target for immediate and short-term actions. Further, media, educators and educational institutions, and scientists and other professional groups and their organizations similarly comprise an important group. Finally, but no less important, industrial, business, and agricultural sectors are key target groups. In particular, attention should be given to the bulk of the rural population.

7.4 MEASURES FOR DEVELOPING INFORMATION

7.4.1 Informational and educational materials must be developed. Materials should be objective and reflect broadly accepted assessments of the science of climate change suitable for use throughout the world. An explanation of the range of uncertainty is essential for clearly understanding the scope of the issue. Because of the dynamic nature of climate issues and because research is constantly producing new knowledge, information and educational materials must be periodically revised and updated. Informational and educational materials should include data on: resource and energy conservation; efficient utilization of energy sources with reduced carbon dioxide emissions; control of air pollutant emissions; and promotion of sound national conservation strategies that highlight the special role of conservation of forests and reforestation/afforestation in climate change issues.

7.4.2 Materials should provide consistent, reliable, and realistic information on climate change and its consequences in simple and attractive formats. Such materials should also include special attention to the mechanics of the atmosphere and the interaction of human activities with it.

7.4.3 Materials should be appropriate for use by technology-based learning media ranging from basic radio, television, film, and video through very sophisticated levels of computer-interactive systems. Information in the form of basic data can be supplied by scientists and other professionals who develop the scientific base from which extension personnel develop the materials provided to the end

users. This maintains a continual flow of knowledge and information needed to keep the public well informed. It is important to choose instruments, however, that are adapted to the receiver of the message. What is appropriate for one country or region likely will not be for another.

Mass media—in particular, radio, television, oral, and print—must play an important role in developing and providing an in-depth approach to both education and information, ranging from information for schools to detailed publications and radio programming to meet special needs. Attention should also be given to developing forms of communication in more remote areas in developing countries.

7.4.4 Information packets, textbooks, and other educational materials for use in schools, as well as more detailed materials to meet teacher-preparation and trainer needs, must be developed.

7.4.5 Countries with information on the labeling of energy-efficient and environmentally preferred or low emission products should compile such information and make it available to all countries. Such information must be continually updated and provided quickly and effectively to all sectors.

7.5 ACTIONS FOR DISSEMINATION

7.5.1 NATIONAL AND LOCAL ACTIONS

7.5.1.1 National committees or other organizations on climate change should be encouraged or clearinghouses should be established to collect, develop, and disseminate materials on climate change. This could help provide focal points for information on issues such as energy efficiency, energy savings, forestry, agriculture, environmentally sound housing and transportation efficiency. Information exchanges would also provide a means to share technical knowledge and expertise.

7.5.1.2 Information obtained through greenhouse gas emission audits may be used to identify the size of the problem as it relates to each nation. Practices may then be identified which would provide simple and cost-effective actions. Use of available research about attitudes and behavior of populations may be used to define local and national programs, identify primary and secondary target audiences, and enable the setting of realistic goals.

7.5.1.3 It is important to consider the following procedures for the development of action programs at this level:

- Organize symposia, seminars, and workshops. (For example, use professional and scientific societies to develop special programs for engineers, architects, land-use planners, and other designers and policymakers, making them aware of how they can contribute to energy efficiency and conservation, and anticipate likely impacts of climate change.)
- Develop specialized programs to provide training for educators and communicators (for example, media personnel).
- Devise or work through existing programs that will take advantage of local language and local culture. (For example, use public meetings and hearings, community organizations, and radio programs to inform the public.)
- Establish or use existing extension-type service organizations to assist in technology transfer and education.
- Establish local outreach through the development of, or use of, existing support groups.
- Call on the skill, structure, and involvement of local and regional special interest groups, for example, women's organizations and environmental interest groups.
- Use media campaigns to raise awareness about climate change concerns affecting the public and about what individuals themselves can do in their daily lives with regard to them.
- Develop and/or strengthen, in the developing countries, the role of NGOs specializing in the field of environment, particularly in the field of global climate change.

7.5.1.4 It is important to recognize that, irrespective of global climate change problems, certain actions at the local level are environmentally useful. These include education and information on the utilization of available alternative technology and on utilization of all resources as efficiently as possible, recycling waste where economically practicable, encouraging more efficient use of all forms of energy, increasing tree planting, and minimizing loss of biomass.

7.5.2   INTERNATIONAL ACTIONS

7.5.2.1 International organizations (UNESCO, UNEP, WMO, etc.) should utilize IPCC reports in developing and providing to all countries understandable guidelines for future program actions.

7.5.2.2 An international institution should be developed or an existing international institution should serve as a clearinghouse for existing and newly developed informational and educational materials. Focal points within existing networks should be established for the development or coordination of materials specific to global climate change issues.

7.5.2.3 Upon completion of IPCC reports, seminars should be developed that focus on providing informational programs and educational materials to developing countries.

7.5.2.4 Information activities should pay special attention to the issues of geographic isolation, cultural and linguistic diversity, and the absence of effective mass media within certain regions and countries throughout the world.

7.5.3   IMMEDIATE RESPONSES

While a number of actions are immediately feasible at the international, national, and local levels, some specific examples are provided below to illustrate the inter-relationship between components discussed earlier and available steps.

7.5.3.1 Develop strategies at the national, regional, and international level for the production, communication, and release of the IPCC reports. Participation by invited experts from the developing countries at this stage in particular, and in the IPCC activities in general, is highly desirable; funding for such participation should be ensured.

7.5.3.2 Upon completion of the IPCC reports, or earlier as also proposed by the IPCC Special Committee, a series of short-term seminars should be targeted to inform high-priority decision makers, world leaders, scientists and other professionals, education and media developers, both in the public and private sectors, of the causes and effects of climate change, and the related international efforts already under way. This should include specially formatted publications, news releases, and related media events. Existing institutions (international organizations and NGOs) may be approached to determine interest in implementing such a seminar series.

7.5.3.3 Establish multi-disciplinary national committees, or other forms of national coordinating mechanisms, on climate change, to build a partnership for implementing educational and informational strategies nationally and to establish appropriate networks with international and re-gional organizations. Through such mechanisms, or otherwise, establish focal points in each country. This and other measures would help improve distribution channels for dissemination of up-to-date information.

7.5.3.4 Solicit the support of international and regional organizations and NGOs for curriculum development by devising model courses of studies at various levels for primary grades through universities. These could be expected to include educational material (trainer and student texts) tailored at the national or local level, including appropriate specific examples of what individuals, organizations, and institutions can themselves do.

7.5.3.5 Solicit the support of the international and regional organizations and NGOs in implementing educational and informational strategies in general.

The immediate responses should take special account of the needs of the developing countries. In this connection, it may be noted that a Special Committee on the Participation of the Developing Countries has been established by IPCC for such purposes.

8

Technology Development and Transfer

COORDINATORS

K. Madhava Sarma (India)

K. Haraguchi (Japan)

CONTENTS

TECHNOLOGY DEVELOPMENT AND TRANSFER

The RSWG called for greater cooperation among all nations to improve and transfer existing technologies, develop and transfer new technologies, and improve the exchange of technological information to limit or adapt to climate change.

Many countries noted the need to tailor technologies to the domestic technological and human resource capabilities of developing countries and to better develop those capabilities.

Impediments to technology transfer and mechanisms to overcome those impediments were discussed at length.

8.1 TECHNOLOGICAL DEVELOPMENT

A. OVERVIEW

1) Technological development, including improvement and reassessment of existing technologies, is urgently needed to:
   (a) limit or reduce anthropogenic greenhouse gas emissions;
   (b) absorb greenhouse gases—i.e., protect and increase greenhouse gas sinks;
   (c) adapt human activities to the impacts of climate change; and
   (d) monitor, detect, and predict climate change and its impacts.

2) Technological actions designed to limit or adapt to climate change must be founded on a sound scientific basis, and must be consistent with the concept of sustainable development.

3) Criteria
   (a) In selecting technologies, priority should be given to those technologies that provide significant economic and social benefits in addition to benefits in limiting or adapting to climate change.
   (b) Appropriate criteria for selecting technologies (and for technology transfer measures) also include economic efficiency, taking into account all the external costs and benefits. Account must also be taken of suitability to local needs and conditions, additional benefits, ease of administration, information needs, legal and institutional constraints, national security (including defense, economy, energy, and food), and acceptability to the public.

4) (a) Technological development will have to be pursued in a wide range of sectors and activities such as energy (including noncommercial sources), industry, agriculture, and transport and management of natural resources.
   (b) Adequate and trained human resources in all the countries are a prerequisite for development and transfer of technologies.
   (c) The growth of industrial and agricultural activities is one of the main anthropogenic components of the greenhouse effect. Technological advances to limit or adapt to climate change are critically important to provide a sound basis of sustainable development.

B. MEASURES FOR PROMOTION OF TECHNOLOGICAL DEVELOPMENT

1) Policies
   (a) Appropriate pricing policy that takes into account external (including environmental) costs and benefits, while recognizing

that it may not be easy to assess them accurately, is one of the key factors in determining the rate and direction of technological development.

(b) In sectors where price signals do not elicit an adequate response, appropriate policies and incentives will be needed.

2) Information

(a) There is a need for international information exchange to enable decision makers to continuously monitor development of technologies to limit or adapt to climate change. In this connection, it is necessary to ask where and how existing information systems are inadequate.

(b) Governments should work cooperatively to conduct periodic assessments of the state of these technologies.

(c) There is a need to develop methodologies to assess the social, environmental, and economic consequences of actions or inactions on climate change.

(d) There is an urgent need for economic analyses and supporting scientific information to determine external costs and benefits referred to B.1.(a) above.

3) Support of Technological Development by Governments

Governments may need to support the development of technologies where such development by the private sector is not feasible.

4) Collaborative Efforts

(a) International collaborative efforts can accelerate development of technologies and offer substantial potential savings for all countries. Such efforts as already exist in bilateral and multilateral contexts, including those in several UN and other international organizations, need to be strengthened and expanded.

(b) Such collaborative efforts as those existing in some countries between the private sector and government to develop technologies could be expanded further to include developing countries and international organizations.

(c) Industrialized countries should take every feasible measure to assist and cooperate with (help) developing countries to acquire and apply these technologies effectively. Developing countries should be encouraged and assisted to develop their own technologies to achieve self-reliance.

C. AREAS FOR TECHNOLOGICAL DEVELOPMENT

The following areas have been suggested for consideration for technological development:

1) Limitation

(a) Limit or reduce CO_2 emissions globally.

(b) Absorb emitted CO_2 (by growing biomass, by work on chemical methods).

(c) Limit or reduce emissions (use) of ozone-depleting substances with greenhouse potential (under the Montreal Protocol to the Vienna Convention).

(d) Limit or reduce CH_4 emissions globally.

(e) Limit or reduce N_2O emissions globally.

(f) Limit or reduce O_3 emissions globally.

2) Adaptation

(a) Manage coastal resources—e.g., erosion or flooding of coastal cities, ports, transport routes, agriculture, estuarine salinization from storm/tidal surges.

(b) Manage water resources—e.g., water supply, flood control (groundwater, catchments, irrigation, river navigation, etc.).

(c) Ensure adequate agricultural response (availability of temperature- and dryness-tolerant varieties; correct rotation).

(d) Ensure adequate forestry response (availability of temperature, dryness- and fire-tolerant varieties; correct rotation).

(e) Ensure adequate fishery response (changes in ocean currents, fishing grounds).

(f) Revise soil mechanics criteria (permafrost, soil moisture, pore-water pressures).

(g) Respond to changes in duration of ice-closure of ports, rivers, coastal seas.

3) Monitoring and Scientific Analyses

(a) Global monitoring and related modeling

of atmospheric and ocean phenomena to try to detect warming signals, improve predictive capacity, assess climate sensitivity, and estimate delays in reaching equilibrium conditions.

(b) Monitor build-up of GHGs in atmosphere.

(c) Monitor sea level changes.

(d) Monitor forests.

(e) Systematic integrated monitoring of human-induced changes in ecosystems, land use, and ocean resources to determine their contribution to climatic change.

(f) Investigate biogeochemistry of past and ongoing climate-change related phenomena for better understanding of climate change mechanisms and impacts.

(The RSWG recommended that the question of biodiversity and climate change be examined by Working Group II.)

8.2 TECHNOLOGY TRANSFER

A. OVERVIEW

1) As the greenhouse gas emissions in developing countries are increasing with their population and economic growth, rapid transfer, on a preferential basis, to developing countries, of technologies that help to monitor, limit, or adapt to climate change, without hindering their economic development, is an urgent requirement.

2) It is important to understand the needs, opportunities, and constraints of the recipients when undertaking technology transfer.

3) Many of the available technologies or those being developed in industrialized countries will have to be adapted to meet the conditions or needs of the developing countries.

4) Substantial efforts to transfer technologies to developing countries are now occurring under

existing bilateral and multilateral arrangements. These should be strengthened and expanded.

B. FACTORS THAT IMPEDE EFFECTIVE TRANSFER OF TECHNOLOGIES AND SUGGESTED SOLUTIONS

1) A consensus has emerged on some of the factors that impede effective transfer of technologies that assist to limit or adapt to climate change and on the steps needed to address them. Such factors include:

(a) lack of necessary institutions and trained human resources in developing countries;

(b) social factors that inhibit change from established ways;

(c) lack of resources for purchase, operation, and maintenance of new technologies; and

(d) higher initial capital costs in the case of some technology options.

The suggested solutions to address these factors are: better utilization of existing multilateral and bilateral development institutions to finance transfer of technologies; strengthening or creation of the necessary institutions; training of human resources; and introduction and operation of new technologies and establishment of new funding mechanisms, as appropriate, to address these issues. The issue of funding mechanisms is discussed in the paper on Financial Measures and the report of the Special Committee. The governments, it is suggested, can provide grants, loans, and loan guarantees to overcome obstacles of high initial costs.

For many countries, in particular developing countries, addressing this matter is of high priority.

2) Introduction of new technologies is sometimes delayed by decisions of governments aimed at prevention of closure of existing industries that might cause economic and social disruption. The national governments and the international community should constantly

strive to remove the difficulties and to facilitate the introduction of new environmentally benign technologies, without hindering sustainable development.

3) There are two other factors that impede transfer of technologies that assist to limit or adapt to climate change. These are:
   (a) existence of legal barriers and restrictive trade practices which impede transfer of technologies; and
   (b) constraints arising from property rights involved in the development of technologies.

There is, however, a divergence of opinion on the solutions needed to address these factors. While some hold that a strict protection of property rights, patents, etc., will promote effective development and transfer of technology, others hold that the level of such protection should conform with overall economic and development policies of the recipient countries. There are differences of opinion on the legal barriers and the methods to remove restrictive trade practices. It is noted that these issues are under in-depth discussion in other forums such as GATT and UNCTAD, but no agreement has emerged.

C.   ISSUES TO BE FOLLOWED UP

Developing countries are of the view that transfer of technologies on a non-commercial basis is necessary and that specific bilateral and multilateral arrangements should be established to promote this. Some other countries where technologies are not owned by the government believe that transfer of technologies would be a function of commercial negotiations. It has not been possible to bridge the difference on views on these questions. It is extremely important to reach early international agreement on these issues, and other questions raised above, in order to promote effective flow of technologies to monitor, limit, or adapt to climate change.

8.3   SUGGESTIONS FOR PROGRAMMES

The following illustrative programmes are suggested to facilitate technology development and transfer:

1) *Global energy programme.* Designate experts on energy programming from respective countries who are to cooperate among themselves to formulate a coherent global energy programme, taking into account energy requirements and environmental objectives.

2) *Technology research centers.* Establish or expand regional or national technology research centers to develop technologies for basic infrastructure, energy efficiency, and alternative energy sources, agriculture, forestry, and water resources.

3) *Energy efficiency standards.* Establish standards on the energy efficiency of imported and locally manufactured technologies.

4) *Energy services.* Establish a policy to encourage private sector involvement in energy conservation and supply.

5) *Access to CFC substitutes.* Initiate a programme to promote CFC substitutes as well as provide assistance to developing countries in the acquisition and manufacture of the substitutes.

6) *Extension services.* Organize a network of extension services to provide assistance to government agencies, communities, and the private sector in designing and implementing mitigation and adaptation projects.

7) *Forestation/Afforestation programme.* New or expanded reforestation or forestation programmes could be linked to offset programmes for fossil-fuel-fired power plants, biomass energy projects, or a timber products industry as a means of obtaining financing and management for the programme. The social and economic incentives needed to en-

courage long-term planting and maintenance should be identified.

8) *Technology advisory committees.* Establish technology advisory committees comprising representatives from government and business to recommend approaches for increasing export, joint venture, and other business opportunities related to global climate change limitation and adaptation efforts and to advise governments on ways to increase access to and encourage the development of commercial sources of new technologies.

9) *Technology research and development.* Promote research and development on technologies to: (1) monitor and detect climate change; (2) limit or adapt to climate change; and (3) predict regional and local change and to assess possible effects.

10) *Technology conferences.* Conduct conferences to transfer existing technologies to: (1) monitor and detect climate change; (2) limit or adapt to climate change; and (3) predict regional and local changes and to assess possible effects.

11) *Pilot transfer programmes.* Conduct pilot programmes to transfer technologies in selected countries and sectors.

12) *International guidelines for technology transfer.* Introduce international guidelines for technology transfer.

13) *International clearinghouse.* Establish a clearinghouse to match needs with the skills and technology.

14) *An inventory of research organizations.* Compile an inventory of research organizations.

15) *Training and education programme.* Organize a programme for training and education of personnel to develop and use new technologies.

9

Economic (Market) Measures

COORDINATORS
J. Tilley (Australia)
J. Gilbert (New Zealand)

CONTENTS

ECONOMIC (MARKET) MEASURES

EXECUTIVE SUMMARY

This paper provides a preliminary assessment of the applicability of economic instruments to limit emissions of greenhouse gases, encourage sinks and facilitate adaptation to climate change. The instruments examined include emissions charges, tradeable emission permits, subsidies, and sanctions.

The paper provides a "top-down" assessment of these measures, which, in conjunction with the more detailed work of the RSWG Subgroups, will comprise a set of ideas helpful to countries in preparing and assessing possible means to address climate change.

These measures have the potential to provide the signals necessary for the more environmentally sensitive operation of markets.

It is considered that economic instruments offer the possibility of achieving environmental improvements at lower cost than regulatory instruments. However, it is unlikely that economic instruments will be applicable to all circumstances. Some combination of economic and regulatory measures will most likely be appropriate. The provision of information and technical assistance are also seen as valuable complementary instruments.

Three factors are considered as potential barriers to the operation of markets and/or the achievement of environmental objectives through market mechanisms. These are: *information problems*, which can often cause markets to produce less effective or unfavorable environmental outcomes; *existing measures and institutions*, which can encourage people to act in environmentally damaging ways; and *balancing competing* social, environmental, and economic *objectives*. Initial response strategies may therefore be to address information problems directly and to review existing measures which may be barriers.

A number of evaluation criteria exist for assessment of instruments that might be adopted at the national and international levels. At the national level, these criteria relate to the need for the measures to: be compatible with sustainable economic development; be cost effective; comprehensively consider all significant sources and sinks; be compatible with international trade rules; be capable of amendment in light of new information; not hinder operation of other measures; be administratively practical overall; and consider distributional and information issues.

Most of these criteria obviously also apply at the international level, where there is also a need for the measures to: be compatible with and support principles of international technology transfer and financial support discussed in other topic papers; take into account the special needs of the developing countries; recognize global climate objectives; and recognize acceptable inter-country arrangements that maintain an overall greenhouse contribution within the sum of their individual obligations.

A general advantage of market based *economic instruments* is that they encourage limitations or reductions in emissions by those who can achieve them at least cost. They also provide an ongoing incentive for industry and individual consumers to apply the most efficient limitation/reduction measures through, for example, more efficient and cleaner technologies. Such incentives may be lacking in the case of regulations.

Regulations, however, are the customary means of controlling pollution in both market and centrally planned economies. A major advantage of

233

regulations is that they create certainty as to desired outcomes, whereas major disadvantages are that they may fail to encourage innovation, introduce inflexibilities in meeting objectives, and offer few incentives to reduce emissions below specified levels.

Tradeable emission permits are based on the concept that emission entitlements or rights to pollute are provided to emitting sources, subject to an overall limit on total emissions, and allow the permit holders to trade or sell their entitlements to another party on the open market.

While there are potential benefits of national or international tradeable emission permit systems, particularly their economic efficiency and cost effectiveness, many contributors had difficulty with the creation of "rights to pollute" and the administrative and monitoring requirements. A number of issues need further examination, such as: the political problem created by a "right to pollute"; the criteria used to determine the initial allocation of emission entitlements; the special situation of developing countries; the potential scope and size of a trading market; and the feasibility of the administrative structure that would be required to implement such a programme.

Emission charges provide a means of encouraging the limitation or reduction of emissions to a socially desirable level. They also provide an ongoing incentive for efficient means of limiting or reducing emissions and they could provide a funding base for further pollution abatement, research, and administration. However, with such measures, it may be difficult to assess the optimal rate of taxation and detailed (and possibly expensive) knowledge is required about likely market reactions to different tax levels. The basis for the tax and means of collection also need to be considered.

Domestic use of *subsidies* (e.g., direct grants, low interest rate financing, loan guarantees, tax deferrals, tax credits) is another possible economic instrument. While external public benefits can justify the use of such subsidies, difficulties such as their expense, need for careful design, need for review, and international trade aspects, mean that subsidies need to be considered with great care. They may also create disincentives for further innovation and development of appropriate technologies.

Trade or financial *sanctions* established under an international convention would need to be consistent with existing trade agreements and many contributors expressed considerable reservations about applying them in a complex situation such as greenhouse gases where monitoring diverse sources would be difficult.

The above instruments raise a number of complex practical problems, particularly in relation to their implementation, and it is evident that further work is required in all countries, and in ongoing IPCC work, to fully evaluate the practicality of such measures and costs and benefits associated with different mechanisms. It is appreciated that each instrument assessed has a role in meeting greenhouse emission objectives, but the suitability of particular instruments is dependent on the particular circumstances and at this stage no measure is considered universally superior to all other available mechanisms.

ECONOMIC MEASURES AS A RESPONSE TO CLIMATE CHANGE

9.1 INTRODUCTION

Over the last few decades there has been an increased concentration of greenhouse gases in the atmosphere. It is clear that this trend will continue in the years to come unless something is done.

While considerable uncertainty surrounds the effect of greenhouse gas emissions, this uncertainty does not justify a lack of action. It is prudent to take reasonable steps at this stage to avoid the risk of catastrophic consequences as a result of changes to the world climate in the future.

This paper provides a preliminary assessment of the applicability of economic instruments to the problem of climate change. The instruments are examined both in terms of their ability to limit net emissions of greenhouse gases and their usefulness in facilitating adaptation to climate change. The following economic mechanisms are examined (both in national and international variants): emissions charges, tradeable emission permits, subsidies, and sanctions.

The paper provides a "top-down" assessment of these economic measures. It is hoped that this, in addition to the more detailed ("bottom-up") work of the RSWG Subgroups, will comprise a set of ideas helpful to countries in preparing and assessing possible means to meet national goals and international commitments relating to climate change.

The analysis is conducted very much at a preliminary level. The paper aims to do no more than provide the basis for the further, more detailed work that remains to be done. This should be carried forward in the light of further work being undertaken in the context of RSWG and the product from the other Working Groups of IPCC and other organizations and individuals currently examining these issues.

The ideas in this document do not necessarily represent the official views of contributing countries or organizations.

9.2 OVERVIEW

While markets will of themselves adjust to take account of the impacts of climate change (for example, through consumer support of environmentally sound goods) this adjustment will not sufficiently address the problems of climate change and careful intervention by national governments is required. These interventions should aim to overcome existing social, institutional or other barriers to the maintenance of environmental quality and provide the signals necessary for the more environmentally sensitive operation of markets.

Economic instruments, through their encouragement of flexible selection of abatement measures, may offer the possibility of achieving environmental improvements at lower cost than regulatory instruments. They also offer the opportunity for countries to introduce limitation/reduction and adaptation measures while more detailed regulatory or other measures are being formulated and implemented. However, it is unlikely that, in the final analysis, economic instruments will be applicable to all circumstances. Some combination of economic and regulatory measures will most likely be appropriate. Provision of information and technical assistance are seen as a valuable complementary instrument.

Experience to date, in addressing issues relating to reduction in CFC production and emissions in the context of the Montreal Protocol, suggests that solutions at the national level will need to draw heavily on the widest variety of economic, regulatory, and other response measures. Differing combinations of instruments are likely to be used by individual countries in order to best meet national goals and international commitments.

Most contributors saw market-based instruments (as well as planning mechanisms in centrally planned economies) as having a role in achieving objectives for limiting or reducing emissions of greenhouse gases, rather than in determining what the optimal level of emissions ought to be. Although it was recognized that the purpose of policy measures is to enhance overall welfare, this requires a balancing of the costs of climate change with the costs of policy measures implemented to reach emission reduction targets. Whether or not a strict cost/benefit approach is applied, it seems clear that all costs must be taken into account in setting policy objectives and determining the mix of options for implementation.

Finally, the nature of the greenhouse issue highlights the importance of genuine cooperation by sovereign countries in implementing national programmes of economic and other measures. This will be as important as bilateral, regional, and international activities enforced through conventions and protocols. Clearly, the two approaches would need to complement each other.

9.3   KEY CONSIDERATIONS

Market-based measures could effectively and efficiently help to reduce emissions of greenhouse gases by operating through environmentally adjusted market forces. These should be framed to reflect fully the social costs of environmental resource use in light of available scientific knowledge. This is consistent with the principle of sustainable development.

Three factors were identified as potential barriers to the operation of markets and/or the achievement of environmental objectives through market mechanisms.

1) Information problems.

2) Counterproductive operation of existing measures and institutions.

3) Other competing objectives.

An understanding of these considerations is necessary in assessing the likely success of a response strategy that utilizes market mechanisms.

9.3.1   INFORMATION

Problems of information can often be the cause of markets producing less effective or even unfavorable environmental outcomes. These problems should be borne in mind in the development of economic instruments.

There are three types of information problems.

1) The information does not exist.

2) The information is not disseminated.

3) The decision makers (government, industry, or individual consumers) do not have the ability or the means to use the information.

If information problems are the cause of economic activity producing unsatisfactory environmental outcomes, it would be appropriate for the policy response to target directly the information problems. This might be through a public education campaign. This is further referred to in the IPCC topic paper on public education and information measures. Better protection of intellectual property so as to encourage research into pollution abatement technology, or other appropriate policy responses, may also be required. This is discussed in more detail in the IPCC topic paper on technology development and transfer measures.

It should also be noted that the different types of economic instruments have different information requirements. The choice of instrument should have regard to the type of information available.

9.3.2 EXISTING MEASURES AND INSTITUTIONS

Environmental problems can often be attributed to existing measures and institutions that encourage people to act in a way that is not consistent with the preservation of environmental quality. Some of these measures and institutions were adopted without consideration of their contribution to increased emissions of greenhouse gases. Countries should be encouraged to reassess these measures (with, of course, due consideration to other economic and social objectives).

Where counterproductive measures (including subsidies, sanctions, regulations, and technical standards) are the problem, it will be appropriate for the policy response to be directed at re-examining them so that they will be conducive to the production of environmentally acceptable outcomes. Removal of these distortions could also provide immediate direct benefits through improved economic efficiency in addition to reducing the risk of climate change.

Governments also have the potential to contribute to the control of greenhouse gas emissions by attending to their public sectors. Options include: improving the practices of state-owned enterprises, and the environmentally and economically sound management of public lands.

9.3.3 OTHER COMPETING OBJECTIVES

There will always be competition among different social, environmental, and economic objectives. Inevitably, this will affect the use of economic measures to limit or reduce greenhouse gas emissions. In a world of limited resources, resources must be allocated between competing objectives and social welfare will not be maximized by pursuing any objective to the exclusion of others. It is recognized that, in balancing competing objectives, there will be trade-offs and total elimination of all environmental problems is unlikely to be achieved. Of course, decision makers and the public have a need to know the extent of environmental damage and what trade-offs would be involved in achieving environmental objectives at the expense of other ob-

jectives, such as economic growth. This should form the basis for any effective and responsible action to protect the environment. However, this is not to say that action on major environmental issues should always be delayed because insufficient information is available.

9.4 EVALUATION CRITERIA

The economic instruments addressed in this document may be considered as options for any national programme established to respond to the problem of climate change or as options for an international institutional structure. Since the form of the instrument may vary according to different circumstances, it is useful to distinguish criteria for the assessment of instruments or policy measures adopted at the national level from those relevant to an assessment of international instruments or policy measures.

The following criteria can be used in the evaluation of a *national* policy measure applicable to industry as well as individuals:

1) The measure should be compatible with principles of sustainable economic development.

2) The measure should be cost effective in limiting specified emissions in accordance with national goals or any international agreements and balanced so as to optimize the net social benefits from the use of resources in pursuit of competing social, environmental, and economic goals.

3) The measure should be comprehensive in terms of taking into account all significant sources and sinks of the specified emissions.

4) The measure should seek to distribute costs toward sources of the specified emissions and to ensure there are no unnecessary cost burdens on the encouragement of sinks.

5) The measure must be compatible with international trade rules and principles and should seek to avoid distortions or restrictions to trade.

6) The measure should be able to be amended or adjusted efficiently and in a predictable manner in response to new information in the fields of science, impacts, technology, and economics, and any international obligations that might emerge.

7) The measure should not hinder the operation of other measures (particularly technology development and information exchange) that facilitate the achievement of objectives in relation to specified emissions.

8) Accurate and reliable information must be available at a national level about sources and sinks of specified emissions.

9) The measure must be administratively practical and effective in terms of application, monitoring and enforcement.

The following criteria should be applicable to any *international* institutional measure incorporated into any convention or protocols:

1) The international policy regime should be compatible with principles of sustainable economic development.

2) The international measure should be cost effective (i.e., designed to achieve the pollution control goals of any agreement at the lowest economic and social cost).

3) The measure should be comprehensive in terms of taking into account all significant sources and sinks of the specified emissions.

4) The international measure should take appropriate advantage of economic measures as well as regulatory measures to encourage utilization of alternative, less polluting technologies and sinks.

5) It should be able to be amended or adjusted, in a predictable manner, in response to new information and any assessments of science, impacts, technology, and economics.

6) The international measure should be equitable in terms of distribution of costs and obligations across national, intergenerational, and income class categories and in terms of access to decision-making processes. The implementation of such measures should take into account the circumstances that most emissions affecting the atmosphere at present originate in industrialized countries where the scope for change is greatest and that under present conditions emissions from developing countries are growing and may need to grow in order to meet their development requirements and thus, over time, are likely to represent an increasingly significant percentage of global emissions.

7) It should respect established international trade rules, with the aim of not distorting or restricting world trade.

8) The international policy measure should be administratively practical and effective in terms of application, monitoring, and enforcement.

9) The measure should be compatible with support principles of international technology transfer and financial support discussed in other topic papers.

10) The measure should take into account the special needs of the developing countries.

11) The international policy regime should recognize that the interests of the international community relate to global climate objectives, not the means by which they are achieved.

12) The international measure should recognize any arrangements between individual countries that maintain an overall greenhouse contribution within the sum of their individual obligations as an acceptable compliance strategy.

Some contributors thought that further investigation dealing with economic responses to climate change could build on the above. This would present a strategy to deal with conflicting or controversial criteria and guide resolution of most potential conflicts.

9.5 INSTRUMENTS EXAMINED

9.5.1 PREAMBLE

Economic instruments offer the possibility of minimizing the total social costs of achieving national goals and international commitments relating to climate change. They seek to meet limitation/reduction or adaptation objectives by adjusting or harnessing market forces so that they take account of environmental costs.

However, the use of economic instruments raises moral concerns in the minds of some contributors relating to the fact that some of the economic instruments imply "paying for the right to pollute" and that economic instruments introduce a profit motive element into the achievement of public goals of environmental protection. Others have responded that regulations allow for the same rights to pollute, the only difference being in the transferability of the right, and that regulation does not banish the profit motive, since profit opportunities in regulatory systems are often directly dependent on securing favorable regulatory treatment. Indeed, the misuse of resources in lobbying for favorable regulatory treatment represents a considerable drawback of the regulatory approach.

Section 9.5.2, below, looks at measures that might be used in centrally planned economies to reduce emissions and then examines the role of regulations in market economies.

Sections 9.5.3–9.5.6 examine the following economic instruments that might be considered as alternatives or supplements to the regulations. The instruments examined are:

9.5.3 Tradeable emission permits
9.5.4 Emissions charges
9.5.5 Subsidies
9.5.6 Sanctions

It will be up to each country to select the mix of measures most appropriate to their particular circumstances in light of the suggested criteria for evaluating measures outlined in Section 9.4 and any steps (such as international consultation) necessary to avoid negative impacts on world trade. More work is required to identify the circumstances under which each measure is most appropriate.

9.5.2 REGULATIONS

9.5.2.1 *Regulations in Centrally Planned Economies*

Because of the differences between the social and economic structures in different countries, different combinations of instruments or at least different emphases in their use will be required. In the particular case of centrally planned economies, the central plan encompasses a comprehensive set of regulations and is the major instrument for achieving sustainable limitations or reductions in emissions.

Central planning provides possibilities for the government to coordinate and direct all efforts to limit or reduce emissions. Possibilities include: planning emissions reductions/limitations by those enterprises that can achieve them best at least cost so as to minimize the financial burden to society as a whole; provision of loans on favorable terms for environmentally sound investments; and alterations in price regulations to enhance the introduction of new environmentally acceptable technologies and activities.

9.5.2.2 *Regulations in Market Economies*

Governments in market economies also have the option to impose direct controls to limit or reduce greenhouse gas emissions. Regulatory controls typically seek to prescribe acceptable standards or to restrict or limit certain activities. They normally do not seek to influence market behavior through price signals. Such controls have been the customary means of controlling pollution. Examples include emission controls for motor vehicles or the installation of scrubbers on power plants.

Direct controls have the advantage of creating certainty as to the desired actions and outcomes. They may also be useful where it may not be possible to frame appropriate market measures. Their implementation, however, may:

- fail to encourage innovation, as there may be no incentive to further limit or reduce emissions;
- introduce inflexibilities in meeting emission ob-

jectives, thereby imposing unnecessary economic costs; and

- require significant monitoring and/or administration to ensure that the desired outcomes are achieved (although the same may be true for market measures).

Some contributors thought that regulations were one area where there is extensive experience with controlling pollution. Further work on determining the potential for increased use of environmental regulations and analysis of their advantages and disadvantages, including areas where regulation might be more effective than economic measures, could be of use.

9.5.3 A SYSTEM OF TRADEABLE EMISSION PERMITS

9.5.3.1 *Method of Operation*

An emission permit system is based on the concept that emission entitlements (like coupons) or rights to pollute are provided to emitting sources (which can be as large as countries), subject to an overall limit on total emissions. The sum of entitlements is equal to an emissions budget set at the national level through regulation or at the international level through international agreements. A "tradeable" emission permit system extends the emission permit system by allowing the permit holders to trade or sell their entitlements to another party on the open market. It is supplementary to a regulatory programme in that it operates within overall limits on emissions of the greenhouse gas.

There has been to date only limited experience with tradeable emission permits at the national level, and no experience at the international level. One example that has been implemented in the United States is its programme to phase down the use of lead in gasoline. Under this programme, the United States government set a target for the usage of lead in gasoline through a regulatory process and used a trading system that allowed gasoline refiners to allocate the lead budget in an economically efficient manner. Although this can be viewed as recognizing a "right to pollute," it is only a right to pollute up to a level determined to be environmentally acceptable by the U.S. government.

Many options exist for determining the *initial* allocation of entitlements under a tradeable emission permit system. For example, at the national level, entitlements could be allocated to existing producers in proportion to their existing emission level or they could be auctioned to the highest bidder or sold by tender.

At the international level, a tradeable emission permit system could be based on the same concepts as those described above. As a first step, an overall emission budget could be set. Global emission budgets and *initial* emission entitlements for each country could be determined in a number of possible ways: the entitlements could be allocated by the level of emissions in a base year, by population size of a country, by a target unit of emissions per unit of GNP, by a target unit of fossil fuel use per unit of population or GNP, etc.

Once the initial entitlements to each country have been determined, countries would of course have to allocate their overall budget to the sources in their country. This allocation procedure could follow whatever lines are deemed most appropriate, given their social and economic needs. However, at the international level, a tradeable permit system might operate like international share, currency, or commodity markets. Governments or individual entitlement holders could be free to buy or sell, brokers could operate, and the spot price of an entitlement would float to a current market valuation of a marginal unit of emissions. Making the market freely accessible and issuing entitlements in small divisible units should go a considerable distance in preventing any nation or firm from cornering the market.

At either the national or international level, a further variation of this system is to allow emission entitlements to be granted for the creation of sinks. An international policy regime that focuses on global climate objectives, rather than the means by which they are achieved, would treat sink creation and emissions reduction as one-for-one substitutes.

9.5.3.2 *Benefits and Difficulties of a National Tradeable Permit Programme*

The following potential benefits have been suggested:

- it reduces the cost of meeting a total emission target set by regulation;

- it provides economic incentives to develop and use cost-effective, energy-efficient industrial processes, consumer products, and emission control technologies;
- it works within a system of agreed emission budgets;
- it provides an efficient mechanism for addressing the trade-offs between sources and sinks for the specified emissions; and
- it provides the opportunity for low-income rights holders to sell rights to others in exchange for compensation of greater value.

The following potential difficulties have been suggested:

- it creates a "right to pollute" that can be purchased by those with the highest level of income;
- it may not be politically acceptable in many countries;
- it requires administrative structures that generally do not exist today; and
- it requires extensive monitoring and record-keeping to track the movement of entitlements and overall compliance with the total emissions budget.

9.5.3.3 Benefits and Difficulties of an International Tradeable Permit Programme

Generally an international programme will carry with it the same benefits and difficulties as a national programme. The potential cost savings may be higher but the level of the administrative and enforcement complexity could be even more severe. An international body of some kind would be required to identify whether the allowable limit of pollution was being exceeded. In addition, political concerns about the "right to pollute" and the ability of wealthy nations to procure those rights would be greater. Also, there is a lack of experience of these programmes at the international level.

Issues Needing Further Examination

Major concerns were expressed by contributors over the use of tradeable emission permits, some thinking that an international system of tradeable permits was not advisable. While it was agreed that studies of tradeable permits should continue, these countries asked that their major misgivings should be taken into consideration.

There are a number of other issues associated with a tradeable emission permit programme that need further exploration, including: the political problem created by a "right to pollute"; the criteria used to determine the initial allocation of emission entitlements; the special situation of developing countries; the potential scope and size of a trading market; and the feasibility of the administrative structure that would be required to implement such a programme, including the conditions necessary for them to be feasible, such as identifiable emission point sources, standard metering practices, availability of comprehensive and up-to-date information on options, and a market in which to trade permits.

The contribution from the Environmental Defense Fund discussed in some detail the practical arrangements that might be adopted in the implementation of a system of tradeable emission rights, both at a national level and internationally. This and other such work should be consulted in the further development of proposals that might utilize this instrument.

Some contributors thought that an alternative approach to a system of emission charges would be to have the polluter pay by conducting research designed to reduce emissions—e.g., through an appropriate tax credit system.

9.5.4 A SYSTEM OF EMISSION CHARGES

Emission charges are levies imposed in relation to the level of emissions. They provide a means of encouraging the limitation or reduction of emissions to a level that is socially desirable. They also provide an ongoing incentive for the parties concerned to implement efficient means of limiting or reducing emissions by, for example, implementing energy efficiency measures. For governments, major attractions of taxes may be the revenue they generate. This revenue could provide a funding base for further pollution abatement, research, and administration. It could also enable other taxes to be lowered, budget deficits to be reduced or government expenditures to be increased in other fields.

A carbon tax on burning fossil fuels is one example of an emission charge. Emissions of carbon dioxide are a major contributor to the greenhouse effect and most of the carbon dioxide comes from the burning of fossil fuels.

A tax on fossil fuels, levied in relation to its carbon content although not a direct tax on emissions, could in some circumstances reduce emissions of carbon dioxide into the atmosphere. For the tax to be effective and efficient, its level should not be influenced by the funding requirements of other programmes; its level should be the same across all countries; and it should apply uniformly to all forms of fossil fuels. It would also be necessary to take account of emissions of other greenhouse gases over the full fuel cycle.

The view was expressed that an "energy tax" that does not vary with carbon content could also be employed to encourage overall energy efficiency. Such a tax would not necessarily lead to fuel switching and its revenues could also be used to fund programmes for promoting environmentally favorable energy technologies.

Recent work by the OECD has shown that the use of taxes to fight pollution has been growing in popularity amongst OECD countries, although the experience to date shows that the primary function of pollution taxes (which in the climate change context can be emission taxes) has been to raise the revenue required for pollution management. The taxes have not generally been set high enough to influence the behavior of polluters.

There are some problems associated with the use of emissions charges. These will require careful consideration in the development of response strategies that might make use of this option.

First, it is difficult to assess the optimal rate at which the tax should be applied in order to meet national goals, where these can be defined in terms of technical levels of emissions. Unlike regulations or the system of tradeable emission rights, the level of emissions is not set directly but attained through the market responding to the taxes. In order for national goals to be attained, some quite detailed knowledge is required about how the market is likely to react to different taxation levels. This information could be difficult and expensive to acquire. The cost-effectiveness studies currently being prepared by the RSWG Energy and Industry Subgroup may help with information on this matter.

Second, there is the difficulty of deciding on the basis for the taxes and how they would be collected. (This criticism also applies to a number of other economic and regulatory instruments.)

The implementation of a coherent emission tax regime at the international level poses further practical difficulties. In particular, factors such as varying local price elasticities and tax structures, exchange rate fluctuations, etc., mean that it would be extremely difficult to derive with any confidence a uniform tax that would lead to the required limitation or reduction in emissions, where there is sufficient technical knowledge to enable an emission level to be specified.

A reasonable and practical option would be to give each country flexibility to set its own national charge rates at a level that would achieve its national goals. However, some contributors felt that an international levy system should not be discarded from consideration in this context.

It is evident that the question of emission charges raises many complex and difficult issues. Careful and substantive analysis of the short and long run environmental and economic impacts of such measures is needed.

9.5.5 SUBSIDIES

The use of subsidies to limit or reduce greenhouse gas emissions by developing countries is discussed in the topic paper on financial mechanisms. This section looks at the domestic use of subsidies.

Subsidies and government financial assistance (hereinafter referred to as "subsidies") are aimed at assisting environmentally sound goods and actions by lowering their costs. Various forms of subsidies have been used, among other things, to encourage the use of energy efficient equipment and to encourage the use of non-fossil energy sources. These include direct grants, low interest rate financing, loan guarantees, tax deferrals (e.g., accelerated depreciation), tax credits, etc.

Because of the external public benefits of developing environmentally sound technologies, subsidies could be used to encourage the development and greater use of such technologies.

A number of difficulties are associated with the use of subsidies.

- In general, subsidizing environmentally sound activities tends to be less efficient than applying charges or fees directed toward the emissions that are to be reduced.
- It is difficult to assess the optimal rate of subsidization.
- They can be expensive.
- More important, unless carefully designed, they could fail to be effective. For example, a subsidy on competing energy sources would not only encourage a move to alternative fuels, but could also encourage a net increase in energy consumption and CO_2 emissions, unless accompanied by economic or regulatory measures.
- Overly generous subsidies or their continual use without review can act as a disincentive for investigation and research into alternative means of meeting the same objective and thus act to lock pollution abatement into outdated or inefficient technology.
- Unqualified use of subsidies could hinder the full direction of costs to sources of the specified emissions, and might cause trade distortions.
- Although some food and agricultural subsidies may lead to some greenhouse gas emissions (e.g., methane) and their removal would result in overall social benefit, this can also produce significant local or regional social costs which need to be considered.

The question of subsidies may prove to be controversial from the point of view of international trade. Environmental subsidies have been the subject of some discussion by members of GATT but there has been no agreement as to what differentiates an environmental subsidy from other subsidies.

9.5.6 SANCTIONS

A final type of economic instrument is the use of economic sanctions for the enforcement of international agreements. This would require an international convention to establish a system of agreed trade or financial sanctions to be imposed on countries that did not adhere to agreed targets. The aim here would be to discourage non-complying countries from deriving benefits without taking action.

Discussion of trade sanctions has already taken place under the Montreal Protocol in relation to products containing CFCs and halons. Sanctions might range from the imposition of import taxes on the offending products to outright bans until the exporter conforms with agreed international standards. The mechanisms for operating such a system would need to be made consistent with existing trade agreements such as GATT.

However, many contributors expressed considerable reservations about applying this measure to greenhouse gases other than CFCs and halons because of the complexity of the situation. In particular, it would be difficult to monitor the many diverse sources of greenhouse gases. Moreover, it was felt that sanctions could appear arbitrary, since it could be difficult to determine accurately whether particular countries exceeded internationally agreed levels or not. This could create confusion and resentment as to why they were subject to the sanctions. Some contributors objected to the concept of sanctions because of the risk that they could be used as a pretext to impose new non-tariff barriers on the exports by developing countries.

9.6 CONCLUSIONS

The following are the significant conclusions of the country contributions and the subsequent discussions at WGIII meetings.

1) Economic instruments offer the potential to achieve national goals and international commitments for limitation/reduction of greenhouse gases and adaptation to climate change, at minimum social cost.

2) The type of instrument chosen will depend to a great extent on the context in which it is to be applied, the economy of the country concerned, and the type of greenhouse gas to be controlled. It is likely that the instruments chosen would include a combination of regulations, economic signals, and economic instruments, plus the provision of information and technical assistance to improve the opera-

tion of the new environmentally adjusted market forces.

3) Economic instruments that may be relevant include: a system of tradeable emission permits; emission charges; sanctions; subsidies to encourage the implementation of measures to limit/reduce emissions; and reduction of existing government interventions in other areas, such as transport, energy, food, and agriculture, that inadvertently encourage emissions of greenhouse gases. Indeed, an international system of tradeable emission permits or, alternatively, a system of international emission charges could offer the potential of serving as a cost-efficient main instrument for achieving a defined target for the reduction of greenhouse gas emissions.

4) A number of contributors expressed major concerns about the use of tradeable emissions permits or sanctions. While it was agreed that studies should be continued on these instruments, contributors requested that their misgivings should be taken into consideration.

5) The instruments considered in this topic paper raise a number of complex practical problems, relating in particular to the implementation of these measures. These should be investigated in respect of the criteria outlined in Section 4 of this paper, the output of the IPCC Subgroups, and studies being carried out by international organizations, national governments and individuals. It is evident that further work is required regarding implications for economic instruments for developing countries.

10

Financial Mechanisms

COORDINATORS
J. Oppenaeu (France)
P. Vellinga (Netherlands)
A. Ibrahim (Egypt)

CONTENTS

FINANCIAL MECHANISMS

10.1 INTRODUCTION

10.1.1 The financial approach presented in this document is based on discussions within the Intergovernmental Panel on Climate Change (IPCC) Response Strategies Working Group (RSWG). A workshop on "Implementation Measures" including "Financial Measures" was held in Geneva, October 2–6, 1989. Prior to that workshop, views on this topic were submitted by Australia, Canada, France, India, Japan, Malta, the Netherlands, New Zealand, Norway, and the United Kingdom. Views were also submitted by the following Nongovernmental Organizations (NGOs): the Natural Resources Defense Council (NRDC) of the United States, and the Stockholm Environment Institute (SEI) of Sweden. Since then, additional contributions on the topic were sent to the coordinators by Canada, Japan, Norway, Switzerland, and the United Kingdom, as well as by NRDC. The document is intended to reflect a wide variety of ideas and opinions and should lead to effective action to tackle the problem of global climate change.

10.1.2 Scientific analysis on the issue is not yet finalized. The nature and the scale of the economic and technical measures which are needed will determine the scope and conception of finance. Response measures will, in part, be determined as a result of ongoing deliberations within the IPCC. The financial options listed here are discussed with a view to defining a sufficiently broad, flexible, and cooperative framework for action.

10.1.3 Although the scope of the discussed mechanisms is not limited to developing countries, this paper pays special attention to the mechanisms that should help countries where the protection of the atmosphere would entail a special or abnormal burden, in particular due to their level of development or their effective responsibility in the deterioration of the atmosphere.

10.1.4 The objective is to encourage adoption by industrialized and developing countries alike, of strategies to adapt their economy to control, limit, or reduce greenhouse gas emissions and at the same time increase the capacity of the natural environment to absorb these gases.

10.2 GENERAL CONSIDERATIONS

10.2.1 The following considerations require special attention. The first tangible results in limiting climate change by these actions will not appear for some time. We are thus engaged in a long-term process.

10.2.2 Environmental problems caused by global warming can often not be disassociated from other environmental problems or from other development factors. This interdependence of environment and development must be recognized and taken into account in the actions of the international community.

10.2.3 The programmes to be undertaken will require actions in both industrialized and developing countries, under the following broad categories: limitation/reduction measures, adaptation measures, research and development, transfer of technologies, public awareness, and education.

10.2.4   The special needs of developing countries, including their vulnerability to problems posed by climate change and their lack of financial resources, must be recognized and cooperation and assistance tailored to meet their individual needs. Financing requirements might be considerable. Substantial economic gains, however, could be achieved, for example, through measures that result in energy conservation and efficiency. Economic and technological measures are described in more detail in a separate implementation paper.

10.2.5   The presence of a positive international economic environment, including further reduction of trade barriers, will also be important to support economic growth of developing countries, and to generate resources that can be applied toward pressing needs.

10.2.6   The scale and diversity of the environmental requirements in developing countries, together with their development problems, favor an approach that ensures mobilizing a wide variety of international organizations, possibly including new ones. It will also be important to ensure the full participation of existing regional organizations.

10.2.7   Priority should be given to those financial measures and policies that can have an early impact in limiting or reducing emissions of greenhouse gases and that make economic sense in their own right.

10.3   GUIDING PRINCIPLES FOR COOPERATION

10.3.1   The following principles derived from, *inter alia*, the UNGA Declaration 43/53 and the Noordwijk Declaration on Climate Change should guide the financial approach. They do not, of course, predetermine legal issues arising from a future climate convention. These issues are discussed in a separate implementation paper as well.

1)   Industrialized and developing countries have a common responsibility and need to limit and/or adapt to climate change. The interdependence of the causes of global warming and the interweaving of the necessary economic solutions require a collaborative approach, with due respect to national sovereignty. This concept is based upon the recognition that:

2)   The atmosphere and climate are vital to all human beings.

3)   It is important for every person and country to contribute to limiting climate change caused by human activities.

4)   In pursuing economic growth as a legitimate means of improving standards of living, countries need to integrate environmental considerations with development policies in order to ensure sustainable development.

5)   The elements outlined above apply to industrialized countries as well as to developing countries. All countries are likely to face increased costs associated with preventive and adaptive measures necessary to address climate change.

10.3.2   Industrialized countries have specific responsibilities on two levels.

1)   Major part of emissions affecting the atmosphere at present originate in industrialized countries where the scope for change is greatest. They should therefore adopt domestic measures to limit climate change and take the lead in setting an example by adapting their own economies in line with future agreements to limit or reduce emissions. The field of action involves faster suppression of the production and consumption of chlorofluorocarbons, energy savings or recourse to non-fossil energy sources, a rational use of forestry resources to commence active reforestation, limiting, or reducing CO_2 emissions, etc.

2)   They should help developing countries participate in international action without compromising their development, by contributing additional financial resources, by appropriate

transfer of technology, by engaging in close cooperation concerning scientific observation, by analysis and research, and finally by means of technical cooperation and assistance geared to forestalling and managing ecological problems. Additional resources should over time be mobilized to help developing countries take the necessary measures that are compatible with their development requirements, and thus compensate for the additional specific effort they would be ready to make. Furthermore, they should show scientific leadership in addressing climate change issues including systematic observations, modeling, prediction, and assessment.

10.3.3 Emissions from developing countries are growing and may need to grow in order to meet their development requirements and thus, over time, are likely to represent an increasingly significant percentage of global emissions. Developing countries should take measures to adapt their economies. They should undertake, according to their individual circumstances, to set out on a course to make a rational use of forestry resources to commence active reforestation and to promote energy savings or recourse to non-fossil energy sources. International cooperation and urgency plead in favor of the adoption of such programs, which are likewise applicable in industrialized countries. The immediate development needs of developing countries would under present conditions require an increase in their present level of CO_2 emissions. The developing countries should attempt to minimize such increases in emissions.

10.3.4 Multilateral development banks, bilateral assistance programs, and United Nations development and scientific and technological organizations should be used in their respective fields of competence. A number of delegations have observed that it is necessary that these institutions expand their fields of competence into new areas that are responsive to environmental necessity. What is justified is a reinforcement of external support for revised internal priorities, policies, and measures, including internal resource allocation, which are in the country's own interest. In addition, regional coopera-

tion bodies should adopt the required initiatives and implement the appropriate solidarity action at their level. A concerted effort by all of the above-mentioned institutions to formulate coordinated, integrated strategies that assure that all development assistance investments, whether undertaken with existing or additional funds, would further the goal of climate stability.

10.4 USES OF FINANCIAL RESOURCES

10.4.1 The Working Group reached a general consensus on priority use for financial resources channeled to developing countries. It was accepted that some activities contributed both to limiting/reducing greenhouse gas emissions and promoting economic development, and that early priority should be attached to these.

10.4.2 It was noted that some activities would require a much greater scale of resources than others, and that considerations of cost effectiveness should apply in deciding particular priorities. It was further noted that there was already a significant amount of activity being funded by existing assistance organizations in some of the priority areas.

10.4.3 Among the highest priority areas for cooperation and assistance are those enumerated below.

1) Efficient use of energy, including appropriate use of technologies, increasing the use of non-fossil fuels, and switching to fossil fuels with lower GHG emission rates.

2) Forestry—to promote the rational management of forest resources and reforestation.

3) Agriculture—to limit or reduce greenhouse gas emissions from agricultural activities.

4) Measures taken under the Montreal Protocol to help developing countries in the develop-

ment and introduction of alternatives to CFCs and halons.

5) Cooperation with and assistance to developing countries for their full participation, as recommended by the Special Committee on the participation of developing countries in the meetings of the IPCC; a simple system of mobilization and coordination (with a leading role for coordinating the participation of developing countries by the IPCC Secretariat) should be elaborated.

6) Cooperation and assistance, including transfer of expertise such as that required to develop and operate forest management programs and large-scale reforestation programs, to help developing countries plan how to tackle problems posed by climate change, including cooperation and assistance in the design of policies and programmes. This holds for adaptation as well as for limitation/reduction-oriented policy planning. It was noted that small investments in policy or programme design could yield large returns.

7) Research and development programmes — including monitoring climate change and research on technologies that might be used to reduce/limit emissions of greenhouse gases and develop adaptive strategies.

8) Technology transfer arrangements, including the training of experts.

9) Public awareness and education.

10.4.4 Some delegations also suggested that financial resources might be used for the following purposes:

1) Cooperation with developing countries, which in future abstain from activities that produce greenhouse gases. In this context, regional banks, which have so far not been actively involved in the IPCC process, should play a coordinating role by introducing environmentally sound criteria in their projects. Furthermore, cooperation with and assistance to developing countries to support their activities that reduce net emissions of greenhouse gases or that introduce alternative technolo-

gies (i.e., those that emit low levels or no greenhouse gases) should be considered.

2) Assistance to developing countries in adjusting to the problems presented by climate change. It was noted that there might be a parallel with support for current economic or structural adjustment programmes.

3) Funding to support the developing countries in fulfilling their obligations under a climate convention, and subsequent related protocols, thereby ensuring widespread implementation of the required measures to limit, reduce, and, as far as possible, prevent climate change.

10.5 SOURCES OF FINANCIAL RESOURCES

10.5.1 The Working Group noted that the question of ways by which financial resources could be *generated* to assist developing countries was separate from the issue of how to *channel* those resources. It was agreed that all avenues should be considered. Among the options for generating official resources were mentioned:

1) General taxation or other revenues not necessarily related to climate change or the natural environment which might be allocated for specific budgetary purposes.

2) Specific taxes or levies related to the emission of greenhouse gases, including taxes or levies on the use of fossil fuels.

10.5.2 Other creative proposals for generating resources have included:

1) Undisbursed official resources, which might result from savings on government energy bills and lower levels of military expenditures.

2) Levying a fixed percentage tax on travel tickets issued worldwide.

3) A world climate lottery.

4) Once any future agreement or protocol on limiting/reducing greenhouse gas emissions is in place, it may be that one source of resources would be levies on or payments from countries unable to meet their treaty or other obligations.

10.5.3 Some delegations stressed the importance of assuring the additionality of resources. It would not always be possible to identify precisely all the resources thus allocated. However, whenever such identification is possible, separate statistics should be kept. The statistics relating to official development assistance of industrialized countries should be adapted so as to record the special effort that these countries would be prepared to make.

10.5.4 Some delegations expressed serious difficulties with proposals for prior dedication of resources, such as all or part of any revenues from taxes, levies, or fees on greenhouse gas emissions, but agreed that these should be studied among future options.

10.6 INSTITUTIONAL MECHANISMS

10.6.1 The Working Group agreed that a progressive approach should be adopted to meet the need for financial assistance and cooperation and the mechanisms to be utilized or developed. This approach would parallel that being pursued in the implementation of the Montreal Protocol on the phasing out of CFCs and halons.

10.6.2 There was agreement that resources should flow in a timely, coordinated, and cost-effective manner throughout the response tracks described below.

1) One track builds on work under way or planned in existing institutions. In this regard, the World Bank, a number of the regional development banks, other multilateral organizations, including the UN specialized agencies such as UNDP, and the bilateral assistance agencies have initiated efforts to incorporate global climate change issues into their programs. The September 25th Communiqué of the Development Committee and the work plans of the new OECD Development Assistance Committee Group on Development Assistance and the Environment are particularly noteworthy. Each organization is expected to define coherent guidelines for action, and to review the need for additional resources and expanded programmes. Bilateral donors should further integrate and reinforce the environmental component in their assistance programmes, and develop co-financing arrangements with multilateral institutions: the deployment of *existing* development assistance monies should, therefore, be re-examined. In this regard, the Working Group welcomed recent initiatives to increase allocations of bilateral funds to support environmental programmes. The second track is described below.

2) A number of delegations from both industrialized and developing countries strongly urged that the parallel creation of new mechanisms or facilities, like a new international Fund, is already justified. They stressed that such new instruments should be directly related to a future climate convention or subsequent protocols. It was added that such new instruments could be located within the World Bank system (with new rules), or elsewhere. With respect to this, WMO, UNDP, and UNEP were mentioned. It was also noted that the Global Environmental Facility proposed by the World Bank in collaboration with UNEP and UNDP was welcomed. Moreover, some suggested that these instruments should receive guidance from an executive body to be created under a climate convention. Other delegations suggested and it was agreed that proposals for new instruments should be clarified. Some delegations stated that full assessment of existing mechanisms or facilities should be made before studying the need for new mechanisms or a new international Fund.

10.6.3   While generation of funds might be distinguished from allocation of funds, additional resources could be generated by a new mechanism from among the options described above (see section on sources of funds) but allocated to existing institutions to administer programmes.

10.6.4   The issue of additionality was discussed in some detail. A number of delegations both from industrialized and developing countries felt strongly that funds raised should be clearly additional and distinct from those provided for development assistance, and that they should be provided through separate windows or facilities to make clear that they would be provided under different terms than normal official development assistance. A number of industrialized countries indicated that they could not, at present, offer significant additional resources. This approach does not preclude the creation of a body, for consultations among the main multilateral partners, similar to the consultative group on international agricultural research. Furthermore, it was noted that unless serious attempts are made to minimize or eliminate contributions to atmospheric burdens of greenhouse gases from *existing* development assistance portfolios, it would be difficult or impossible to justify *additional* funds for this purpose.

10.6.5   Some delegations expressed the strong view that supportive economic measures would be important in the context of assistance provided. Others expressed reservations about the appropriateness of certain types of economic policy dialogue in this respect.

10.6.6   With regard to the two tracks outlined above, the group agreed that individual developing countries should initiate studies on their current and projected emissions levels and assistance needs in limiting/reducing them, across the three most important sectors—energy, forestry, and agriculture. A number of delegations advocated the involvement of the World Bank from the start. Each RSWG Subgroup should take these into consideration in its deliberations. Donors were urged to provide resources for such studies, which should be completed as expeditiously as possible. In advance of such studies donors and recipients should agree on the terms of reference. The Working Group noted that this approach paralleled that currently under way to implement the Montreal Protocol in which an accelerated effort is under way to complete assessments and country studies (supported by external resources) of the increased needs for cooperation and assistance to make possible a phase-out of production and use of CFCs.

10.6.7   While recognizing the complexity of such studies, the Working Group expressed the hope that such assessments and strategy plans would proceed expeditiously to permit early consideration of the magnitude of financing requirements, which might be encountered by developing countries. Some delegations strongly expressed the view that such information is indispensable to assess the capabilities of existing institutions, once strengthened and reoriented, to meet the financing needs identified, and to assess the need and scope for new mechanisms.

10.7   FUTURE WORK PROGRAMME

10.7.1   The Working Group agreed that further progress was needed in a number of areas, particularly in assessing the magnitude of the financing needed for developing countries, the need for new mechanisms, and the contribution of the private sector.

10.7.2   Individual country studies of current and projected emissions of greenhouse gases and plans for limiting or reducing emissions are needed to permit the assessment of the magnitudes involved. It was noted that the application of efficient technologies can have a major bearing on future capital requirements.

10.7.3   Such information is fundamental to determining the future demands on existing institutions and the potential need for new mechanisms. There was consensus on the need to further study the context of a new mechanism, in the concept of a future climate convention or its protocols.

10.7.4 The Working Group further noted the important contribution that the private sector might make in cooperating with developing countries to respond to climate change. In particular, the private sector can facilitate effective technology transfer, and, through foreign direct investment, support economic growth and provide additional resources. The potential for co-financing and other forms of public and private sector collaboration should be explored.

10.8 CONCLUDING REMARKS

The framework for action that has just been presented, and the financial mechanisms that could accompany it, are of a progressive nature. The first far-reaching measures (for example, research, swifter reduction of the emissions of chlorofluorocarbons, reinforcement of the scientific observations system, technical assistance for developing countries, support for forestry) are already under way and should be strengthened expeditiously.

11

Legal and Institutional Mechanisms

COORDINATORS
R. Rochon (Canada)
D. Attard (Malta)
R. Beetham (U.K.)

CONTENTS

LEGAL AND INSTITUTIONAL MECHANISMS

EXECUTIVE SUMMARY

1) The coordinators' report has as its primary objective the compilation of elements that might be included in a future framework Convention on Climate Change, and a discussion of the issues that are likely to arise in the context of developing those elements.

2) There is a general view that while existing legal instruments and institutions with a bearing on climate should be fully utilized and further strengthened, they are insufficient alone to meet the challenge. A very broad international consensus has therefore emerged in the IPCC, confirmed notably at the 44th United Nations General Assembly, on the need for a framework Convention on Climate Change. Such a Convention should generally follow the format of the Vienna Convention for the Protection of the Ozone Layer, in laying down, as a minimum, general principles and obligations. It should further be framed in such a way as to gain the adherence of the largest possible number and most suitably balanced spread of countries while permitting timely action to be taken; it should contain provision for separate annexes/protocols to deal with specific obligations. As part of the commitment of the parties to action on greenhouse gas emissions and the adverse effects of global warming, the Convention would also address the particular financial needs of the developing countries, the question of the access to and transfer of technology, and institutional requirements.

3) The paper points out a number of issues to be

decided in the negotiation of a Convention. In general these are:
- the political imperative of striking the correct balances: on the one hand, between the arguments for a far-reaching, action-oriented Convention and the need for urgent adoption of such a Convention so as to begin tackling the problem of climate change; and, on the other hand, between the cost of inaction and the lack of scientific certainty;
- the extent to which specific obligations, particularly on the control of emissions of carbon dioxide and other greenhouse gases, should be included in the Convention itself or be the subject of separate protocol(s):
- the timing of negotiation of such protocol(s) in relation to the negotiations on the Convention.

4) In particular, within the Convention the following specific issues will need to be addressed:
- *financial needs of developing countries.* The need for additional resources for developing countries and the manner in which this should be addressed, particularly in terms of the nature, size, and conditions of the funding, even if detailed arrangements form the subject of a separate protocol, will have to be considered by the negotiating parties;
- *development and transfer of technology.* The basis on which the promotion of the development and transfer of technology and provision of technical assistance to develop-

ing countries should take place will need to be elaborated, taking into account considerations such as terms of transfer, assured access, intellectual property rights, and the environmental soundness of such technology;

- *institutions*. Views differ substantially on the role and powers of the institutions to be created by the Convention, particularly in exercising supervision and control over the obligations undertaken.

5) The inclusion of any particular element in the paper does not imply consensus with respect to that element, or the agreement of any particular

government to include that element in a Convention.

6) The coordinators have not sought to make a value judgment in listing and summarizing in the attached paper the elements proposed for inclusion in a framework Convention: their text seeks merely to assist the future negotiators in their task. They note, however, that a readiness to address the foregoing fundamental problems in a realistic manner will be a prerequisite for ensuring the success of the negotiations and the support of a sufficiently wide and representative spread of nations.

POSSIBLE ELEMENTS FOR INCLUSION IN A FRAMEWORK CONVENTION ON CLIMATE CHANGE

11.1 PREAMBLE

In keeping with common treaty practice including the format of the Vienna Convention, the Climate Change Convention would contain a preamble that might seek to address some or all of the following items:

- a description of the problem and reasons for action (need for timely and effective response without awaiting absolute scientific certainty);
- reference to relevant international legal instruments (such as the Vienna Convention and Montreal Protocol) and declarations (such as UNGA Resolution 43/53 and Principle 21 of the Stockholm Declaration);
- recognition that climate change is a common concern of mankind, affects humanity as a whole, and should be approached within a global framework, without prejudice to the sovereignty of states over the airspace superadjacent to their territory as recognized under international law;
- recognition of the need for an environment of a quality that permits a life of dignity and well-being for present and future generations;
- reference to the balance between the sovereign right of states to exploit natural resources and the concomitant duty to protect and conserve climate for the benefit of mankind, in a manner not to diminish either;

- endorsement and elaboration of the concept of sustainable development;
- recognition of the need to improve scientific knowledge (e.g., through systematic observation) and to study the social and economic impacts of climate change, respecting national sovereignty;
- recognition of the importance of the development and transfer of technology and of the circumstances and needs, particularly financial, of developing countries; need for regulatory, supportive, and adjustment measures to take into account different levels of development and thus differing needs of countries;
- recognition of the responsibility of all countries to make efforts at the national, regional, and global levels to limit or reduce greenhouse gas emissions and prevent activities that could adversely affect climate, while bearing in mind that:
  - most emissions affecting the atmosphere at present originate in industrialized countries where the scope for change is greatest;
  - implementation may take place in different time frames for different categories of countries and may be qualified by the means at the disposal of individual countries and their scientific and technical capabilities;
  - emissions from developing countries are growing and may need to grow in order to meet their development requirements and thus, over time, are likely to represent an in-

creasingly significant percentage of global emissions;

- recognition of the need to develop strategies to absorb greenhouse gases, i.e., protect and increase greenhouse gas sinks; to limit or reduce anthropogenic greenhouse gas emissions; and to adapt human activities to the impacts of climate change.

Other key issues that will have to be addressed during the development of the preambular language include:

- Should mankind's interest in a viable environment be characterized as a fundamental right?
- Is there an entitlement not to be subjected, directly or indirectly, to the adverse effects of climate change?
- Should there be a reference to the precautionary principle?
- In view of the inter-relationship among all greenhouse gases, their sources and sinks, should they be treated collectively?
- Should countries be permitted to meet their aggregate global climate objectives through joint arrangements?
- Should reference be made to weather modification agreements such as the ENMOD treaty as relevant legal instruments?
- Is there a common interest of mankind in the development and application of technologies to protect and preserve climate?
- Does the concept of sustainable development exclude or include the imposition of new conditionality in the provision of financial assistance to developing countries, and does it imply a link between the protection and preservation of the environment, including climate change, and economic development so that both are to be secured in a coherent and consistent manner?
- Should the preamble address the particular problems of countries with an agricultural system vulnerable to climate change and with limited access to capital and technologies, recognizing the link with sustainable development?
- Is there a minimum standard of living that is a prerequisite to adopting response strategies to address climate change?

11.2 DEFINITIONS

As is the practice, definitions will need to be elaborated in a specific article on definitions. The terms that will need to be defined will depend on the purpose of the Convention and thus the language used by the negotiating parties.

11.3 GENERAL OBLIGATIONS

Following the format of such treaties as the Vienna Convention, an article would set out the general obligations agreed to by the parties to the Convention. Such obligations may relate to, for example:

- the adoption of appropriate measures to protect against the adverse effects of climate change, to limit, reduce, adapt to, and, as far as possible, prevent climate change in accordance with the means at the disposal of individual countries and their scientific and technical capabilities; and to avoid creating other environmental problems in taking such measures;
- the protection, stabilization, and improvement of the composition of the atmosphere in order to conserve climate for the benefit of present and future generations;
- taking steps having the effect of limiting climate change but that are already justified on other grounds;
- the use of climate for peaceful purposes only, in a spirit of good neighborliness;
- cooperation by means of research, systematic observation, and information exchange in order to understand better and assess the effects of human activities on the climate and the potential adverse environmental and socio-economic impacts that could result from climate change, respecting national sovereignty;
- the encouragement of the development and transfer of relevant technologies, as well as the provision of technical and financial assistance, taking into account the particular needs of developing countries to enable them to fulfill their obligations;

- cooperation in the formulation and harmonization of policies and strategies directed at limiting, reducing, adapting to, and, as far as possible, preventing climate change;
- cooperation in the adoption of appropriate legal or administrative measures to address climate change;
- provision for bilateral, multilateral, and regional agreements or arrangements not incompatible with the Convention and any annex/ protocol, including opportunities for groups of countries to fulfill the requirements on a regional or subregional basis;
- cooperation with competent international organizations effectively to meet the objectives of the Convention;
- the encouragement of and cooperation in the promotion of public education and awareness of the environmental and socio-economic impacts of greenhouse gas emissions and of climate change;
- the strengthening or modification if necessary of existing legal and institutional instruments and arrangements relating to climate change; and
- a provision on funding mechanisms.

Other key issues that will have to be addressed in the process of elaborating this article include the following:

- Should there be a provision setting any specific goals with respect to levels of emissions (global or national) or atmospheric concentrations of greenhouse gases while ensuring stable development of the world economy, particularly stabilization by industrialized countries, as a first step, and later reduction of CO_2 emissions and emissions of other greenhouse gases not controlled by the Montreal Protocol? Such provision would not exclude the application of more stringent national or regional emission goals than those that may be provided for in the Convention and/or any annex/protocol.
- In light of the preambular language, should there be a provision recognizing that implementation of obligations may take place in different time frames for different categories of countries and/or may be qualified by the means

at the disposal of individual countries and their scientific and technical capabilities?
- Should there be a commitment to formulate appropriate measures, such as annexes, protocols or other legal instruments, and, if so, should such formulation be on a sound scientific basis or on the basis of the best available scientific knowledge?
- In addressing the transfer of technology, particularly to developing countries, what should be the terms of such transfers (i.e., commercial versus non-commercial, preferential versus non-preferential, the relationship between transfers and the protection of intellectual property rights)?
- Should funding mechanisms be limited to making full use of existing mechanisms or also entail new and additional resources and mechanisms?
- Should provision be made for environmental impact assessments of planned activities that are likely to cause significant climate change as well as for prior notice of such activities?
- What should be the basis of emission goals (e.g., total emission levels, per capita emissions, emissions per GNP, emissions per energy use, climatic conditions, past performance, geographic characteristics, fossil fuel resource base, carbon intensity per unit of energy, energy intensity per GNP, socio-economic costs and benefits, or other equitable considerations)?
- Should the particular problem of sea level rise be specifically addressed?
- Is there a link between nuclear stockpiles and climate change?

11.4 INSTITUTIONS

It has been the general practice under international environmental agreements to establish various institutional mechanisms. The parties to a Climate Change Convention might, therefore, wish to make provision for a Conference of the Parties, an Executive Organ, and a Secretariat.

The Conference of the Parties may, among other things: keep under continuous review the implementation of the Convention and take appropriate decisions to this end; review current scientific information; and promote harmonization of policies and strategies directed at limiting, reducing, adapting to, and, as far as possible, preventing climate change.

Questions that will arise in developing provisions for appropriate institutional mechanisms include:

- Should any of the Convention's institutions (e.g., the Conference of the Parties and/or the Executive Organ) have the ability to take decisions *inter alia* on response strategies or functions in respect of surveillance, verification and compliance that would be binding on all the parties, and, if so, should such an institution represent all of the parties or be composed of a limited number of parties (e.g., based on equitable geographic representation)?
- What should be the role of the Secretariat?
- What should be the decision-making procedures, including voting requirements (e.g., consensus, majority)?
- If a trust fund or other financial mechanism were established under the Convention, how should it be administered?
- Should scientific and/or other bodies be established on a permanent or ad hoc basis, to provide advice and make recommendations to the Conference of the Parties concerning research activities and measures to deal with climate change?
- Should the composition of the above bodies reflect equitable climatic or geographic representation?
- Should there be a provision for working groups (e.g., on scientific matters as well as on socio-economic impacts and response strategies)?
- Is there a need for innovative approaches to institutional mechanisms in the light of the nature of the climate change issue?
- What should be the role of non-governmental organizations?

11.5 RESEARCH, SYSTEMATIC OBSERVATIONS, AND ANALYSIS

It would appear to follow general practice to include provision for cooperation in research and systematic monitoring. In terms of research, each party might be called upon to undertake, initiate, and/or cooperate in, directly or through international bodies, the conduct of research on and analysis of:

- physical and chemical processes that may affect climate;
- substances, practices, processes, and activities that could modify the climate;
- techniques for monitoring and measuring greenhouse gas emission rates and their uptake by sinks;
- improved climate models, particularly for regional climates;
- environmental, social, and economic effects that could result from modifications of climate;
- alternative substances, technologies, and practices;
- environmental, social, and economic effects of response strategies;
- human activities affecting climate;
- coastal areas, with particular reference to sea level rise;
- water resources; and
- energy efficiency.

The parties might also be called upon to cooperate in establishing and improving, directly or through competent international bodies, and taking fully into account national legislation and relevant ongoing activities at the national, regional, and international levels, joint or complementary programmes for systematic monitoring and analysis of climate, including a possible worldwide system; and cooperate in ensuring the collection, validation and transmission of research, observational data and analysis through appropriate data centers.

Other issues that will arise in developing this provision include:

- Should consideration be given to the establishment of panels of experts or of an independent scientific board responsible for the coordination of data collection from the above areas of research and analysis and for periodic assessment of the data?
- Should provision be made for on-site inspection?
- Should there be provision for open and non-discriminatory access to meteorological data developed by all countries?
- Should a specific research fund be established?

11.6 INFORMATION EXCHANGE AND REPORTING

Precedents would suggest the inclusion of a provision for the transmission of information through the Secretariat to the Conference of the Parties on measures adopted by them in implementation of the Convention and of protocols to which they are party. In an annex to the Vienna Convention, the types of information exchanged are specified and include scientific, technical, socio-economic, commercial and legal information.

For the purposes of elaborating this provision, issues having to be addressed by the negotiating parties include those below.

- Is there a need for the elaboration of a comprehensive international research programme in order to facilitate cooperation in the exchange of scientific, technological, and other information on climate change?
- Should parties be obliged to report on measures they have adopted for the implementation of the Convention, with the possible inclusion of regular reporting on a comparable basis of their emissions of greenhouse gases?
- Should each party additionally be called upon to develop a national inventory of emissions, strategies, and available technologies for addressing climate change? If so, the Convention might also call for the exchange of information on such inventories, strategies, and technologies.

11.7 DEVELOPMENT AND TRANSFER OF TECHNOLOGY

While the issue of technology has been addressed in the section on General Obligations, it might be considered desirable to include separate provisions on technology transfer and technical cooperation. Such provisions could call upon the parties to promote the development and transfer of technology and technical cooperation, taking into account particularly the needs of developing countries, to enable them to take measures to protect against the adverse effects of climate change, to limit, reduce, and, as far as possible, prevent climate change, or to adapt to it.

Another issue that will arise is: should special terms be attached to climate-related transfers of technology (such as a preferential and/or non-commercial basis and assured access to, and transfer of, environmentally sound technologies on favorable terms to developing countries), taking into consideration the protection of intellectual property rights?

11.8 SETTLEMENT OF DISPUTES

It would be usual international practice to include a provision on the settlement of disputes that may arise concerning the interpretation or application of the Convention and/or any annex/protocol. Provisions similar to those in the Vienna Convention for the Protection of the Ozone Layer might be employed, i.e., voluntary resort to arbitration or the International Court of Justice (with a binding award) or, if neither of those options is elected, mandatory resort to conciliation (with a recommendatory award).

It would be the usual international practice to include clauses on the following topics:

- amendment of the Convention;
- status, adoption and amendment of annexes;
- adoption and entry into force of, and amendments to, protocols;
- signature;
- ratification;
- accession;
- right to vote;
- relationship between the Convention and any protocol(s);
- entry into force;
- reservations;
- withdrawal;
- depositary; and
- authentic texts.

11.9 ANNEXES AND PROTOCOLS

The negotiating parties may wish the Convention to provide for the possibility of annexes and/or protocols. Annexes might be concluded as integral parts of the Convention, while protocols might be concluded subsequently (as in the case of the Montreal Protocol to the Vienna Convention on Protection of the Ozone Layer). While it is recognized that the Convention is to be all-encompassing, the negotiating parties will have to decide whether greenhouse gases, their sources and sinks, are to be dealt with: individually, in groups, or comprehensively; in annexes or protocols to the Convention. The following, among others, might also be considered as possible subjects for annexes or protocols to the Convention:

- agricultural practices;
- forest management;
- funding mechanisms;
- research and systematic observations;
- energy conservation and alternative sources of energy;
- liability and compensation;
- international emissions trading;
- international taxation system; and
- development and transfer of climate change-related technologies.

Issues that will arise in connection with the development of annexes and protocols include:

- timing, i.e., negotiating parties advocating a more action-oriented Convention may seek to include specific obligations in annexes as opposed to subsequent protocols and/or negotiate one or more protocols in parallel with the Convention negotiations;
- sequence, i.e., if there is to be a series of protocols, in what order should they be taken up?

List of Acronyms
and Chemical Symbols

AAGR Average Annual Growth Rate

AFOS Agriculture, Forestry and Other Human Activities Subgroup of IPCC Working Group III

ASF Atmospheric Stabilization Framework

BaU "Business as Usual" Scenario. Same as Scenario A of Working Group III

Bt Billion (1000 million or 10^7) tonnes

BTC Billion (1000 million or 10^7) tonnes Carbon

CFCs Chlorofluorocarbons

CH$_4$ Methane

CI Carbon Intensity in kilogram carbon per gigajoule

CO Carbon monoxide

CO$_2$ Carbon dioxide

CZMS Coastal Zone Management Subgroup

EIS Energy and Industry Subgroup of Working Group III

Gg Gigagram (10^9 grams)

GHG Greenhouse Gas

GDP Gross Domestic Product

GNP Gross National Product

GtC Gigatonnes (10^9 tonnes) Carbon

ha hectare

HCFC Hydrochlorofluorocarbon

HFC Hydrofluorocarbon

IMAGE Integrated Model for the Assessment of the Greenhouse Effect

IOC Intergovernmental Oceanographic Commission of UNESCO

IPCC Intergovernmental Panel on Climate Change

ICSU International Council of Scientific Unions

ITTO International Tropical Timber Organization

Mt Megatonnes (10^6 tonnes)

N$_2$O Nitrous oxide

NGOs Non-Governmental Organizations

NO$_x$ Nitrogen oxides

O$_3$ Ozone

OECD Organization for Economic Cooperation and Development

pa per annum

PC per capita carbon emissions in tonne carbon

PgC Petagrams Carbon

ppm part per million

RSWG Response Strategies Working Group of IPCC Working Group III

SO$_x$ Sulphur oxides

TC Tonne Carbon

TC-GJ Tonne Carbon per GigaJoule

TFAP Tropical Forestry Action Plan

Tg   Teragrams (10^{12} grams)

TgC   Teragram Carbon

TgCH$_4$ Teragram Methane

Tg N   Teragram Nitrogen

UN   United Nations

UNGA   United Nations General Assembly

UNDP   United Nations Development Programme

UNEP   United Nations Environment Programme

UNESCO   United Nations Educational, Scientific and Cultural Organization

VOCs   Volatile Organic Compounds

WMO   World Meteorological Organization

Also Available from Island Press

Ancient Forests of the Pacific Northwest
By Elliott A. Norse

Balancing on the Brink of Extinction: The Endangered Species Act and Lessons for the Future
Edited by Kathryn A. Kohm

Better Trout Habitat: A Guide to Stream Restoration and Management
By Christopher J. Hunter

Beyond 40 Percent: Record-Setting Recycling and Composting Programs
The Institute for Local Self-Reliance

The Challenge of Global Warming
Edited by Dean Edwin Abrahamson

Coastal Alert: Ecosystems, Energy, and Offshore Oil Drilling
By Dwight Holing

The Complete Guide to Environmental Careers
The CEIP Fund

Economics of Protected Areas
By John A. Dixon and Paul B. Sherman

Environmental Agenda for the Future
Edited by Robert Cahn

Environmental Disputes: Community Involvement in Conflict Resolution
By James E. Crowfoot and Julia M. Wondolleck

Fighting Toxics: A Manual for Protecting Your Family, Community, and Workplace
Edited by Gary Cohen and John O'Connor

Forests and Forestry in China: Changing Patterns of Resource Development
By S. D. Richardson

From *The Land*
Edited and compiled by Nancy P. Pittman

The Global Citizen
By Donella Meadows

Hazardous Waste from Small Quantity Generators
By Seymour I. Schwartz and Wendy B. Pratt

Holistic Resource Management Workbook
By Allan Savory

In Praise of Nature
Edited and with essays by Stephanie Mills

The Living Ocean: Understanding and Protecting Marine Biodiversity
By Boyce Thorne Miller and John Catena

Natural Resources for the 21st Century
Edited by R. Neil Sampson and Dwight Hair

The New York Environment Book
By Eric A. Goldstein and Mark A. Izeman

Overtapped Oasis: Reform or Revolution for Western Water
By Marc Reisner and Sarah Bates

Permaculture: A Practical Guide for a Sustainable Future
By Bill Mollison

Plastics: America's Packaging Dilemma
By Nancy A. Wolf and Ellen D. Feldman

The Poisoned Well: New Strategies for Groundwater Protection
Edited by Eric Jorgensen

Race to Save the Tropics: Ecology and Economics for a Sustainable Future
Edited by Robert Goodland

Recycling and Incineration: Evaluating the Choices
By Richard A. Denison and John Ruston

Reforming The Forest Service
By Randal O'Toole

The Rising Tide: Global Warming and World Sea Levels
By Lynne T. Edgerton

Saving the Tropical Forests
By Judith Gradwohl and Russell Greenberg

Trees, Why Do You Wait?
By Richard Critchfield

War on Waste: Can America Win Its Battle With Garbage?
By Louis Blumberg and Robert Gottlieb

Western Water Made Simple
From *High Country News*

Wetland Creation and Restoration: The Status of the Science
Edited by Mary E. Kentula and Jon A. Kusler

Wildlife and Habitats in Managed Landscapes
Edited by Jon E. Rodiek and Eric G. Bolen

For a complete catalog of Island Press publications, please write:
  Island Press
  Box 7
  Covelo, CA 95428

or call: 1-800-828-1302